Lecture Notes in Computer Scie

T0238871

Commenced Publication in 1973
Founding and Former Series Editors:
Gerhard Goos, Juris Hartmanis, and Jan van Leeuwen

For further volumes:
http://www.springer.com/series/7411

Paola Flocchini · Jie Gao
Evangelos Kranakis
Friedhelm Meyer auf der Heide (Eds.)

Algorithms for Sensor Systems

9th International Symposium on Algorithms
and Experiments for Sensor Systems,
Wireless Networks and Distributed Robotics
ALGOSENSORS 2013, Sophia Antipolis
France, September 5–6, 2013
Revised Selected Papers

 Springer

Editors
Paola Flocchini
University of Ottawa
Ottawa, ON
Canada

Jie Gao
Stony Brook University
Stonybrook, NY
USA

Evangelos Kranakis
School of Computer Science
Carleton University
Ottawa, ON
Canada

Friedhelm Meyer auf der Heide
University of Paderborn
Paderborn
Germany

ISSN 0302-9743   ISSN 1611-3349   (electronic)
ISBN 978-3-642-45345-8   ISBN 978-3-642-45346-5   (eBook)
DOI 10.1007/978-3-642-45346-5
Springer Heidelberg New York Dordrecht London

Library of Congress Control Number: 2013956805

CR Subject Classification (1998): F.2, C.2, G.1, I.2.9

LNCS Sublibrary: SL5 – Computer Communication Networks and Telecommunications

Printed on acid-free paper

Springer is part of Springer Science+Business Media (www.springer.com)

Preface

ALGOSENSORS, the International Symposium on Algorithms and Experiments for Sensor Systems, Wireless Networks and Distributed Robotics, is an international forum dedicated to the algorithmic aspects of wireless networks, static or mobile. The 9th edition of ALGOSENSORS was held during September 5–6 in Sophia Antipolis, France, within the ALGO annual event.

Originally focused solely on sensor networks, ALGOSENSORS now covers more broadly algorithmic issues arising in all wireless networks of computational entities, including sensor networks, sensor-actuator networks, and systems of autonomous mobile robots. In particular, it focuses on the design and analysis of discrete and distributed algorithms, on models of computation and complexity, on experimental analysis, in the context of wireless networks, sensor networks, and robotic networks and on all foundational and algorithmic aspects of the research in these areas.

This year papers were solicited in three tracks: Sensor Network Algorithms (Track A), Wireless Networks and Distributed Robotics Algorithms (Track B), and Experimental Algorithms (Track C).

In response to the call for papers, 30 submissions were received, out of which 19 papers were accepted after a rigorous reviewing process by the (joint) Program Committee, which involved at least three reviewers per paper. In addition to the technical papers, the program included two invited presentations, the keynote talk by Magnús M. Halldórsson (Reykjavik University), and the opening talk by Giuseppe Prencipe (University of Pisa). This volume contains the technical papers as well as summaries of the two keynote talks.

We would like to thank the Program Committee members, as well as the external reviewers, for their fundamental contribution in selecting the best papers resulting in a strong program. We would also like to warmly thank the ALGO/ESA 2013 organizers for kindly accepting the proposal of the Steering Committee to co-locate ALGOSENSORS with some of the leading events on algorithms in Europe.

October 2013

Paola Flocchini
Jie Gao
Evangelos Kranakis
Friedhelm Meyer auf der Heide

Organization

Program Committee

Christian Blum	Ikerbasque and University of the Basque Country, Spain
Prosenjit Bose	Carleton University, Canada
Ioannis Chatzigiannakis	University of Patras, Greece
Shantanu Das	Aix-Marseille University, France
Shlomi Dolev	Ben-Gurion University of the Negev, Israel
Alon Efrat	University of Arizona, USA
Sandor Fekete	Braunschweig University of Technology, Germany
Paola Flocchini	University of Ottawa (Program Chair), Canada
Hannes Frey	University of Koblenz-Landau, Germany
Jie Gao	Stony Brook (Co-chair Track A), USA
Maria Gradinariu	University of Paris 6, France
Himanshu Gupta	Stony Brook University, USA
Qiangsheng Hua	Tsinghua University, China
Taisuke Izumi	Nagoya Institute of Technology, Japan
Ralf Klasing	LaBRI - CNRS, France
Dariusz Kowalski	University of Liverpool, UK
Evangelos Kranakis	Carleton University (Co-chair Track B), Canada
Danny Krizanc	Wesleyan University, USA
Alexander Kroeller	Braunschweig University of Technology, Germany
Pierre Leone	University of Geneva, Switzerland
Xiangyang Li	Illinois Institute of Technology, USA
Mingyan Liu	University of Michigan, USA
Tzvi Lotker	Ben-Gurion University, Israel
Friedhelm Meyer Auf der Heide	Paderborn University (Co-chair Track C), Germany
Lata Narayanan	Concordia University, Canada
Alfredo Navarra	University of Perugia, Italy
Amiya Nayak	University of Ottawa, Canada
Dennis Pfisterer	University of Lübeck, Germany
Giuseppe Prencipe	University of Pisa, Italy
Michael Rabbat	McGill University, Canada
Rik Sarkar	University of Edinburgh, UK
Elad Michael Schiller	Chalmers University of Technology, Sweden
Christian Schindelhauer	University of Freiburg, Germany

Subhash Suri   University of California, Santa Barbara, USA
My Thai   University of Florida, USA
Andrea Vitaletti   Sapienza University, Italy
Masafumi Yamashita   Kyushu University, Japan

Additional Reviewers

Akribopoulos, Orestis
Amaxilatis, Dimitrios
Azimi, Navid
Barenboim, Leonid
Chalopin, Jérémie
Chatzis, Konstantinos
D'Angelo, Gianlorenzo
D'Emidio, Mattia
Lamani, Anissa
Landsiedel, Olaf
Li, Ximing
Morales, Oscar
Nguyen, Dung
Viglietta, Giovanni
Yu, Dongxiao
Zhang, Le

Contents

Modeling Reality Algorithmically: The Case of Wireless Communication

Magnús M. Halldórsson(✉)

ICE-TCS, School of Computer Science, Reykjavik University, 101 Reykjavik, Iceland
mmh@ru.is

1 Algorithmic Models and Their Properties

Computation is increasingly being viewed as the 21st century approach to modeling the world. Classical sciences have become increasingly more computer-driven, necessitating a computational perspective. The equation – bastion of 20th century science – is becoming supplanted by the algorithm. To properly address real-world phenomena, we need models appropriate for algorithmic approaches.

This note contains the author's reflections on the choice and design of models, particularly those capturing aspects of the measurable world. What it is that we look for in models and the essential properties that we seek. We do this in the context of wireless networking, but hope that some of the lessons have wider relevance.

We postulate that algorithmic models must satisfy four properties to be truly useful.

Fidelity. A model must be a fair representation of reality. Whereas physics has the advantage that its relatively simple laws hold with extremely high precision, the settings for most fields of study are inherently noisy, making perfect models a pipe dream. Instead, abstractions are an intrinsic part of most models, where the intent is to factor out unimportant ideosyncracies. On the other hand, if essential features are eliminated, the model fails its primary task: to faithfully represent reality.

Simplicity. Overly complex models generally result in limited usage. The utility of such models for algorithmic design is necessarily limited, as it complicates all the efforts involved. Occam's razor suggests that models should be as simple as possible, but also no simpler than that, paraphrasing an aphorism attributed to Einstein. Simplicity also has implications for analysis.

Analyzability. In order to truly understand real-world phenomena, we need to be able to analyze them and study, both individually and in relation to other phenomena. A model with characteristics that defy analysis may allow for uniformed heuristic use, but will hamper our understanding of the intrinsic properties of the concept at hand.

Generality. Finally, we seek explanations of general utility with wide applicability. There is always the danger to introduce context-specific attributes to strengthen the predictive power of the model, but the more we do so, the less useful the model is as a means to explain general properties.

P. Flocchini et al. (Eds.): ALGOSENSORS 2013, LNCS 8243, pp. 1–5, 2014.
DOI: 10.1007/978-3-642-45346-5_1, © Springer-Verlag Berlin Heidelberg 2014

It may be helpful to consider some examples.

Integer Linear Programming. ILP is an extremely general tool to represent combinatorial problems, allowing for generic solution methods. Depending on the context, it can be very faithful to the real phenomena. The utilities for algorithm design and analysis depends a lot on the specific domain, and can range from very high to minimal.

Maxwell's equations. The equations for electrodynamics that underlie electrical and communication technologies are both very general and highly precise, omitting only the quantum effects that are usually immeasurably small. However, when examined at the scale of wireless networks, the details involved are overwhelming, rendering them unusable for all but exceptional settings of algorithms and analysis.

In general, the utility of a model may depend on the issue/problem under consideration.

2   Selected Wireless Models

A fair number of algorithmic models has been proposed for wireless networks. Let us consider the more prominent models, explore the problems that they address well and examine the issues they raise and their weaknesses. Usually, the distributed setting is assumed, but one can also evaluate them with respect to centralized algorithms. Let n denote the number of wireless transceiver nodes.

Radio model. In the earliest and the most basic model, a wireless transmission is successful if exactly one transmitter is transmitting, in which case all the other nodes receive the message.

A core problem addressed in this model, which has been extended to other models, is *leader election*: the nodes should agree on a single node as a leader. With this primitive, many other issues are simplified. One surprising result due to Willard is that this can be achieved in $O(\log \log n)$ steps [15], when the nodes have *collision detection*, i.e., can distinguish silence from the case when two or more nodes are attempting to transmit.

The key limitation of this model is the assumption that all nodes are within communication range. It is also pessimistic in that it does not allow for any spatial reuse of the wireless channel.

General graphs. In the (general) graph model, the graph represents which pairs of nodes can communicate (and interfere) with each other. In this sense, the radio model corresponds to the clique graph.

The prototypical problem addressed in this model is the *broadcast* problem: how to transmit a message from a given source to all other nodes in the graph. A celebrated result of Bar-Yehuda, Goldreich and Itai [2] shows that this can be achieved in $O(D \log n)$ time steps with a randomized distributed algorithm, where D is the diameter of the graph. This is essentially optimal for a distributed algorithm, and within logarithmic factor of the best possible by a centralized algorithm.

One downside of the model is that many problems become hard to solve even approximately. For instance, the coloring problem, which captures core questions regarding scheduling wireless communications, has no sublinear approximation algorithm [6].

Disc graphs. Communication occurs in the physical world, which is three-dimensional, and distances do matter. A natural approach to limit complexity is therefore to embed the nodes in a Euclidean space and assume that nodes can connect if they are sufficiently close. In the basic setting, nodes can communicate (and interfere) if and only if they are within a fixed distance apart, giving rise to *unit disc graphs* (UDG). UDGs have been the source of a large amount of interesting theory, with the early paper of Clark et al. [3] cited over 1000 times.

Many variations and extensions exist, such as allowing for differing power/radii of the nodes (disc graphs) or different ranges for interference than for communication (protocol model). All disc graphs, however, make strong assumptions: the world is flat (i.e., planar), radio transmission ranges are circular, and reception is symmetric. More generally, all graph-based models assume that reception quality is a binary and that interference is a pairwise relationship. Numerous empirical results (see, e.g., [10]) have shown these to be simplistic.

Physical model. The model of choice in engineering circles has been the *physical model*, where the radio signal is assumed to be a decaying function of distance. Here, interference is no longer binary but additive, with successful reception achieved if the total amount of interference is sufficiently small relative to the strength of the intended signal (i.e., high enough signal-to-interference-and-noise-ratio, SINR).

The standard assumption is that signal decays as a polynomial function of distance, known as *geometric pathloss*. Namely, if the signal travels distance d from a sender transmitting with power P, it will be received with strength P/d^α, where α is an absolute constant depending on the setting, understood to be in the range $[2,6]$.

The physical model was mostly ignored by algorithm theory for a long time, assumed to be too complicated and hard to analyze (failing our Simplicity and Analyzability axioms). Recent years have, however, seen great improvements in our understanding of the model and increase in analytic results.

One problem for which results in the physical model are qualitatively different from those in other models is the *aggregation* problem (and the related connectivity problem): Compute aggregation statistic (say, the minimum) of a set of values, where each wireless node contains only a single value. In any disc or graph-based model, the worst-case round complexity is necessarily linear. Surprisingly, perhaps, Moscibroda and Wattenhofer showed that in the physical model it is only poly-logarithmic [13]. In fact, the worst-case bound is only $O(\log n)$ [7].

Even though the physical model adds several attributes of realism, it still has issues regarding fidelity. Geometric path loss means assuming that antennas are omnidirectional and that signal decays smoothly as a function of distance. Real environments have obstacles and walls that can reflect, scatter and damper signals, and the mere appearance of floors or a ground introduces multi-path effects that are beyond the pure geometric path loss.

3   Future Directions

The preceding models have now all been fairly well studied. Each has aspects that fit certain problems better than others, allowing us to draw distinct lessons. None, however, captures all the important aspects of real environments. Experimental evidence has indeed found that wireless reception is tricky, defying simplistic characterizations [1]. We point out a few additional approaches that have been considered.

If the assumption of geometric path loss is jettisoned from the physical model, we are left with the *abstract SINR* model. This is extremely general, with general graphs being a special case. Thus, scheduling-type problems become highly inapproximable. Still, it may be instructive to consider this general model further, identifying other types of restricted instances or parameterized properties that allow us to recover the Analyzability axiom. The *inductive independence* or *maximum average affectance* property of [9] is one such candidate.

Temporal variability in wireless signal reception has been captured in the recent *dual graph* model [12]. It extends the (general) graphs model by allowing for both reliable and adversarially chosen unreliable links. Whether this exact definition is the best one remains to be seen.

Random artifacts appear to be unavoidable in real networks, at least at a low level. Different versions are known as "fading", "shadowing", or "Gaussian" noise. One of the more common ones, Rayleigh fading, has been analyzed in conjunction with the physical model [4] under the assumption of full independence. Correlations and other variations await further study.

Our coverage is by no means exhaustive. Among exciting recent directions are the Abstract MAC Layer [11], multi-channel models (e.g., [5]), jamming resistance [14], and MIMO extensions of the physical model (e.g., [8]).

We believe that the time is ripe for tackling the challenge of faithfully modeling real wireless environments, while obeying the other axioms of simplicity, generality and analyzability. A natural direction would be to meld some of the recent variations with the classical models and assess the resulting model according to these criteria.

Acknowledgements. I thank the attendees of WRAWN (Workshop on Realistic models for Algorithms in Wireless Networks) 2013 for stimulating presentations and discussions that motivated these reflections. Any misrepresentations or omissions are, of course, mine only.

References

1. Baccour, N., Koubaa, A., Mottola, L., Zuniga, M.A., Youssef, H., Boano, C.A., Alves, M.: Radio link quality estimation in wireless sensor networks: a survey. ACM Trans. Sens. Netw. (TOSN) **8**(4), 34 (2012)
2. Bar-Yehuda, R., Goldreich, O., Itai, A.: On the time-complexity of broadcast in radio networks: an exponential gap between determinism and randomization. In: PODC, pp. 98–108 (1987)
3. Clark, B.N., Colbourn, C.J., Johnson, D.S.: Unit disk graphs. Discrete Math. **86**(1), 165–177 (1990)
4. Dams, J., Hoefer, M., Kesselheim, T.: Scheduling in wireless networks with Rayleigh-fading interference. In: SPAA, pp. 327–335 (2012)
5. Dolev, S., Gilbert, S., Guerraoui, R., Newport, C.: Gossiping in a multi-channel radio network. In: Pelc, A. (ed.) DISC 2007. LNCS, vol. 4731, pp. 208–222. Springer, Heidelberg (2007)
6. Feige, U., Kilian, J.: Zero knowledge and the chromatic number. In: CCC, pp. 278–287. IEEE (1996)
7. Halldórsson, M.M., Mitra, P.: Wireless connectivity and capacity. In: SODA (2012)
8. Janson, T., Schindelhauer, C.: Broadcasting in logarithmic time for ad hoc network nodes on a line using mimo. In: SPAA, pp. 63–72. ACM (2013)
9. Kesselheim, T., Vöcking, B.: Distributed contention resolution in wireless networks. In: DISC, pp. 163–178, August 2010
10. Kotz, D., Newport, C., Gray, R.S., Liu, J., Yuan, Y., Elliott, C.: Experimental evaluation of wireless simulation assumptions. In: MSWiM, pp. 78–82. ACM (2004)
11. Kuhn, F., Lynch, N., Newport, C.: The abstract MAC layer. Distrib. Comput. **24**(3–4), 187–206 (2011)
12. Kuhn, F., Lynch, N., Newport, C., Oshman, R., Richa, A.: Broadcasting in unreliable radio networks. In: PODC, pp. 336–345. ACM (2010)
13. Moscibroda, T., Wattenhofer, R.: The complexity of connectivity in wireless networks. In: INFOCOM (2006)
14. Ogierman, A., Richa, A., Scheideler, C., Schmid, S., Zhang, J.: Competitive MAC under Adversarial SINR. ArXiv e-print 1307.7231, July 2013
15. Willard, D.E.: Log-logarithmic selection resolution protocols in a multiple access channel. SIAM J. Comput. **15**(2), 468–477 (1986)

Autonomous Mobile Robots: A Distributed Computing Perspective

Giuseppe Prencipe[⊠]

Dipartimento di Informatica, Università di Pisa, Pisa, Italy
`prencipe@di.unipi.it`

Abstract. The distributed coordination and control of a team of autonomous mobile robots is a problem widely studied in a variety of fields, such as engineering, artificial intelligence, artificial life, robotics. Generally, in these areas, the problem is studied mostly from an empirical point of view.

Recently, the study of what can be computed by such team of robots has become increasingly popular in theoretical computer science and especially in distributed computing, where it is now an integral part of the investigations on computability by mobile entities [28]. In this paper we describe the current investigations on the algorithmic limitations of what autonomous mobile robots can do with respect to different coordination problems, and overview the main research topics that are gaining attention in this area.

1 Introduction

For the last twenty years, the major trend in robotic research, both from engineering and behavioral viewpoints, has been to move away from the design and deployment of few, rather complex, usually expensive, application-specific robots. In fact, the interest has shifted towards the design and use of a large number of "generic" robots which are very simple, with very limited capabilities and, thus, relatively inexpensive, but capable, together, of performing rather complex tasks.

The advantages of such an approach are clear and many, including: reduced costs; ease of system expandability which in turns allows for incremental and on-demand deployment; simple and affordable fault-tolerance capabilities; reusability of the robots in different applications [26, 49].

One of the first studies conducted in this direction in the AI community is that of Matarić [44]. The main idea in Matarić's work is that "interactions between individual agents need not to be complex to produce complex global consequences".

Other investigations in the AI community include the study of [4] on stigmergy communication and on the use a set of simple robots that operate completely autonomously and independently to collect pucks spread over a square

---

This research is supported in part by MIUR of Italy under project ARS TechnoMedia.

P. Flocchini et al. (Eds.): ALGOSENSORS 2013, LNCS 8243, pp. 6–21, 2014.
DOI: 10.1007/978-3-642-45346-5_2, © Springer-Verlag Berlin Heidelberg 2014

arena in a single cluster; the ALLIANCE architecture and the studies on selfish behavior of cooperative robots in animal societies by Parker [49]; the formation and navigation problems in multi-robot teams in the context of primitive animal behavior in pattern formation by Balch and Arkin [3]; and the experiments in cooperative cleaning behavior of Jung et al. [38].

Alternative approaches to the problem of studying multi-robot systems, can be found in the CEBOT system of Fucuda, Kawaguchi et al. [32,41], in the planner-based architecture of Noreils [47], in the information requirements theory of Donald et al. [26] (see [7] for a survey), in the Swarm Intelligence of Beni and Hackwood [5], in the Self-Assembly Machine ("fructum") of Murata et al. [46], etc.

The common feature of all these approaches is that they do not deal with formal correctness of the solutions, that are only analyzed empirically. In all these investigations, algorithmic aspects were somehow implicitly an issue, but clearly not a major concern, let alone the focus, of the study. An investigation with an algorithmic flavor has been undertaken within the AI community by Durfee [27], who argues in favor of limiting the knowledge that an intelligent robot must possess in order to be able to coordinate its behavior with others.

More recently, the study of teams of autonomous mobile robots has gained attention also in distributed computing area, keeping pace with the trend originally started in robotics and AI. However, here the problem has been tackled from a different perspective: from a *computational* point of view. In other words, the focus is to understand the relationship between the capabilities of the robots and the solvability of the tasks they are given. In these studies, the impact of the *knowledge* of the environment is analyzed: can the robots form an arbitrary geometric pattern if they have a *compass*? Can they gather in a point? Which information each robot must have about its fellows in order for them to collectively achieve their goal? The goal is to look for the minimum power to give to the robots so that they can solve a given task; hence, to formally analyze the strengths and weaknesses of the distributed coordination and control.

In this paper we describe the current investigations on the interplay between robots capabilities, computability, and algorithmic solutions of coordination problems by autonomous mobile robots.

2   Modeling Autonomous Mobile Robots

The considered computational universe is a 2-dimensional plane populated by a set of n autonomous mobile robots, denoted by $r_1, \ldots, r_n$, that are modeled as devices with computational capabilities which are able to freely move on a two-dimensional plane.

The Robots and Their Behavior. A robot is a computational unit capable of sensing the positions of other robots in its surrounding, performing local computations on the sensed data, and moving towards the computed destination. The local computation is done according to a deterministic algorithm that takes in input the sensed data (i.e., the robots' positions), and returns a destination point towards which the executing robot moves. All the robots execute the same algorithm. The local view of each robot includes a unit of length, an origin, and a Cartesian coordinate system defined by the *directions* of two coordinate axes, identified as the X and Y axis, together with their *orientations*, identified as the positive and negative sides of the axes. Each robot cyclically performs the following operations: **(i)** *Look*: The robot observes the world by activating its sensors which will return a *snapshot* of the positions of all other robots within the visibility range with respect to its local coordinate system. Each robot is viewed as a point, hence its position in the plane is given by its coordinates, and the result of the snapshot (hence, of the observation) is just a set of coordinates in its local coordinate system: this set forms the *view of the world* of r. **(ii)** *Compute*: The robot performs a *local computation* according to a deterministic algorithm $\mathcal{A}$ (we also say that the robot *executes* $\mathcal{A}$). The algorithm is the same for all robots, and the result of the *Compute* state is a *destination point*. **(iii)** *Move*: If the destination point is the current location of r, r performs a *null movement* (i.e., it does not move); otherwise it moves towards the computed destination but it can stop anytime during its movement.[1]

The robots are completely *autonomous*: no central control is needed. Furthermore they are *anonymous*, meaning that they are a priori indistinguishable by their appearance, and they do not (need to) have any kind of identifiers that can be used during the computation.

Moreover, the robots are *silent*: there are no explicit direct means of communication, and any communication occurs in a totally implicit manner. Specifically, it happens by means of observing the robots' positions in the plane, and taking a deterministic decision accordingly. In other words, the only mean for a robot to send information to some other robot is to move and let the others observe (reminiscent of bees in a bee dance).

Activation and Operation Schedule. With respect to the *activation* schedule of the robots and of the timing of the *operations* within their cycles, there are two main models, asynchronous and semi-synchronous.

In the asynchronous model (ASYNC), no assumptions on the cycle time of each robot, and on the time each robot takes to execute each state of a given cycle are made [29]. It is only assumed that each cycle is completed in finite time, and that the distance traveled in a cycle is finite. Moreover, the robots do not need to have a common notion of time, and each robot can execute its actions at unpredictable time instants.

More precisely, there are only two limiting assumptions. The first one refers to space; namely, the distance traveled by a robot during a computational cycle. **(A1)** *The distance traveled by a robot r in a move is not infinite. Furthermore, there exists an arbitrarily small constant $\delta_r > 0$, such that if the destination point*

---

[1] e.g. because of limits to the robot's motorial capabilities.

is closer than δ_r, r will reach it; otherwise, r will move towards it of at least δ_r. As no other assumptions on space exist, the distance traveled by a robot in a cycle is unpredictable.

The second limiting assumption is on the length of a cycle. **(A2)** *The amount of time required by a robot r to complete a computational cycle is not infinite. Furthermore, there exists a constant $\varepsilon_r > 0$ such that the cycle will require at least ε_r time.* As no other assumption on time exists, the resulting system is *fully asynchronous* and the duration of each activity (or inactivity) is unpredictable; this setting is usually denoted by ASYNC.

There are two important consequences: First, since the time that passes after a robot starts observing the positions of all others and before it starts moving is arbitrary, but finite, the actual move of a robot may be based on a situation that was observed arbitrarily far in the past, and therefore it may be totally different from the current situation. Second, since movements can take a finite but unpredictable amount of time, and different robots might be in different states of their cycles at a given time instant, it is possible that a robot can be seen *while* it is moving by other robots that are observing.[2]

In the *semi-synchronous* (SSYNC) model, the activations of the robots is logically divided into global rounds; in each round, one or more robots are activated and obtain the same snapshot; based on that snapshot, they compute and perform their move [57].

In particular, there is a global clock tick reaching all robots simultaneously, and a robot's cycle is an instantaneous event that starts at a clock tick and ends by the next. The only unpredictability is given by the fact that at each clock tick, every robot is either *active* or *inactive*, and only active robots perform their cycle. The unpredictability is restricted by the fact that at least one robot is active at every time instant, and every robot becomes active at infinitely many unpredictable time instants. A very special case is when every robot is active at every clock tick; in this case the robots are *fully synchronized* (this specific setting is usually denoted by FSYNC).

In this setting, at any given time, all active robots are executing the same cycle state; thus no robot will look while another is moving. In other words, a robot observes other robots only when they are stationary. This implies that the computation is always performed based on accurate information about the current configuration. Furthermore, since no robot can be seen *while* it is moving, the movement can be considered *instantaneous*. An additional consequence of atomicity and synchronization is that, for them to hold, the maximum distance that a robot can move in one cycle is bounded.

Capabilities. Different settings arise from different assumptions that are made on the robots' capabilities, and on the amount of information that they share and use during the accomplishment of the assigned task. In particular,

---

[2] Note that this does not mean that the observing robot can distinguish a moving robot from a non moving one.

- **Visibility.** The robots may be able to sense the complete plane or just a portion of it. We will refer to the first case as the *Unlimited Visibility* case. In contrast, if each robot can sense only up to a distance $V > 0$ from it, we are in the *Limited Visibility* case. In the following, we will say also that the robots have unlimited/limited visibility.
- **Geometric Agreement.** Each robot r has its own unit of length, and a *local compass* defining a local Cartesian coordinate system defined by the *directions* of two coordinate axes, identified as the X and Y axis, together with their *orientations*, identified as the positive and negative sides of the axes. This local coordinate system is self-centric, i.e. the origin is the position of the observing robot. Depending on the level of consistency among the robots on the direction and orientation of the axes of their local compasses, different classes of *global* geometric agreement can be identified: *total agreement* (or *consistent compass*), when the robots agree on the direction and orientation of both axes; *partial agreement* (or *one axis*) when all robots agree on the direction and orientation of only one axis; *chirality* when the robots agree on the orientation of the axes (i.e., clockwise); and *no agreement* (or *disorientation*), where no consistency among the local coordinate systems is known to exist.
- **Memory.** The robots can access local memory to store different amount of information regarding the positions in the plane of their fellows. In the *oblivious* model, all the information contained in the workspace is *cleared* at the end of each cycle. In the *non-oblivious* (or *persistent memory*) model, part (or all) of the local memory is *legacy*: unless explicitly erased by the robot, it will persist throughout the robot's cycles. In this model, an important parameter is the *size* of the persistent workspace. One extreme is the *unbounded memory* case, where no information is ever erased; hence robots can remember all past computations and actions. On the opposite side is the case when the size of the persistent workspace is *constant*; in this case, the entities are just Finite-State Machines, and are called *finite-state* robots.

Let us stress that the only means for the robots to coordinate is the observation of the others' positions and their change through time. For oblivious robots, even this form of communication is impossible, since there is no memory of previous positions.

3   Static Problems

Pattern Formation. The PATTERN FORMATION problem is one of the most important coordination problem and has been extensively investigated in the literature (e.g., see [10,56,57,60]). The problem is practically important, because, if the robots can form a given pattern, they can agree on their respective roles in a subsequent, coordinated action.

In its most general definition, the robots are required to form an *arbitrary* pattern. The geometric pattern to be formed is a set of points (given by their Cartesian coordinates) in the plane, and it is initially known by all the robots in the system.

The robots are said to *form the pattern* if, at the end of the computation, the positions of the robots coincide, in everybody's local view, with the points of the pattern. The formed pattern may be *translated*, *rotated*, *scaled*, and *flipped* into its mirror position with respect to the initial pattern. Initially the robots are in arbitrary positions, with the only requirement that no two robots are in the same position, and that, of course, the number of points prescribed in the pattern and the number of robots are the same.

The basic research questions are which patterns can be formed, and how they can be formed. Many proposed procedures do not terminate and never form the desired pattern: the robots just converge towards it; such procedures are said to *converge*.

There exists solution to solve this problem in both ASYNC (e.g., [31]) and SSYNC (e.g., [57]), by always considering robots with unlimited visibility. In all the solutions, the kind of patterns that can be formed by the robots depends on the level of agreement the robots have on their local coordinate systems.

Several studies also investigated on the formation of specific patterns, such as lines and circles. In the LINE FORMATION problem, the robots are required to place themselves on a line, whose position is not prescribed in advance (if $n = 2$, then a line is always formed). In [15], this problem has been tackled by studying an apparently totally different problem: the *spreading*. In this problem, the robots, that at the beginning are arbitrarily placed on the plane, are required to evenly spread within the perimeter of a given region. In their work, the authors focus on the one-dimensional case: in this case, the robots have to form a line, and place themselves uniformly on it. A very interesting aspect of the study, is that [15] addresses the issue of *local algorithms*: each robots decides where to move based on the positions of its close neighbors. In particular, in the case of the line, the protocol is quite simple: each robot r observes its left and right neighbor. If r does not see any robot, it simply does not move; otherwise, it moves to the median point between its two neighbors. The authors prove its convergence in SSYNC. Furthermore, if each robot knows the exact number of robots at each of its sides, it is possible to achieve the spreading in one dimension in a finite number of cycles.

In the CIRCLE FORMATION problem, the robots want to place themselves on the plane to form a non degenerated circle of a given diameter.[3] One of the first discussion on circle formation by a group of mobile entities was by Debest [20], who introduced it as an illustration of self-stabilizing distributed problems, but did not provide an algorithm. This problem was first studied by Sugihara and Suzuki [56]. They presented an heuristic distributed protocol,

---

[3] If the diameter is not fixed a priori, the problem becomes trivial, even in ASYNC: each robot computes the smallest circle enclosing all the robots' positions and moves on the circumference of such a circle.

successively improved by Tanaka [58], that allowed the robots to form an approximation of a circle (more similar to a Reuleaux triangle) having a given diameter D. A variant of this problem is the UNIFORM CIRCLE FORMATION problem: the n robots on the plane must be arranged at regular intervals on the boundary of a circle. This kind of formation can be usefully deployed in surveillance tasks: the robots are placed on the border of the area (or around the target) to surveil (e.g., see [34]). Both problems have since been extensively investigated in SSYNC and ASYNC [8, 21–24, 39, 52, 58].

Gathering. In the GATHERING problem, the robots, initially placed in arbitrary and distinct positions, are required to gather in a single location within finite time. This problem is also called *point formation, homing,* or *rendezvous.* A problem closely related to GATHERING is that of CONVERGENCE, where the robots need to be arbitrarily close to a common location, without the requirement of ever reaching it.

In spite of their apparent simplicity, these problems have been investigated extensively both in SSYNC and in ASYNC under a variety of assumptions on the robots' capabilities: in fact, several factors render this problem difficult to solve. First of all, some basic results about GATHERING: It is possible in FSYNC, with an algorithm that exploits the properties of the center of gravity of the team [13]; it is impossible without additional assumptions in SSYNC, hence in ASYNC [51, 57], and trivially achievable even in ASYNC with totally agreement on the coordinate systems (gather at the position occupied by the rightmost and topmost robot).

Rendezvous. When the system contains only *two* robots, the GATHERING problem is very special, and it is often called RENDEZVOUS. We have just stated that, with a common coordinate system, there is an easy solution to GATHERING, and hence to RENDEZVOUS even in ASYNC. In absence of a common coordinate system the problem is not solvable even in SSYNC. Hence, with $n = 2$, the focus is on gathering in FSYNC, and on the CONVERGENCE problem.[4]

The RENDEZVOUS has been extensively studied by assuming different level of agreement on the compass systems of the robots. In particular, the problem is solvable in ASYNC when the robots agree on chirality, but the axis are however tilted up to a $\phi < \frac{\pi}{2}$ degrees [37], and the tilt is fixed. If the robots still agree on chirality, but the tilt of their compasses might be variable, rendezvous can be achieved in SSYNC with *fully variable* compasses if and only if $\phi < \frac{\pi}{4}$, and in ASYNC with *semi-variable* compasses[5] if and only if $\phi < \frac{\pi}{6}$ [37].

---

[4] Notice that RENDEZVOUS has a trivial solution in FSYNC: a robot moves to the halfway point to the other robot. In both SSYNC [57] and ASYNC [13], this *move-to-half* strategy guarantees only *convergence.*

[5] The tilted compasses are said to be *fully variable* if the actual tilt of each compass may vary at any time (but always with no more than ϕ from the global coordinate system); they are *semi-variable* if the tilt of each compass may vary (but no more than ϕ) between successive cycles, but it does not change during a cycle.

Gathering and Convergence. The GATHERING problem has been extensively investigated both experimentally and theoretically in the *unlimited visibility* setting, that is assuming that the entities are capable to sense the entire space. As stated above, when no additional assumptions are made in the model, there is no deterministic solution to the GATHERING problem in SSYNC. However, CONVERGENCE is possible even in ASYNC: The robots get closer to a gathering point, but never reach it in finite time. One quite simple and effective convergence solution in ASYNC exploits the *Center of Gravity* of the robots [13]. With the strongest assumption of *unlimited mobility* (all robots always reach their destinations when performing a *Move*), convergence time in ASYNC can be improved [17].

Thus, the GATHERING problem has a solution only adding additional assumptions. The most common assumption is that of *multiplicity detection*: a robot is able to detect whether a point on the plane is occupied by more than one robot. With this assumption, there exists solutions in both SSYNC [57] and ASYNC [12]. Another capability that has also been considered is a stronger form of multiplicity detection, where robots can detect the exact number of robots located at a given position [25]. Adding this capability, it is impossible to solve the problem for all possible initial configurations containing an even number of robots; however the robots can gather from an arbitrary configuration with n robots, when n is odd. In this case, *initial* configurations include also configurations containing more than one robot on the same point. Note that, since this algorithm is correct starting from all possible configurations provided n is odd (even the ones containing more than one robot), it is truly self-stabilizing.

In contrast, the multiplicity detection is not used in the solution described in [11]; however, it is assumed that the robots can rely on an unlimited amount of memory: the robots are said to be *non-oblivious*. In other words, the robots have the capability to store the results of all computations since the beginning, and freely access to these data and use them for future computations.

Furthermore, in SSYNC agreement on chirality and unlimited mobility suffice for making the problem solvable, even with variable tilted compasses, if the tilt of the local compasses is $\phi < \frac{\pi}{4}$ [36];

A different setting that has been studied is when robots have *limited visibility*: in this scenario, an obvious necessary condition to solve the problem, is that at the beginning of the computation the *visibility graph* (having the robots as nodes and an edge (r_i, r_j) if r_i and r_j are within viewing distance) is connected [2,30]. In [2] the proposed protocol solves the CONVERGENCE problem. In [30], the authors present an algorithm that let the robots to gather in a finite number of cycles. However, in this case the robots can rely on the presence of a common coordinate system: that is, they share a compass.

With limited visibility, the CONVERGENCE problem has been studied in FSYNC when the robots operate in a *non-convex* region (of which they have no map) [33]; in ASYNC with a limited form of asynchrony [42], where the time spent by a robot in the *Look*, and *Compute* states is bounded by a globally predefined amount, while the time spent in the *Move* state is bounded by a locally predefined quantity (not necessarily the same for each robot); and in ASYNC under a *1-fair scheduler* [40]: Between two successive activations of each robot *r*, all other robots have been activated at most once (as a consequence, from the moment *r* observes the current situation to the moment it finishes its movement, no other robot performs more than one *Look*).

The GATHERING problem has been also investigated in the context of robots *failures*. In this context, the goal is for the non-faulty robots to gather regardless of the action taken by the faulty ones. Two types of robot faults were investigated by Peleg et al. [1]: *crash* failure, in which the robot stops any activity and will no longer execute any computational cycle; and the *byzantine* failure, in which the robot acts arbitrarily and possibly maliciously.

In [14] it is analyzed the case of systems where the robots have inaccuracies in sensing the positions of other robots, in computing the next destination point, and in moving towards the computed destination. The authors provide a set of limitations on the amount of inaccuracies allowing convergence; hence, they present an algorithm for convergence under bounded measurement, movement and calculation errors. In [43], the case of *radial errors* has also been considered.

Finally, beside the inaccuracies in the compasses that have already been cited above (*tilted* compasses), with *eventually consistent compasses* (i.e., transient errors on the compasses), the GATHERING problem has also been studied in SSYNC, with robots that agree on chirality: in this case, it has been proven that the robots can gather in finite time [53].

Near-Gathering A problem that is very close to the CONVERGENCE problem is NEAR-GATHERING, where a set of robots with limited visibility, at the beginning arbitrarily placed in the plane on distinct positions, are required to get close enough to each other, without any collisions. In particular, in finite time, the robots are required to move within distance ε from each other for some predefined ε. This problem is particularly useful to overcome the limitations introduced by having robots with limited sensing capabilities: in fact, once they are *close enough*, all robots can see each other, hence they can operate as they had unlimited visibility power. This problem has been recently solved in ASYNC for robots with consistent compass [48].

4   Dynamic Problems: Flocking and Capture

In this set of problems, the robots dynamically move, and there is really no ending in the robots' tasks. Let us consider the FLOCKING problem first: There are mainly two versions of this problem. In the first one, there are two kinds of robots in the environment: the *leader L*, and the *followers* (this scenario is also called *guided flocking*). The leader acts independently from the others, and it can be assumed that it is driven by an human pilot. The followers are required to follow the leader wherever it goes (*following*), while keeping a formation they are given in input (*flocking*). In this context, a formation is simply a pattern described as a set of points in the plane, and all the robots have the same formation in input.

In [35], an algorithm solving this problem in ASYNC has been tested by using computer simulation; the algorithm assumes no agreement. All the experiments demonstrated that the algorithm is well behaved, and in all cases the followers were able to assume the desired formation and to maintain it while following the leader along its route. Moreover, the obliviousness of the algorithm contributes to this result, since the followers do not base their computation on past leader's positions.

In the second version of the problem, also known as *homogeneous flocking*, there is no exogenous source (i.e., no guide) and every robot knows the trajectory: The path along which the flock has to move is known in advance to every robot (e.g., [6, 54, 55]).

Finally, if the leader is considered an "enemy" or "intruder", and the pattern surrounds it, the problem is known as CAPTURE (or *intruder*). A protocol that assumes no agreement and solves the problem in ASYNC has been presented in [34]. The proposed algorithm exhibits remarkable robustness, and numeric simulations indicate that the intruder is efficiently captured in a relatively short time and kept surrounded after that, as desired. Furthermore, the solution is self-stabilizing. In particular, any external intervention (e.g., if one or more of the cops are stopped, slowed down, knocked out, or simply faulty) does not prevent the completion of the task.

5   New Directions

Computing with Colors. A new direction of investigation that just started being explored is the introduction in the model of some form of direct communication. The first attempt in this direction is in [19], where the robots make *visible* to their fellows $O(1)$ persistent bits [19]: Each robot is equipped with a light bulb that can display a constant numbers of different colors; the colors are visible to all other robots, and are persistent, that is, the light bulbs are not automatically reset at the end of each cycle. Thus, they can be used to remember states and to communicate. Apart from these lights, the robots are oblivious in all other respects.

Studies in this direction just started, and here is a brief summary of the major results obtained so far.

Colored ASYNC *versus* SSYNC. The presence of lights with visible colors is undoubtably a very powerful computational tool even if just constant in number. Indeed, it can overcome the limitations of ASYNC making the robots strictly more powerful than traditional SSYNC robots, as we see in the following. In fact, it has been shown that asynchronous robots with lights are *at least as powerful as* semi-synchronous ones: the proof consists of a protocol that allows to execute any semi-synchronous algorithm in an asynchronous setting, each robot using a light with a constant number of colours [19].

There are problems that robots cannot solve *without* visible bits, even if they are semi-synchronous, but can be solved with $O(1)$ visible bits even if the robots are asynchronous [19]. One such a problem is *rendezvous*, i.e., the gathering of two robots; from previous Sect. 3, we know that this problem is not solvable in SSYNC. However, this problem can be solved if the robots have $O(1)$ colors.

Hence, these two results lead to conclude that asynchronous robots endowed with $O(1)$ *visible lights* are strictly more powerful than semi-synchronous robots without any light [19].

Colored ASYNC *versus* FSYNC. The relationship between FSYNC and *Colored* ASYNC is less understood. What is known is that asynchronous robots, if empowered with both a constant number of visible lights and the ability to remember a single snapshot from the past, become at least as powerful as traditional fully synchronous robots [19].

Interestingly, there are problems that can be solved in ASYNC with three colours and one past snapshot, but are not solvable in FSYNC without additional information. This is the case, for example, of the BLINKING problem, which requires $n > 2$ robots to perform subtasks T_1 and T_2 repeatedly in alternation. In T_1, the robots must form a circle, i.e. each robot lies on a distinct point on the same circle $\mathcal{C}$ of radius $Rad > 0$; while in T_2, the robots must gather at a single point.

The presence of a problem not solvable in FSYNC but solvable in ASYNC with lights and one past snapshot, leads to the following conclusion: Asynchronous robots, endowed with $O(1)$ *visible lights* and able to remember a single *snapshot*, are strictly more powerful than fully-synchronous oblivious robots without any lights [19].

This is to be contrasted with the fact that, without lights, ASYNC robots are not even as powerful as SSYNC, even if they remember an unlimited number of previous snapshots [50].

Solid Robots. In the standard model, the robots are viewed as points, i.e., they are *dimensionless*. An interesting variant of the model is to consider entities that occupy a physical space of some size; that is, the entities have a solid dimension. These robots, called *solid* (or *fat*), are assumed to have a common unit distance and are viewed as circular disks of a given diameter. The disks of two robots can touch but cannot overlap. Moreover, it is assumed that, if during its movement a robot collide with another, its movement stops (*fail-stop collision*).

The robots' visibility might be affected by their solid dimension. If so, two robots r_1 and r_2 can see each other if there exist points x and y in the visibility radius of r_1 and r_2 respectively, such that the segment $[xy]$ does not contain any point of any other robot. Note that if a robot r_1 can see robot r_2, it can see some non-zero arc of its bounding circle and thus it can always compute its centre. Otherwise, if no visibility obstruction occurs, the robots are said to be *transparent*.

Very few problems have been investigated for solid robots. One of these is the GATHERING. Obviously, in the case of solid robots, the definition of gathering needs to be modified.

The robots are said to form a connected configuration in the plane if between any two points of any two robots there exists a polygonal line each of whose points belongs to some robot. Gathering is accomplished if the robots form some connected configuration and they are all visible to each other (and thus are aware that a connected configuration is achieved).

Adding a physical dimension to the robots significantly complicates the task, mainly because of the fact that their "body" can obstruct visibility. An example that shows one of the difficulty is given by a team of 4 robots whose centres are situated on two intersecting non-perpendicular lines, one robot in each of the four half- lines. The obvious algorithm that would work if the robots were points would be to have them move towards the intersection of the two lines, which is invariant under straight moves. However, it is easy to see that an adversary might have two robots meet in their move toward the centre, thus obstructing the view to the other two, without forming a connected configuration. In general, the lack of full visibility due to obstruction, prevent the robots from being able to compute easily an invariant point.

For the gathering of solid robots, currently there are only solutions for very small teams; in fact, no gathering algorithm is known for $n > 4$ non-transparent solid robots [18]. Furthermore, these algorithms are not collision-free and they rely on the fail-stop collision assumption to work.

In [9] it is presented an algorithm that works for $n \geq 5$ robots that are solid but transparent. The robots must be initially placed in an asymmetry configuration (so that a leader can be elected) and the desired gathering pattern is a circular layered structure of robots with the elected leader in the center.

In [16] gathering by solid robots is considered in a different setting. Each robot is given in input the position of the gathering point in its own coordinate system. All robots have the same dimension dim, and they are said to be gathered when they form a sphere with minimum radius around the predefined gathering point. Robots have *limited visibility*, large enough to avoid collisions (thus, a visibility radius $V \geq 2 \cdot dim$ is sufficient), and they operate in FSYNC.

Solid robots have been also studied in the context of circle formation, in [56, 58] for robots with unlimited visibility, and in [45] for mobile robots whose vision is not only limited but also *directional*.

Simulation Environments. A promising area of research on these topics is represented by the development of computer simulation environments dedicated to autonomous mobile robots. Several studies can be found in the literature right on this track [2, 34, 35, 56, 58]. All these simulation environments are specifically designed and developed for a particular problem: for instance, the one in [58] for the circle formation; the one in [35] for the flocking problem; the one in [34] for the intruder problem.

Recently, there has been a first attempt in designing a modular simulation environment to test and execute generic protocols for the autonomous mobile robots addressed in this paper: SYCAMORE [59]. In this environment, the protocol of a robot is defined as a plugin given in input to the simulation engine, and it can be easily set to simulate both 2D an 3D scenarios.

6   Conclusions

In this paper, we surveyed a number of recent results on the interplay between robots capabilities and solvability of problems. The goal of these studies is to gain a better understanding of the power of distributed control from an algorithmic point of view. The area is quite young, thus still offers many research quests. First, one outstanding theoretical open problem: no solution is still known for the GATHERING problem where the robots have limited visibility and no agreement; actually, it is not even clear whether the problem is solvable (a similar problem stands for the NEAR-GATHERING).

Then, operating capabilities of our robots are quite limited: New research directions can be taken by expanding the capabilities of the robots, in the attempt of better modeling the real robots. It would be interesting to look at models where the robots have more complex capabilities, e.g.: the robots have some kind of direct communication capabilities (besides the use of lights); the robots are distinct and externally identifiable; etc. Little is known about the solvability of other problems like spreading and exploration (used to build maps of unknown terrains), about the physical aspects of the models, such as those related to energy saving issues, and about the relationships between geometric problems and classical distributed computations. In the area of reliability and fault-tolerance, lightly faulty snapshots, a limited and directional (i.e., not 360°) range of visibility, obstacles that limit the visibility and that moving robots must avoid or push aside, as well as robots that appear and disappear from the scene clearly are all topics that have not yet been studied.

We believe that investigations in these areas will provide useful insights on the ability of weak robots to solve complex tasks.

Acknowledgements. The author would like to thank Paola Flocchini and Nicola Santoro for their help and suggestions in the preparation of this paper.

References

1. Agmon, N., Peleg, D.: Fault-tolerant gathering algorithms for autonomous mobile robots. SIAM J. Comput. **36**, 56–82 (2006)
2. Ando, H., Oasa, Y., Suzuki, I., Yamashita, M.: A distributed memoryless point convergence algorithm for mobile robots with limited visibility. IEEE Trans. Robot. Autom. **15**(5), 818–828 (1999)
3. Balch, T., Arkin, R.C.: Behavior-based formation control for multi-robot teams. IEEE Trans. Robot. Autom. **14**, 926–939 (1998)
4. Beckers, R., Holland, O.E., Deneubourg, J.L.: From local actions to global tasks: stigmergy and collective robotics. In: Artificial Life IV, 4th International Workshop on the Synthesis and Simulation of Living Systems (1994)
5. Beni, G., Hackwood, S.: Coherent swarm motion under distributed control. In: Proceedings of the DARS'92, pp. 39–52 (1992)
6. Canepa, D., Potop-Butucaru, M.G.: Stabilizing flocking via leader election in robot networks. In: Masuzawa, T., Tixeuil, S. (eds.) SSS LNCS. 2007, vol. 4838, pp. 52–66. Springer, Heidelberg (2007)

7. Cao, Y.U., Fukunaga, A.S., Kahng, A.B., Meng, F.: Cooperative mobile robotics: antecedents and directions. In: International Conference on Intelligent Robots and System, pp. 226–234 (1995)
8. Chatzigiannakis, I., Markou, M., Nikoletseas, S.E.: Distributed circle formation for anonymous oblivious robots. In: Ribeiro, C.C., Martins, S.L. (eds.) WEA 2004. LNCS, vol. 3059, pp. 159–174. Springer, Heidelberg (2004)
9. Chaudhuri, S.G., Mukhopadhyaya, K.: Gathering asynchronous transparent fat robots. In: Janowski, T., Mohanty, H. (eds.) ICDCIT 2010. LNCS, vol. 5966, pp. 170–175. Springer, Heidelberg (2010)
10. Chen, Q., Luh, J.Y.S.: Coordination and control of a group of small mobile robots. In: Proceedings of the IEEE International Conference on Robotics and Automation, pp. 2315–2320 (1994)
11. Cieliebak, M., Gathering non-oblivious mobile robots. In: Proceedings of the 6th Latin American Symposium on Theoretical Informatics, pp. 577–588 (2004)
12. Cieliebak, M., Flocchini, P., Prencipe, G., Santoro, N.: Distributed computing by mobile robots: gathering. SIAM J. Comput. **41**(4), 829–879 (2012)
13. Cohen, R., Peleg, D.: Convergence properties of the gravitational algorithm in asynchronous robot systems. SIAM J. Comput. **34**, 1516–1528 (2005)
14. Cohen, R., Peleg, D.: Convergence of autonomous mobile robots with inaccurate sensors and movements. In: Durand, B., Thomas, W. (eds.) STACS 2006. LNCS, vol. 3884, pp. 549–560. Springer, Heidelberg (2006)
15. Cohen, R., Peleg, D.: Local spreading algorithms for autonomous robot systems. Theor. Comput. Sci. **399**, 71–82 (2008)
16. Cord-Landwehr, A., et al.: Collisionless gathering of robots with an extent. In: Černá, I., Gyimóthy, T., Hromkovič, J., Jefferey, K., Královič, R., Vukolić, M., Wolf, S. (eds.) SOFSEM 2011. LNCS, vol. 6543, pp. 178–189. Springer, Heidelberg (2011)
17. Cord-Landwehr, A., et al.: A new approach for analyzing convergence algorithms for mobile robots. In: Aceto, L., Henzinger, M., Sgall, J. (eds.) ICALP 2011, Part II. LNCS, vol. 6756, pp. 650–661. Springer, Heidelberg (2011)
18. Czyzowicz, J., Gasieniec, L., Pelc, A.: Gathering few fat mobile robots in the plane. Theor. Comput. Sci. **410**(6–7), 481–499 (2009)
19. Das, S., Flocchini, P., Prencipe, G., Santoro, N., Yamashita. M.: The power of lights: synchronizing asynchronous robots using visible bits. In: Proceedings of the 32nd International Conference on Distributed Computing Systems (ICDCS), pp. 506–515 (2012)
20. Debest, X.A.: Remark about self-stabilizing systems. Comm. ACM **238**(2), 115–117 (1995)
21. Défago, X., Konagaya, A.: Circle formation for oblivious anonymous mobile robots with no common sense of orientation. In: Workshop on Principles of Mobile Computing, pp. 97–104 (2002)
22. Défago, X., Souissi, S.: Non-uniform circle formation algorithm for oblivious mobile robots with convergence toward uniformity. Theor. Comput. Sci. **396**(1–3), 97–112 (2008)
23. Dieudonné, Y., Labbani-Igbida, O., Petit, F.: Circle formation of weak mobile robots. ACM Trans. Auton. Adapt. Syst. **3**(4), 16:1–16:20 (2008)
24. Dieudonné, Y., Petit, F.: Circle formation of weak robots and Lyndon words. Inf. Process. Lett. **4**(104), 156–162 (2007)
25. Dieudonné, Y., Petit, F.: Self-stabilizing gathering with strong multiplicity detection. Theor. Comput. Sci. **428**(13), 47–57 (2012)

26. Donald, B.R., Jennings, J., Rus, D.: Information invariants for distributed manipulation. Int. J. Robot. Res. **16**(5), 63–73 (1997)
27. Durfee, E.H.: Blissful ignorance: knowing just enough to coordinate well. In: ICMAS, pp. 406–413 (1995)
28. Flocchini, P., Prencipe, G., Santoro, N.: Distributed Computing by Oblivious Mobile Robots. Morgan & Claypool, San Rafeal (2012)
29. Flocchini, P., Prencipe, G., Santoro, N., Widmayer, P.: Hard tasks for weak robots: the role of common knowledge in pattern formation by autonomous mobile robots. In: Proceedings of the 10th International Symposium on Algorithm and Computation, pp. 93–102 (1999)
30. Flocchini, P., Prencipe, G., Santoro, N., Widmayer, P.: Gathering of asynchronous robots with limited visibility. Theor. Comput. Sci. **337**, 147–168 (2005)
31. Flocchini, P., Prencipe, G., Santoro, N., Widmayer, P.: Arbitrary pattern formation by asynchronous oblivious robots. Theor. Comput. Sci. **407**(1–3), 412–447 (2008)
32. Fukuda, T., Kawauchi, Y., Asama, H., Buss. M.: Structure decision method for self organizing robots based on cell structures-CEBOT. In: Proceedings of the IEEE International Conference on Robotics and Automation, vol.2, pp. 695–700 (1989)
33. Ganguli, A., Cortés, J., Bullo, F.: Multirobot rendezvous with visibility sensors in nonconvex environments. IEEE Trans. Robot. **25**(2), 340–352 (2009)
34. Gervasi, V., Prencipe, G.: Robotic Cops: the intruder problem. In: Proceedings of the IEEE Conference on Systems, Man and Cybernetics, pp. 2284–2289 (2003)
35. Gervasi, V., Prencipe, G.: Coordination without communication: the case of the flocking problem. Discrete Appl. Math. **143**, 203–223 (2004)
36. Izumi, T., Katayama, Y., Inuzuka, N., Wada, K.: Gathering autonomous mobile robots with dynamic compasses: an optimal result. In: Pelc, A. (ed.) DISC 2007. LNCS, vol. 4731, pp. 298–312. Springer, Heidelberg (2007)
37. Izumi, T., Souissi, S., Katayama, Y., Inuzuka, N., Defago, X., Wada, K., Yamashita, M.: The gathering problem for two oblivious robots with unreliable compasses. Siam J. Comput. **41**(1), 26–46 (2012)
38. Jung, D., Cheng, G., Zelinsky, A.: Experiments in realising cooperation between autonomous mobile robots. In: ISER (1997)
39. Katreniak. B.: Biangular circle formation by asynchronous mobile robots. In: Proceedings of the 12th International Colloquium on Structural Information and Communication Complexity (2005)
40. Katreniak, B.: Convergence with limited visibility by asynchronous mobile robots. In: Kosowski, A., Yamashita, M. (eds.) SIROCCO 2011. LNCS, vol. 6796, pp. 125–137. Springer, Heidelberg (2011)
41. Kawauchi, Y., Inaba, M., Fukuda, T.: A principle of decision making of cellular robotic system (CEBOT). In: Proceedings of the IEEE Conference on Robotics and Automation, pp. 833–838 (1993)
42. Lin, J., Morse, A.S., Anderson, B.D.O.: The multi-agent rendezvous problem. Part 2: the asynchronous case. SIAM J. Control Optim. **46**(6), 2120–2147 (2007)
43. Martínez, S.: Practical multiagent rendezvous through modified circumcenter algorithms. Automatica **45**(9), 2010–2017 (2009)
44. Matarić, M.J.: Interaction and intelligent behavior. Ph.D. thesis. MIT, May 1994
45. Miyamae, T., Ichikawa, S., Hara, F.: Emergent approach to circle formation by multiple autonomous modular robots. J. Robot. Mechatron. **21**(1), 3–11 (2009)
46. Murata, S., Kurokawa, H., Kokaji, S.: Self-assembling machine. In: Proceedings of the IEEE Conference on Robotics and Automation, pp. 441–448 (1994)
47. Noreils, F.R.: Toward a robot architecture integrating cooperation between mobile robots: application to indoor environment. Int. J. Robot. Res. **12**, 79–98 (1993)

48. Pagli, L., Prencipe, G., Viglietta, G.: Getting close without touching. In: Even, G., Halldórsson, M. (eds.) SIROCCO 2012. LNCS, vol. 7355, pp. 315–326. Springer, Heidelberg (2012)
49. Parker, L.E.: On the design of behavior-based multi-robot teams. J. Adv. Robot. **10**(6), 547–578 (1996)
50. Prencipe, G.: The effect of synchronicity on the behavior of autonomous mobile robots. Theory Comput. Syst. **38**, 539–558 (2005)
51. Prencipe, G.: Impossibility of gathering by a set of autonomous mobile robots. Theor. Comput. Sci. **384**(2–3), 222–231 (2007)
52. Samia, S., Défago, X., Katayama, T.: Convergence of a uniform circle formation algorithm for distributed autonomous mobile robots. In: Journés Scientifiques Francophones (JSF), Tokio, Japan (2004)
53. Souissi, S., Défago, X., Yamashita, M.: Using eventually consistent compasses to gather memory-less mobile robots with limited visibility. ACM Trans. Auton. Adapt. Syst. **4**(1), 1–27 (2009)
54. Souissi, S., Yang, Y., Défago. X.: Fault-tolerant flocking in a k-bounded asynchronous system. In: Proceedings of 12th International Conference on Principles of Distributed Systems (OPODIS), pp. 145–163 (2008)
55. Souissi, S., Yang, Y., Défago, X., Takizawa, M.: Fault-tolerant flocking for a group of autonomous mobile robots. J. Syst. Softw. **84**, 29–36 (2011)
56. Sugihara, K., Suzuki, I.: Distributed algorithms for formation of geometric patterns with many mobile robots. J. Robot. Syst. **13**, 127–139 (1996)
57. Suzuki, I., Yamashita, M.: Distributed anonymous mobile robots: formation of geometric patterns. Siam J. Comput. **28**(4), 1347–1363 (1999)
58. Tanaka, O.: Forming a circle by distributed anonymous mobile robots. Technical report, Department of Electrical Engineering, Hiroshima University, Hiroshima, Japan (1992)
59. Volpi, V.: Sycamore: a 2D–3D simulation environment for autonomous mobile robots algorithms. https://code.google.com/p/sycamore/
60. Wang, P.K.C.: Navigation strategies for multiple autonomous mobile robots moving in formation. J. Robot. Syst. **8**(2), 177–195 (1991)

Token Dissemination
in Geometric Dynamic Networks

Sebastian Abshoff$^{(\boxtimes)}$, Markus Benter, Andreas Cord-Landwehr,
Manuel Malatyali, and Friedhelm Meyer auf der Heide

Heinz Nixdorf Institute & Computer Science Department, University of Paderborn,
Fürstenallee 11, 33102 Paderborn, Germany
{abshoff,benter,phoenixx,malatya,fmadh}@hni.upb.de

Abstract. We consider the k-token dissemination problem, where k initially arbitrarily distributed tokens have to be disseminated to all nodes in a dynamic network (as introduced by Kuhn et al. STOC 2010). In contrast to general dynamic networks, our dynamic networks are unit disk graphs, i.e., nodes are embedded into the Euclidean plane and two nodes are connected if and only if their distance is at most R. Our worst-case adversary is allowed to move the nodes on the plane, but the maximum velocity $v_{\max}$ of each node is limited and the graph must be connected in each round. For this model, we provide almost tight lower and upper bounds for k-token dissemination if nodes are restricted to send only one token per round. It turns out that the maximum velocity $v_{\max}$ is a meaningful parameter to characterize dynamics in our model.

Keywords: Geometric dynamic networks · Token dissemination · Distributed computing

1 Introduction

Dynamic networks appear in many scenarios like peer-to-peer networks, mobile wireless ad-hoc networks or swarms of mobile robots. The dynamics in such models is diverse and different. Kuhn et al. [10] have introduced a very general model with the aim of understanding limitations and possibilities when coping with dynamics in networks, independent of specific application. In this paper, we look at special dynamics motivated by agents that move in the Euclidean plane and that are able to communicate with nearby agents only. More particularly, we look at dynamic unit disk graphs as they are often used to model ad-hoc networks or robotic networks. We are mainly interested in exploring the impact of a velocity limit of the agents on the time required to perform fundamental tasks such as token dissemination.

This work was partially supported by the German Research Foundation (DFG) within the Collaborative Research Centre "On-The-Fly Computing" (SFB 901), by the EU within FET project MULTIPLEX under contract no. 317532, and the International Graduate School "Dynamic Intelligent Systems".

P. Flocchini et al. (Eds.): ALGOSENSORS 2013, LNCS 8243, pp. 22–34, 2014.
DOI: 10.1007/978-3-642-45346-5_3, © Springer-Verlag Berlin Heidelberg 2014

The nodes of our geometric dynamic network are embedded into the Euclidean plane and two nodes are connected if and only if their distance is at most a constant R, which models the limited range of a wireless communication device. We consider a worst-case dynamic that is able to move the nodes within this plane. This worst-case dynamic is restricted by a maximal velocity parameter $v_{\max}$ and it must preserve connectivity of the network. In our model, we can prove lower and upper bounds for the k-token dissemination problem, which has also been studied by Kuhn et al. in a general model. In the k-token dissemination problem, k initially arbitrarily distributed tokens have to be disseminated by the nodes of the dynamic network such that each node receives all tokens and also decides that it has received all k tokens since we assume k is not known by the nodes beforehand. Note that solving the all-to-all token dissemination problem, where each node starts with exactly one token, implicitly solves the counting problem if the nodes' unique IDs are considered as tokens. While solving the token dissemination problem, a distributed algorithm must cope with the dynamic of the network, i.e., the changes of edges as induced by a worst-case dynamic that moves the nodes.

In our model, we restrain the dynamic network model by Kuhn et al. by introducing a geometry that gives a natural restriction of the power of a worst-case dynamic by geometric means. From this, we expect new insights into the complexity of distributed computational problems by using different techniques that exploit the geometry of the dynamic network. As a first step, both our lower and upper bounds for the k-token dissemination problem contain the maximal velocity parameter $v_{\max}$, i.e., they are bound by the characteristic value of the network dynamic. More precisely, we define a dynamic unit disk graph with maximal node velocity $v_{\max}$ and communication radius R and require connectivity w.r.t. a unit disk graph with radius 1. Our algorithm terminates after $\mathcal{O}(n(n+k) \cdot \min\{v_{\max}, R\} \cdot R^{-2})$ rounds if $R > 1$. Moreover, we present a lower bound of $\Omega(n \cdot k \cdot \min\{v_{\max}, R\} \cdot R^{-3})$ for randomized knowledge-based token-forwarding algorithms. Note that for $k = \Omega(n)$, the upper bound simplifies to $\mathcal{O}(n \cdot k \cdot \min\{v_{\max}, R\} \cdot R^{-2})$ and the upper and the lower bound become almost tight.

2   The Geometric Dynamic Network Model

In this paper, we consider the following dynamic network model adapted from Kuhn et al. [10, 12, 17]: we assume a dynamic graph with a fixed set V of n nodes, and a discrete, synchronous time model. Each node v is identified by a unique ID, assigned by some injective function id : $V \rightarrow \{1, \ldots, \mathrm{poly}(n)\}$. In round r, the dynamic graph has some edge set E_r, forming the graph $G_r = (V, E_r)$. We assume local broadcast communication, i.e., a message sent by node u in round r is delivered to u's neighbors in round $r+1$. Therefore, when sending a message in round r, a node usually does know to which neighbors the message will be delivered. In this paper, the message each node can send via local broadcast communication is limited to one token per round. Kuhn et al. introduced the concept of T-interval connectivity as a reasonable restriction of the dynamic: For each time interval I of length $T \geq 1$, there must be some stable and connected subgraph in all graphs G_r with $r \in I$. If $T = 1$, this just means the graph must be connected in each round r.

Our modifications address a dynamic model motivated by geometric mobility as it appears, e.g., in swarm robotics. Here, we assume that each node v in each round r has a position $p_r(v) \in \mathbb{R}^2$ (however, our results also hold for $\mathbb{R}^3$). The distance between two nodes u and v in round r is denoted by $d_r(u,v) = |p_r(u) - p_r(v)|$. Then, for each round r, we define G_r as the unit disk graph with communication radius R. We omit the round parameter r when the round is clear from context. In addition, $1 \leq R \leq n$ holds throughout the paper for technical reasons. Furthermore, we assume that the maximum velocity of each node is bounded by a parameter $v_{\max} > 0$, i.e., the position of a node changes at most by a distance $v_{\max}$ from round to round. Such a model was also considered by Bienkowski et al. [1], and it is often (implicitly or explicitly) assumed for designing local strategies for robotic formation problems (for a survey see [11]).

Our results require a somewhat stronger notion of connectivity than in the general model by Kuhn et al.: We demand that the graph in round r is connected even if we restrict the communication radius to 1 instead of R. To distinguish these graphs, we talk about the *communication graph* G_r if radius R is used, and about the *connectivity graph* G_r' if radius 1 is used. Thus, we require that the connectivity graph G_r' is connected in each round r. This geometric model gives rise to another natural restriction of dynamics. A graph G_r is called C-connected if at least C edges have to be removed to transform G_r into a disconnected graph.

The focus of our paper lies on the k-token dissemination problem. In this problem, each node u in the network receives as input $I(u)$ a possibly empty subset of tokens such that $\left| \bigcup_{v \in V} I(v) \right| = k$. Then, the nodes have to disseminate these tokens such that each node eventually knows all k tokens and then explicitly terminates (i.e., it outputs the result and does not send/receive any further messages). Here, k is not known by the nodes beforehand. Additionally, we examine the implications of our results for the problem of counting, which is to determine the exact number of nodes in the network.

We show a result for a restricted class of algorithms that is called *knowledge-based token-forwarding algorithms* (cf. Kuhn et al.): let $A_u(r)$ denote the set of messages node u has received by the beginning of round r including its input $I(u)$. A *token-forwarding algorithm* requires each node to send only one pure token from $A_u(r)$ (without modification and without annotation) or the empty message, and it must not terminate before it has received all k tokens. A token-forwarding algorithm is called *knowledge-based* if the distribution that determines which token is sent by u in round r is a function only of its unique ID $\mathrm{id}(u)$, $A_u(0), \ldots, A_u(r-1)$ and the sequence of u's coin tosses up to round r (including r). Many natural strategies can be found in this class of knowledge-based token-forwarding algorithms, e.g., strategies like sending a known token sampled uniformly at random.

3   Related Work

Dynamic networks, where the set of edges in the network may change arbitrarily and in an adversarial way from round to round as long as the graph is strongly connected in each round, were introduced by Kuhn et al. [10,12,17]. In each round, each node may send a message of size $\mathcal{O}(\log n)$ bits that is delivered to all neighboring nodes in the following round. Computation in their model requires termination. On the one hand, for the k-token dissemination problem in T-interval connected dynamic networks, Kuhn et al. present a deterministic $\mathcal{O}(n(n + k)/T)$ token-forwarding algorithm. This algorithm can be used to obtain an $\mathcal{O}(n^2/T)$ algorithm for the counting problem. On the other hand, they give a $\Omega(nk/T)$ lower bound for the restricted class of knowledge-based token-forwarding algorithms and they provide an $\Omega(n \log k)$ lower bound for deterministic centralized token-forwarding algorithms.

Dutta et al. [5] improved the latter lower bound by Kuhn et al. to $\Omega(nk/\log n + n)$ for any randomized (even centralized) token-forwarding algorithm and showed for a weakly-adaptive adversary that k-token dissemination can be done in $\mathcal{O}((n + k) \log n \log k)$ w.h.p. Furthermore, they provide two polynomial time, randomized and centralized offline algorithms, one returns an $\mathcal{O}(n, \min\{k, \sqrt{k \log n}\})$ schedule w.h.p. and another one an $\mathcal{O}((n + k) \log^2 n)$ schedule w.h.p. if nodes can send a token along each edge per round. Using similar techniques, Haeupler and Kuhn [9] showed lower bounds if nodes are allowed to forward $b \leq k$ tokens or if they are only required to obtain a δ-fraction in T-interval connected dynamic networks and dynamic networks that are c-vertex connected in every round.

O'Dell and Wattenhofer [18] analyzed information dissemination problems in slightly different but worst-case adversarial models. Das Sarma et al. [19] developed randomized token-forwarding algorithms based on random walks on dynamic networks. Here, an oblivious adversary that is not aware of the random choices of the algorithm modifies the network. Haeupler and Karger go beyond the class of token-forwarding algorithms and send linear combinations of tokens. With this technique, they are able to solve the k-token dissemination problem in $\mathcal{O}(nk/\log n)$ rounds w.h.p. [8]. Brandes and Meyer auf der Heide [4] develop algorithms for counting if in addition every edge in the network fails with some probability. Michail et al. [16] studied computation in possibly disconnected dynamic networks and introduced temporal connectivity conditions. The same authors [15] looked into naming and counting in the absence of unique IDs in dynamic networks. Here, naming refers to the problem of generating unique IDs. Interestingly, they introduce a different communication model where the nodes in the network can send different, individual messages to their neighbors but without any information about their states.

The unit disk model has been extensively studied in the area of routing in wireless ad-hoc and sensor networks, in particular, in geographic routing algorithms. Geographic routing takes advantage of the availability of position information to decide which node becomes the next hop. Those algorithms assume that a node can get its own position using a location service such as GPS.

A worst-case optimal and average-case efficient geographic routing algorithm has been proposed by Kuhn et al. [13,14]. However, geographic routing focuses on single source routing while a well-designed token dissemination algorithm has to take congestion into account. For a broad overview about further routing algorithms in this area, we refer to Frey et al. [7].

Gossiping algorithms are a class of algorithms for distributed computation in arbitrary graphs following a simple principle:the nodes are initialized with some values, they continuously exchange these values and calculate new values based on the ones they received together with a problem specific function.

Boyd et al. [2,3] analyzed the mixing time of the averaging problem. Later on, Dimakis et al. [6] showed that the mixing time can be significantly improved in grid graphs and random geometric graphs, two communication models for realistic sensor networks. In a random geometric graph, the n sensor locations are chosen uniformly and independently in the unit square, and each pair of nodes is connected if their Euclidean distance is smaller than some constant transmission radius R. They proposed an algorithm that computes true average to accuracy $1/n^\alpha$ using $\mathcal{O}(n^{1.5}\sqrt{\log n})$ radio transmissions. This reduces the energy consumption by a factor of $\sqrt{n/\log n}$ compared to standard gossip algorithms.

4   Lower Bound on Token Dissemination in Geometric Dynamic Networks

In this chapter, we show that any knowledge-based token-forwarding algorithm needs $\Omega(n \cdot k \cdot \min\{v_{\max}, R\} \cdot R^{-3})$ rounds for solving the k-token dissemination problem in geometric dynamic networks. To do so, we follow a similar analysis like the one by Kuhn et al. for an $\Omega(nk)$ lower bound for dynamic networks with arbitrarily changing edges [10].

For the sake of a simple presentation, we first introduce our construction for the special case $R = 1$, i.e., the communication graph is equal to the connectivity graph. Later, this result will be generalized to the case $R \geq 1$.

Theorem 1. *If $R = 1$, then any knowledge-based token-forwarding algorithm for k-token dissemination requires $\Omega(n \cdot k \cdot \min\{v_{\max}, 1\})$ rounds to succeed with probability $> \frac{1}{2}$.*

Proof. We create the setting as follows: Initially, some node v_0 knows all k tokens and all other nodes do not know any token. As the token-forwarding algorithm is knowledge-based, the probability distribution of the tokens sent by v_0 does not depend on the dynamic graph. Let $r^* := \left\lfloor \frac{(n-4)k}{2L} \right\rfloor - 1$ for $L := \left\lceil \frac{1+\varepsilon}{2v_{\max}} \right\rceil$, which is used as the length of a row of nodes in our construction. ϵ is a suitably small chosen value. Then, by linearity of expectation and Markov's inequality, there is some infrequently sent token t that is sent $< \frac{n-4}{L}$ times by v_0 until round r^* with probability of at least $\frac{1}{2}$. For this, we define a dynamic such that v_0 cannot terminate by round r^* since some node must be unaware of t at this time.

All nodes are positioned as shown in Fig. 1: Except for the four nodes v_0, v_{n-1}, v_{n-2}, and v_{n-3}, all other nodes are assigned to horizontal rows where each

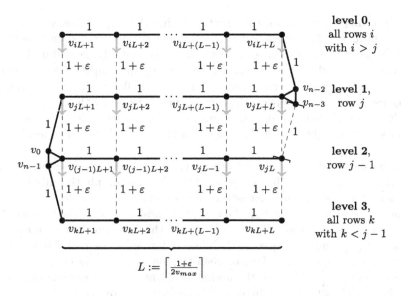

Fig. 1. Construction for $R = 1$ showing the positions of the nodes for a fixed j.

row consists of L nodes that are positioned on four levels. On each level, the nodes have a distance of exactly 1, which maximizes the distance between the nodes such that the row is still connected. The distance between two levels is at least $1 + \epsilon$.

Node v_0 is connected to $v_{(j-1)L+1}$, v_{jL+1}, and v_{n-1}, and positioned such that the distance between v_0 and v_{jL+1} is exactly 1. Analogously, node v_{n-1} is connected to $v_{(j-1)L+1}$, v_{kL+1}, and v_0, and positioned such that the distance between v_n and v_{kL+1} is exactly 1. We will see later that these nodes are essential for preserving connectivity during the movement of the nodes.

Similarly, node v_{n-2} is connected to v_{jL+L}, v_{iL+L}, and v_{n-3}, and positioned such that the distance between v_{n-2} and v_{iL+L} is exactly 1. In contrast to that, node v_{n-3} is only connected to v_{jL+L} and v_{n-2} but the distance between v_{n-3} and v_{jL} is slightly larger than 1. As we will see later, this is important to ensure that the infrequently sent token t cannot be learned by the nodes on level 0. Initially, one row $j = 0$ is at level 1 and all other rows $i > j$ are stacked at level 0. Level 2 and level 3 are not occupied by rows at this time.

When v_0 sends the token t, three rows start moving down. In particular, one row $j + 1$ at level 0, row j at level 1, and row $j - 1$ at level 2 start moving down with maximal relative velocity $2v_{\max}$[1] for the next L rounds until they reach level 1, 2, and 3, respectively. Once a row reaches level 3, it does not move any further and all rows $k < j - 1$ stack again. As soon as rows $j + 1, j, j - 1$ reach the next level, j can be incremented and Fig. 1 shows the current situation. If v_0 again sends token t, the described procedure repeats and three rows move down.

---

[1] To upper bound the worst-case traveling distance for a fixed node pair u and v, we can w.l.o.g. assume that u is static while v moves with velocity of at most $2v_{\max}$.

The crucial property of our construction is that the graph is always connected while the infrequently sent token t never reaches the rows stacked on level 0. Let us first argue why the graph is always connected. Initially, level 0 and level 1 are occupied and the graph is connected. During movement, the location of node v_{n-2} ensures that the row moving between level 0 and level 1 is connected. Analogously, node v_{n-1} connects the row moving in between of level 2 and level 3. The placement of node v_0 and v_{n-3} ensures that the left side of the graph is connected to the right side of the graph via the row in between of level 1 and level 2.

Let us now consider the second property: level 0 never gets the infrequently sent token t. According to the definition of the dynamic graph, the only possibility for level 0 to get t is via the row moving between level 1 and level 2 via node v_{n-3} or v_{n-2}. Since t is sent less than $\lfloor \frac{n-4}{L} \rfloor$ times, the token needs more than L rounds to cross one row. Thus, according to the definition of the movement, the nodes v_{jL+L} and v_{n-3} become disconnected at least one round before t can be sent from v_{jL+L} to v_{n-3}. This is when the row arrives at level 2.

Since there are $\lfloor \frac{n-4}{L} \rfloor$ rows, the nodes on at least one row are unaware of the token. Hence, $r^* = \Omega(n \cdot k \cdot \min\{v_{\max}, 1\})$ rounds are required.   □

Next, we extend this construction for an arbitrary communication radius $R \geq 1$:

Theorem 2. *If $R \geq 1$, then any knowledge-based token-forwarding algorithm for k-token dissemination requires $\Omega(n \cdot k \cdot \min\{v_{\max}, R\} \cdot R^{-3})$ rounds to succeed with probability $> \frac{1}{2}$.*

Proof. As before, we want to find a token that is sent infrequently over some cut. Yet, for $R \geq 2$, the communication graph is $\lfloor R \rfloor$-connected in each round, i.e., there is no single cut vertex as in the construction before for $R = 1$. Therefore, multiple nodes $v_0, v_{n-1}, \ldots, v_{n-\lfloor R \rfloor + 3}$ initially receive all k tokens such that the probability distribution of the tokens sent by them does not depend on the dynamic graph. Define $r^* := \lfloor \frac{(n-cR)k}{2L\lfloor R \rfloor} \rfloor - 1$ for $L := \lceil \frac{3(R+\varepsilon)}{2v_{\max}} \rceil \cdot (R+1)$ and some constant $c \in \mathbb{N}^+$ that is specified later. Then, by linearity of expectation and Markov's inequality, there is some infrequently sent token t which is sent $< \frac{(n-cR)k}{L\lfloor R \rfloor}$ times by all nodes $v_0, v_{n-1}, \ldots, v_{n-\lfloor R \rfloor + 3}$ until round r^* with probability of at least $\frac{1}{2}$. We present a dynamic such that all nodes $v_0, \ldots, v_{\lfloor R \rfloor - 1}$ cannot terminate by round r^* since there still is a node that is unaware of t at this time.

All nodes are positioned as shown in Fig. 2: similar to the construction in Theorem 1, except for $c := 2 \left(\lceil 3(R+\epsilon) \rceil - 2 \right)$ nodes $v_0, v_{n-1}, \ldots, v_{n-2\lfloor 3(R+\epsilon) \rfloor + 3}$, all other nodes are assigned to horizontal rows where each row consists of L nodes that are positioned on six levels. The distance between $v_{n-2\lceil 3(R+\epsilon)\rceil + 3}$ and the position below v_{jL} is slightly greater than one. Initially, one row $j = 0$ is at level 1 and all other rows $i > j$ are stacked at level 0. Levels $2, \ldots, 5$ are not occupied by rows at this time.

When one of the nodes $v_0, v_{n-1}, \ldots, v_{n-\lfloor R \rfloor + 3}$ sends the token t, one row at level 0 and all rows on levels $1, \ldots, 4$ start moving down with maximal relative

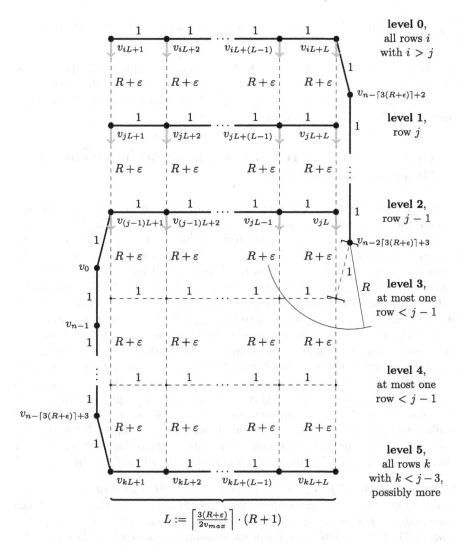

$$L := \left\lceil \frac{3(R+\epsilon)}{2v_{max}} \right\rceil \cdot (R+1)$$

Fig. 2. Construction for $R \geq 1$ showing the positions of the nodes for a fixed j.

velocity $2v_{\max}$ for the next L rounds until they reach the next level. The row from level 0 stops at level 1, but all other rows continue moving until they reach level 5, where they do not move any further and stack again. As soon as row $j+1$ reaches level 0, j can be incremented and Fig. 2 shows the current situation until t is sent again. If any of the nodes $v_0, v_{n-1}, \ldots, v_{n-\lfloor R \rfloor +3}$ sends the token t again, the described procedure repeats and further rows move down.

Observe that the graph is always connected and that token t cannot reach $v_{n-\lceil 3(R+\epsilon) \rceil +3}$ or any node above since it is sent less than $\left\lfloor \frac{(n-cR)k}{L \lfloor R \rfloor} \right\rfloor$ times and the token needs more than L rounds to cross one row. Since there are

$\lfloor \frac{n-cR}{L} \rfloor$ rows, the nodes on at least one row are unaware of the token. Therefore, $r^* = \Omega(n \cdot k \cdot \min\{v_{\max}, R\} \cdot R^{-3})$ rounds are required.   □

Remark 1. Our results implies that the lower bound by Kuhn et al. [10] already holds for a much more restricted model of dynamics: If we choose $R = 1$ and $v_{\max}$ constant, e.g. $v_{\max} = 1$, then we achieve the $\Omega(nk)$ lower bound for knowledge-based token-forwarding algorithms.

5   Upper Bound on Token Dissemination in Geometric Dynamic Networks

In this chapter, we present a k-token dissemination algorithm for geometric dynamic networks with bounded maximum velocity $v_{\max}$. The algorithm is basically an extension of the algorithm by Kuhn et al. which allows to solve k-token dissemination under arbitrary edge dynamics in $\Theta(n(n + k))$ rounds. Under the restriction of $R > 1$, it is possible to speed up the algorithm up to $\Theta(n(n+k) \cdot \min\{v_{\max}, R\} \cdot R^{-1})$. Moreover, if $R \geq 2$, then the $\Theta(R)$-connectivity of the communication graph can be exploited to get another speed-up of $\Theta(R)$, i.e., the algorithm needs $O(n(n + k) \cdot \min\{v_{\max}, R\} \cdot R^{-2})$ rounds in total.

Let us first sketch the dissemination algorithm by Kuhn et al. for $2T$-interval connected graphs. By definition of $2T$-interval connectivity, there is a spanning connected subgraph for at least $2T$ rounds. This subgraph is used to establish a pipelining effect such that at least the T smallest tokens are disseminated to all nodes in $\Theta(n)$ rounds. The algorithm proceeds in $\lfloor \frac{n}{T} \rfloor$ phases, where each phase consists of $2T$ rounds.[2] In each round of each phase, each node sends the smallest token it has not yet sent in this phase. To disseminate k tokens, this procedure can be repeated $\lceil \frac{k}{T} \rceil$ times. We restate the following results are either provided in the paper by Kuhn et al. or that directly follow from their results.

Theorem 3 ([10,17]). *For $T \geq 1$, in a T-interval connected dynamic network with arbitrarily changing edges, the algorithms by Kuhn et al. for k-token dissemination and counting can be sped up by a factor of T, i.e., they need $\Theta(n(n + k) \cdot T^{-1})$ rounds for k-token dissemination and $\Theta(n^2 \cdot T^{-1})$ rounds for counting.*

Theorem 4 ([10,17]). *For $T, C \geq 1$, in a T-interval connected dynamic network with arbitrarily changing edges where the stable subgraph is C-connected, the algorithms by Kuhn et al. for k-token dissemination and counting can be sped up by a factor of $T \cdot C$, i.e., they need $\Theta(n(n + k) \cdot T^{-1} \cdot C^{-1})$ rounds for k-token dissemination and $\Theta(n^2 \cdot T^{-1} \cdot C^{-1})$ rounds for counting.*

Note that it is assumed that T and C are known by the nodes. Furthermore, we would like to stress that it is not enough that $G(r)$ is C-connected in each

---

[2] Note that n is not known by the nodes beforehand but as described by Kuhn et al. [10] it can be determined involving the dissemination procedure itself using different estimates for n.

round r. To make use of the pipelining effect it is also important that the stable subgraph is C-connected.

In the following, we show that our geometric dynamic networks are $\Theta(R \cdot v_{\max}^{-1})$-interval connected if $R > 1$ and that the stable subgraphs are $\Theta(R)$-connected if $R \geq 2$.

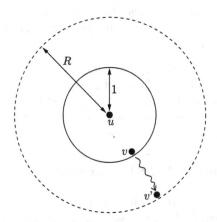

Fig. 3. After v moved to position v', the nodes u and v are still connected.

Lemma 1. *Assume the nodes of a geometric dynamic network move with maximum velocity $v_{\max}$. Then, the geometric dynamic network is $\left\lfloor \frac{(R-1)}{2 \cdot v_{\max}} \right\rfloor + 1$-interval connected.*

Proof. Consider a fixed node pair u and v which is connected in the connectivity graph of round r. Observe that the distance between two nodes can increase by at most $2v_{\max}$ per round. Thus, nodes that are connected in the connectivity graph (radius 1) stay connected in the communication graph (radius R) for at least $\left\lfloor \frac{(R-1)}{2 \cdot v_{\max}} \right\rfloor$ further rounds. This implies the lemma (cf. Fig. 3). $\square$

Lemma 2. *Assume the nodes of a geometric dynamic network move with maximum velocity $v_{\max}$. Then, the geometric dynamic network contains a spanning $\left\lfloor \frac{1}{2}R \right\rfloor$-connected subgraph that is stable for $\left\lfloor \frac{R}{4v_{\max}} \right\rfloor + 1$ rounds.*

Proof. Consider a path of length $\left\lfloor \frac{1}{2}R \right\rfloor$ in the connectivity graph of round r. The nodes on this path form a clique in the communication graph. Following a similar argument as in Lemma 1, these nodes stay connected in the communication graph for further $\left\lfloor \frac{\frac{1}{2}R}{2v_{\max}} \right\rfloor$ rounds. This implies the lemma. $\square$

Theorem 5. *If $R > 1$, then k-token dissemination can be done in $\Theta(n(n+k) \cdot \min\{v_{\max}, R\} \cdot R^{-2})$ rounds and counting can be done in $\Theta(n^2 \cdot \min\{v_{\max}, R\} \cdot R^{-2})$ rounds.*

Proof. If $R > 1$, then according to Lemma 1 the geometric dynamic network is $\Theta(R \cdot v_{\max}^{-1})$-interval connected. Thus, by Theorem 3, the algorithms by Kuhn et al. need $\Theta(n(n+k) \cdot \min\{v_{\max}, R\} \cdot R^{-1})$ rounds for k-token dissemination and $\Theta(n^2 \cdot \min\{v_{\max}, R\} \cdot R^{-1})$ rounds for counting.

If in addition $R \geq 2$, then according to Lemma 2, the communication graph contains a spanning $\Theta(R)$-connected subgraph that is stable for $\Theta(R \cdot v_{\max}^{-1})$ rounds. Thus, by Theorem 4, the algorithms by Kuhn et al. need $\Theta(n(n+k) \cdot \min\{v_{\max}, R\} \cdot R^{-2})$ rounds for k-token dissemination and $\Theta(n^2 \cdot \min\{v_{\max}, R\} \cdot R^{-2})$ rounds for counting. □

Comparing the lower and the upper bound, we can observe that the bounds are almost matching (despite of a factor of R^{-1}) if $k = \Theta(n)$. However, it should be pointed out that the graph model is a bit relaxed by introducing the connectivity graph in addition to the communication graph. It is an interesting question for further research to consider less relaxed models or even matching models.

6   Conclusion and Future Prospects

We showed that the k-token dissemination problem in geometric dynamic networks can be solved asymptotically faster than in traditional dynamic networks. For this, we utilized a communication radius larger than one, which is the connectivity radius. Specifically, by introducing natural conditions, the k-token dissemination problem can be solved in $\mathcal{O}(n(n+k) \cdot \min\{v_{\max}, R\} \cdot R^{-2})$ rounds in our model with $R > 1$ while the lower bound for arbitrary edge dynamics for knowledge-based token-forwarding algorithms is $\Omega(nk)$ [10]. Additionally, these results can also be applied to count the number of nodes of the network in $\mathcal{O}(n^2 \cdot \min\{v_{\max}, R\} \cdot R^{-2})$ rounds if $R > 1$.

Our lower bound shows that even an optimal knowledge-based token-forwarding algorithm needs $\Omega(n \cdot k \cdot \min\{v_{\max}, R\} \cdot R^{-3})$ rounds to disseminate k tokens. For the more interesting case of $k = \Omega(n)$, the upper bound simplifies to $\mathcal{O}(n \cdot k \cdot \min\{v_{\max}, R\} \cdot R^{-2})$ and becomes almost tight. It should be pointed out that our upper bound model is somehow relaxed by introducing the connectivity graph in addition to the communication graph. However, we think that this is a good starting point for analyzing, how we can improve the performance of k-token dissemination, counting, and other related problems by restricting general edge dynamics.

As a next step, it would be interesting to further restrict the model used for the upper bound. Particularly, is it possible to show similar round complexity results with matching models in the upper and lower bound such that we can omit the restriction $R > 1$? To gain more intuition about bounds in the geometric dynamic network model, it is another open question whether global knowledge is an advantage in this model and if so, to what degree. In other words, can a central online algorithm for k-token dissemination that is able to observe the positions of the nodes perform better when facing the network dynamic?

Moreover, more general geometric network models could be of interest. Those are for example asynchronous time models or different graph classes such as disk graphs or quasi unit disk graphs. Yet, one could also think about looking at different and specifically non-geometric restrictions to the network dynamic. A challenging but very interesting question is, whether it is possible to build up a hierarchy of dynamic network restrictions similar to hierarchies known from complexity theory.

References

1. Bienkowski, M., Byrka, J., Korzeniowski, M., Meyer auf der Heide, F.: Optimal algorithms for page migration in dynamic networks. J. Discrete Algorithms **7**(4), 545–569 (2009)
2. Boyd, S., Ghosh, A., Prabhakar, B., Shah, D.: Analysis and optimization of randomized gossip algorithms. In: 43rd IEEE Conference on Decision and Control, 2004. CDC, vol. 5, pp. 5310–5315 (2004)
3. Boyd, S.P., Ghosh, A., Prabhakar, B., Shah, D.: Gossip algorithms: design, analysis and applications. In: INFOCOM, pp. 1653–1664 (2005)
4. Brandes, P., Meyer auf der Heide, F.: Distributed computing in fault-prone dynamic networks. In: TADDS, pp. 9–14 (2012)
5. Dutta, C., Pandurangan, G., Rajaraman, R., Sun, Z., Viola, E.: On the complexity of information spreading in dynamic networks. In: SODA, pp. 717–736 (2013)
6. Dimakis, A.G., Sarwate, A.D., Wainwright, M.J.: Geographic gossip: Efficient averaging for sensor networks. IEEE Trans. Sig. Process. **56**(3), 1205–1216 (2008)
7. Hannes, F., Stefan, R., Ivan, S.: Routing in wireless sensor networks. In: Misra, S., Woungang, I., Misra, S.C. (eds.) Guide to Wireless Sensor Networks, pp. 81–111. Springer, London (2009)
8. Haeupler, B., Karger, D.R.: Faster information dissemination in dynamic networks via network coding. In: PODC, pp. 381–390 (2011)
9. Haeupler, B., Kuhn, F.: Lower bounds on information dissemination in dynamic networks. In: Aguilera, M.K. (ed.) DISC 2012. LNCS, vol. 7611, pp. 166–180. Springer, Heidelberg (2012)
10. Kuhn, F., Lynch, N.A., Oshman, R.: Distributed computation in dynamic networks. In: STOC, pp. 513–522 (2010)
11. Kempkes, B., Meyer auf der Heide, F.: Local, self-organizing strategies for robotic formation problems. In: Erlebach, T., Nikoletseas, S., Orponen, P. (eds.) ALGOSENSORS 2011. LNCS, vol. 7111, pp. 4–12. Springer, Heidelberg (2012)
12. Kuhn, F., Oshman, R.: Dynamic networks: models and algorithms. SIGACT News **42**(1), 82–96 (2011)
13. Kuhn, F., Wattenhofer, R., Zollinger, A.: Worst-case optimal and average-case efficient geometric ad-hoc routing. In: MobiHoc, pp. 267–278 (2003)
14. Kuhn, F., Wattenhofer, R., Zhang, Y., Zollinger, A.: Geometric ad-hoc routing: of theory and practice. In: PODC, pp. 63–72 (2003)
15. Michail, O., Chatzigiannakis, I., Spirakis, P.G.: Brief announcement: naming and counting in anonymous unknown dynamic betworks. In: Aguilera, M.K. (ed.) DISC 2012. LNCS, vol. 7611, pp. 437–438. Springer, Heidelberg (2012)
16. Michail, O., Chatzigiannakis, I., Spirakis, P.G.: Causality, influence, and computation in possibly disconnected synchronous dynamic networks. In: Baldoni, R., Flocchini, P., Binoy, R. (eds.) OPODIS 2012. LNCS, vol. 7702, pp. 269–283. Springer, Heidelberg (2012)

17. Oshman, R.: Distributed computation in wireless and dynamic networks. Ph.D. thesis, Department of Electrical Engineering and Computer Science, Massachusetts Institute of Technology, Cambridge, MA 02139, September 2012
18. O'Dell, R., Wattenhofer, R.: Information dissemination in highly dynamic graphs. In: DIALM-POMC, pp. 104–110 (2005)
19. Das Sarma, A., Molla, A.R., Pandurangan, G.: Fast distributed computation in dynamic networks via random walks. In: Aguilera, M.K. (ed.) DISC 2012. LNCS, vol. 7611, pp. 136–150. Springer, Heidelberg (2012)

The Wake Up Dominating Set Problem

Amir Bannoura$^{(\boxtimes)}$, Christian Ortolf, Christian Schindelhauer, and Leonhard Reindl

Faculty of Engineering, University of Freiburg, Freiburg, Germany
{ortolf,schindel}@informatik.uni-freiburg.de,
{bannoura,reindl}@imtek.de

Abstract. Recently developed wake-up receivers pose a viable alternative for duty-cycling in wireless sensor networks. Here, a special radio signal can wake up close-by nodes. We model the wake-up range by the unit-disk graph. Such wake-up radio signals are very energy expensive and limited in range. Therefore, the number of signals must be minimized. So, we revisit the Connected Dominating Set (CDS) problem for unit-disk graphs and consider an online variant, where starting from an initial node all nodes need to be woken up, while the online algorithm knows only the nodes woken up so far and has no information about the number and location of the sleeping nodes.

We show that in general this problem cannot be solved effectively, since a worst-case setting exists where the competitive ratio, i.e. the number of wake-up signals divided by the size of the minimum CDS, is $\Theta(n)$ for n nodes. For dense random uniform placements, this problem can be solved within a constant factor competitive ratio with high probability, i.e. $1 - n^{-c}$.

For a restricted adversary with a reduced wake-up range of $1 - \epsilon$ we present a deterministic wake-up algorithm with a competitive ratio of $O(\epsilon^{-\frac{1}{2}})$ for the general problem in two dimensions.

In the case of random placement without any explicit position information we present an $O(\log n)$-competitive epidemic algorithm with high probability to wake up all nodes. Simulations show that a simplified version of this oblivious online algorithm already produces reasonable results, that allows its application in the real world.

Keywords: Wake-up receivers · Online algorithm · Connected dominating set · Unit-disk graph · Epidemic algorithms

1 Introduction

Energy is the driving problem of wireless sensor networks (WSNs), since sensor nodes usually operate for long periods and the only source of energy is battery cells which are difficult to be exchanged. The functionality of WSNs can be extended through the use of low power microprocessors, sensors, and radio transceivers. The availability of low power hardware components provided a

P. Flocchini et al. (Eds.): ALGOSENSORS 2013, LNCS 8243, pp. 35–50, 2014.
DOI: 10.1007/978-3-642-45346-5_4, © Springer-Verlag Berlin Heidelberg 2014

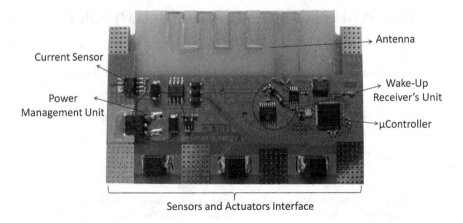

Fig. 1. Wireless sensor node with integrated wake-up receiver

technological break-through of wake-up receivers. These receivers interact only when a special wake-up signal is addressed to them. When a wake-up signal is received, the wake-up receiver triggers an interrupt to wake the sensor node. In addition, any sensor node has the capability of transmitting a wake-up signal to wake up all other close-by sensor nodes. Recent research [9] has decreased the energy consumption of sensor nodes when no activity is required to less than $9\,\mu W$, whereas sensor nodes that are not equiped with wake-up receivers who uses duty cycle would be spending around $51\,mW$ checking the medium from time to time. Figure 1 shows a sample board of a designed wake-up receiver integrated with a wireless sensor node.

Wake-up receivers integrated in sensor nodes constitute a paradigm shift for wireless sensor protocols. In which, sensor nodes interact with the surrounding neighbors only when they are required to receive and send information. The duty cycle process for periodically checking to find out whether messages need to be received or synchronizing with other sensor nodes may not be required anymore.

Despite that this technology provides a new solution for the energy consumption problem, new problems arise. Sensor nodes are required to produce a wake-up signal, these signals are energy expensive compared with the signals that are required for normal data communication. Furthermore, the communication range of a wake-up signal is smaller than the normal data communication range, which requires a multi-hop wake-up signals to wake up sensor nodes that are located in the normal data communication range. Our aim is to minimize the number of wake-up signals transmitted as much as possible and maximize the covered area to reduce the energy required to wake up sensor nodes.

A straight-forward solution is to establish minimum set of sensors which are able to wake up all sensor nodes in case some data need to be collected or distributed. This is the well known Connected Dominating Set (CDS) problem, where one tries to compute the Minimum CDS (MCDS). This problem plays an important role in wireless networks and it is known to be NP-complete.

For simplicity we assume a unit-disk range model and we are interested in computing the minimum dominating connected set in unit disk graphs, i.e. geometric graphs where an undirected edge exists between nodes, if their distance is at most 1.

However, our problem is somehow different. When the sensor nodes are placed, no positions are known before the first wake-up signal. Also, the nodes may be moving, sensor nodes may fail, and possible persistent memory might not be available. All these are reasons to build up a CDS from scratch regularly.

So, we face an online version of the MCDS problem in the context of wake-up receivers. At the beginning, one sensor node wakes up, e.g. because of new sensor data. It sends a wake-up signal and receives responses from all next neighbors. Then, a decision needs to be taken which of the neighbored sensors is allowed to send the next wake-up signal. Since normal data communication consumes only little energy compared to the wake-up signal, we can assume that all active nodes are aware of each other. Furthermore, the information which sensor received a wake-up signal is available to us, even if the sensor has already been woken up. The question is now, can we wake up all nodes with minimal number of wake-up signals. This is what we address as the *wake-up minimum connected dominating set problem in unit disk graphs*. In this variant the positions of the woken up nodes become available as soon as they are awake. For the *wake-up position-aware minimum connected dominating set problem in unit disk graphs* positions are not known at all.

2   Related Work

The new wake-up receivers developed by Gamm et al. in [9] give us an alternative to the concept of duty cycles for awaiting incoming messages in wireless sensor networks.

A perfectly efficient online wake-up would use a minimum connected dominating set of nodes to wake up all the nodes. Finding such a MCDS was already shown to be NP-hard for general graphs, as well as for unit disc graphs [4,13]. For the general (non unit-disk graph) problem no polynomial time approximation exists unless $NP \subseteq DTIME[n^{O(\log \log n)}]$ [10], yet for MCDS with unit disc graphs a PTAS has been presented in [3].

Movement of sensor nodes and maintaining an existing MCDS in their presence was discussed before by Das et al. in [5].

Of course the wake-up problem is an online version of MCDS, because of the differences to the online version presented by Eidenbenz [8] we are referring to it as the wake-up problem.

Eidenbenz models the online problem by node added every round by an adversary, while the online algorithm has to present a CDS, but may never remove nodes once added to the CDS. He shows a competitive ratio of $\Theta(n)$ for the CDS size.

Another online MCDS problem closer related to the wake-up problem is discussed as broadcast problem by Bar-Yehuda et al. [1] and leads to the same lower bound of $\Omega(n)$, so does the reactive routing problem [14]. This motivates why comparing to an adversary with the same radio range is pointless, as asymptotically no online algorithm can beat trivial flooding, i.e. using the whole graph as CDS.

Further research discusses Minimum Routing Cost CDS (MOC-CDS) [6] which requires the hop distance, between any pair of nodes to be minimal. This problem is also NP hard, but for the unit disc graph a PTAS exists [7]. A more generalized version of the problem is called α-MOC-CDS, where the dominating set must have an α-spanner property additionally. For $\alpha = 1$ these problems are the same.

For distributed generation of MCDS approximations are discussed in [15]. Our problem differs, since not all nodes are awake in the beginning, but have to discover all other nodes from the starting node, leading to different time and message complexity.

The motivation for proving an algorithm for a dense random network stems from the requirement for density in a random unit disk graph to guarantee connectivity of the network [11].

In the position oblivious case we use a push-based epidemic rumor spreading algorithm with a simple counter mechanism, similar to the one in [12]. Rumor spreading turns out to be simple and robust. For other applications epidemic algorithms have been already proposed. For a survey of epidemic algorithms in wireless sensor networks we refer to the Chap. 3, Epidemic Models, Algorithms and Protocols in Wireless Sensor and Ad-hoc Networks in [2] by Das and Prabib.

3   Preliminaries

We assume that points are in general positions, i.e. that neither three points are on a line nor four points on a circle in two dimensions. For points in two dimensions or three dimensions the unit-disk graph (UDG) of a given point set V is an undirected graph with edge set $E := \{\{u,v\} \mid u,v \in V : |u,v| \leq 1\}$. Later on we refer also to UDGs with different radius r.

We consider the following problems.

Definition 1. *Given an undirected graph $G = (V,E)$ a connected dominating set (CDS) S has the following properties*

1. *S is connected in G, i.e. for all $u,v \in S$ there exists a path from u to v in G using only nodes of S.*
2. *S is dominating all nodes in V, i.e. for all $u \in V$ there exists a node $v \in S$ such that $\{u,v\} \in E$.*

Definition 2. *The* wake-up position oblivious Minimum Connected Dominating Set Problem in Unit Disk Graphs (Wake-Up-PO-MCDS-UDG) *is to construct a CDS where the algorithm works in rounds and starts with a node s_0.*

1. At the beginning only the nodes $V_1 = \{u \in V : \{u, s_0\} \in E\}$ and edges $E_1 = \{\{u, s_0\} : u \in V_1\}$ are known.
2. In round i a node $s_i \in V_1$ may be selected by the algorithm and then the nodes $V_{i+1} = \{u \in V : \{u, s_i\} \in E\}$ and edges $E_{i+1} = \{\{u, s_i\} : u \in V_1\}$ are added to the knowledge base of the algorithm.

In the wake-up position-aware minimum connected dominating set problem in unit disk graphs also the position of the known nodes is available to the algorithm.

4 Lower Bounds

The problem of computing the minimum connected dominating set for unit-disk graphs (MCD-UD) has been proven to be NP-complete by Lichtenstein [13]. Lower approximation bounds are not known, while the best approximation factor so far has a bound of 3.8 [16].

For the Wake-up version there is a trivial, but hard computational lower bound for the competitive ratio, i.e. the number of nodes of a connected dominating set woken up by an algorithm divided by the number of nodes of the MCDS.

Proposition 1. *The competitive ratio of all deterministic algorithms for Wake-Up-MCD-UD is at least $\frac{n}{2} - \frac{1}{2}$. For probabilistic algorithms the expected competitive ratio is at least $\frac{n}{4}$.*

Proof. We use a variant of the construction presented in [1,8,14], see Fig. 2. The optimal solution uses wake-up calls from the start node s and the node u_i connected to t. Any deterministic algorithm can be fooled to use $n-1$ wake-up calls of nodes $u_1, \ldots, u_{n-2}$ such that the final wake up call reaches t.

If the connected node u_i is chosen randomly, then any randomized algorithm needs in the expectation $1 + \frac{n-2}{2} = \frac{n}{2}$ calls to launch a wake-up at u_i.

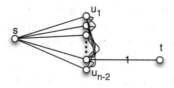

Fig. 2. Lower bound construction with competitive ratio $n/2 - \frac{1}{2}$

5 Algorithms

While the wake-up problem can not be efficiently solved in general, for high node density a straight-forward grid based algorithm already achieves constant approximation ratio.

5.1 A Grid-Based Online Algorithm

We partition the area into a grid of a square size of $\frac{1}{\sqrt{5}}$ for two dimensions and $\frac{1}{\sqrt{6}}$ for three dimensions, see Fig. 7. This grid size guarantees that any node in a cell can reach all nodes in (orthogonally) neighbored cells with a unit-distance wake-up call. We assume that each node is aware of its grid position and let denote $\text{cell}(u)$ the grid cell of u.

The grid based wake-up algorithm 1 chooses a representative for each cell and performs a flooding on the grid structure. In particular, it solves the problem if all grid cells are non-empty. Note that such a $m \times m$ grid can be covered only by a CDS of size of at least $(\frac{m}{\sqrt{5}} - 1)^2 = \frac{1}{5}m^2 - \frac{2}{\sqrt{5}}m + 1$ in two dimensions, while the number of nodes who perform wake up calls is bounded by m^2. Hence, in a square we have a competitive ratio of $5 + o(1)$, and in a cube a competitive ratio of $6 + o(1)$ by an analogous calculation.

Algorithm 1: Grid based wake-up algorithm

Send wake up from s
$G_{\text{done}} \leftarrow \{\text{cell}(s)\}$
$G_{\text{to-do}} \leftarrow \{\text{cell}(u)\ :\ \{u, s\} \in E\} \setminus \{\text{cell}(s)\}$
while $G_{\text{to-do}} \neq \emptyset$ **do**
    Pick a node w such that $\text{cell}(w) \in G_{\text{to-do}}$
    Send wake up from w
    $G_{\text{done}} \leftarrow G_{\text{done}} \cup \{\text{cell}(w)\}$
    $G_{\text{to-do}} \leftarrow G_{\text{to-do}} \cup \{\text{cell}(u)\ :\ \{u, w\} \in E\} \setminus \text{cell}(s) \setminus G_{\text{done}}$

end

If enough nodes are placed randomly, then every cell is occupied with high probability, i.e. $1 - n^{-c}$ for a constant c where $c \in (0, 1)$.

Theorem 1. *If n nodes are place randomly in a $m \times m$-grid with $m \leq \sqrt{\frac{n}{c \ln n}}$ for some constant c. Then the grid based wake-up algorithm computes a CDS with a constant competitive ratio with high probability, i.e. $1 - \frac{1}{n^{c+1}}$ for any $c > 1$.*

Proof. A node is placed in one of the m^2 cells with probability $\frac{1}{m^2} \geq \frac{c \ln n}{n}$. Therefore, the chance that it is not placed in a cell is $1 - \frac{c \ln n}{n}$. So, the probability that a cell is empty can be upperbounded as follows.

$$\left(1 - \frac{c \ln n}{n}\right)^n \leq e^{-c \ln n} = n^{-c} \tag{1}$$

We have used $(1 - 1/m)^m \leq 1/e$ for $m > 0$. By the union bound the probability that any of cell is empty is therefore at most n^{-c+1}. $\qquad\square$

Note that when the node density is decreased only by a constant factor, that the unit disk graphs becomes disconnected [11] and there is no solution for CDS.

5.2   A Competitive Algorithm with Respect to a Weaker Adversary

We have seen that for non-randomized placement the wake-up variant cannot compete with the offline version, which can be seen as an adversary which places the sleeping nodes in the area outside of the wake up signals. However, if we compare the wake-up algorithm with a weaker adversary, we can show some interesting results.

For this we consider two unit-disk graphs with radius 1 and $1 - \epsilon$ for some $\epsilon \in (0, 1)$. The wake-up algorithm solves the wake-up problem for CDS in unit disk graphs, i.e. with radius 1. We compare its performance to the size of the minimum connected dominating set of the unit disk graph with radius $1 - \epsilon$ of the same event. In this way, counter-examples cannot occur as shown in Fig. 2.

Algorithm 2: $(1 - \epsilon)$ cover two hop wake-up grid algorithm

---

Send wake up from s
$G_{\text{done}} \leftarrow \{\text{cell}(s)\}$
$G_{\text{to-do}} \leftarrow \{\text{cell}(u) \ : \ \{u, s\} \in E\} \setminus \{\text{cell}(s)\}$
while $G_{\text{to-do}} \neq \emptyset$ **do**
> Pick a node w such that $\text{cell}(w) \in G_{\text{to-do}}$
> Send wake up from w
> Compute the $\left\lceil \frac{32\pi}{\sqrt{\epsilon}} \right\rceil$-coverage boundary set S of the neighbored nodes of w
> **forall the** $v \in S$ **do**
>> $G_{\text{done}} \leftarrow G_{\text{done}} \cup \{\text{cell}(v)\}$
>> $G_{\text{to-do}} \leftarrow G_{\text{to-do}} \cup \{\text{cell}(v) \ : \ \{v, w\} \in E\} \setminus G_{\text{done}}$
>
> **end**

end

Theorem 2. *In two dimensions there is a wake-up algorithm which produces at most $O(\epsilon^{-1/2})$ more wake-up calls than the number of nodes of the CDS of the $1 - \epsilon$ unit disk graph for $\epsilon \in (0, 1)$.*

Proof. The key idea is for a node u to cover the two-hop neighborhood of the $1 - \epsilon$ unit disk graph with a set of nodes in the neighborhood of the unit-disk graph. We prove that the size of this set of this $\left\lceil \frac{32\pi}{\sqrt{\epsilon}} \right\rceil$-coverage boundary set nodes is bounded by $O(\frac{1}{\sqrt{\epsilon}})$.

Using this observation we use a grid based approach like in [14]. If in two dimensions we choose the grid size of $\frac{1}{\sqrt{2}}(1 - \epsilon)$, then any node in a cell of size u can reach with two hops all nodes which any node of its cell can reach in one hop, since the diagonal of the cell is $1 - \epsilon$. So, it suffices to broadcast a message in a grid, which can be done in constant factor overhead.

It remains to construct the coverage of the two-hop neighborhood of the $1 - \epsilon$ unit disk graphs with $O(\sqrt{\epsilon})$ nodes. For this, we need to investigate some properties of the two-hop neighborhood.

Definition 3. *The two-hop covering node set of a unit-disk-graph of radius r starting with node s is the set of nodes $S \subseteq V$ with the following properties:*

1. *All nodes of S are within distance r of s.*
2. *Each node of S is necessary, i.e. for all nodes of $u \in S$ there exists a point p within distance $[r, 2r]$ from s with $|u, p| \leq r$ and $|v, p| > r$ for all $v \in S \setminus \{u\}$.*

We call the nodes of $S = \{c_1, \ldots, c_m\}$ the *cover nodes*. The *outer ring* is the disk of radius $2r$ without the disk of radius r with center s. The *coverage area* is the union of all disks of radius r and center points of S. The *coverage boundary* is the boundary of this coverage area. By definition it consists of arcs with radius r and center points of S. When two points u, v of S have distance r to the same point w the coverage boundary, we call this point a *boundary point*, see Fig. 3. Now the following geometric observations can be made.

Lemma 1

1. *Given a point p of the coverage boundary, which is not a boundary point, and the tangent T of the boundary, then the angle δ between (p, s) and T is in the range $\delta \in [\frac{1}{6}\pi, \frac{5}{6}\pi]$.*
2. *The length of the coverage boundary is at most $8\pi r$.*

Proof. The first statement comes from a simple geometric observation which is based on possible placement of nodes c_i, see Fig. 4.

For the second statement, note that each distance d traveled on the boundary region corresponds to an angle difference of at least $\frac{d}{z} \frac{1}{\sin \delta}$, where z is the distance to s which is in the range $[r, 2r]$. The angle of the tangent is δ. Since $\frac{1}{2} \leq \sin \delta \leq 1$ and since the total angle difference is bounded by 2π, it follows that the maximum length of the boundary is bounded by $8\pi r$.   □

We neglect the case, where the coverage boundary intersects with the inner ring. It is straight-forward that the following claims also hold for this case.

Using the observation of Lemma 1 we can order all cover points according to their direction seen from s as $c_1, \ldots, c_m$. Only neighbored nodes share a boundary point. We name the angles according to Fig. 3. We denote b_i as the boundary point between c_i and c_{i+1}, and b_m as the boundary node between c_1 and c_m. Let $\alpha_i = \angle c_i b_i c_{i+1}$, $\beta_i = \angle b_{i-1} c_i b_i$ and $\gamma_i = \pi - \angle c_{i-1} c_i c_{i+1}$.

From the definition of the angles we derive the following equalities for all $i \in [1, m]$:

$$\gamma_i = \beta_i - \frac{\alpha_i + \alpha_{i-1}}{2}, \tag{2}$$

$$\sum_{i=1}^{m} \gamma_i = 2\pi. \tag{3}$$

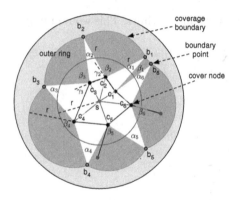 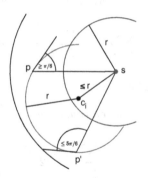

Fig. 3. Definition of coverage boundary, boundary points $b_1, \ldots, b_m$ and cover nodes $S = \{c_1, \ldots, c_m\}$ labeling

Fig. 4. Angle property of the boundary region

Since the boundary region is defined by the arcs of angles $\alpha_1, \ldots, \alpha_n$ with radius r now Lemma 1 implies

$$\sum_{i=1}^{m} \beta_i \le 8\pi \; .$$

So,

$$\sum_{i=1}^{m} \alpha_i = \left(\sum_{i=1}^{m} \alpha_i + \gamma_i \right) - 2\pi = \left(\sum_{i=1}^{m} \beta_i \right) - 2\pi \le 6\pi$$

While the cover points do not necessarily form a convex hull, its form is quite well behaved. For a large number of cover points we need to find groups of near points, i.e. with small angles α_i and β_i.

Lemma 2. *Given m boundary points then, there exist at least $k \le m$ interval indices $i_1, \ldots, i_k$ such that for all $\nu \in \{1, \ldots k\}$:*

$$\sum_{j=i_\nu}^{i_{\nu+1}-1} \alpha_j \le \frac{12\pi}{k} \quad and \quad \sum_{j=i_\nu+1}^{i_{\nu+1}-1} \beta_j \le \frac{16\pi}{k}$$

Proof. Start with $i_1 = 1$. Now for $\nu = 1, 2, \ldots$ choose the largest q such that

$$\sum_{j=i_\nu}^{q} \alpha_j \le \frac{12\pi}{k} \quad and \quad \sum_{j=i_\nu+1}^{q} \beta_j \le \frac{16\pi}{k}$$

and set $i_{\nu+1} := q + 1$.

By definition

$$\sum_{j=i_\nu}^{i_{\nu+1}} \alpha_j > \frac{12\pi}{k} \quad or \quad \sum_{j=i_\nu}^{i_{\nu+1}} \beta_j > \frac{16\pi}{k}$$

One of this property must be violated more than $k/2$ times. This would imply

$$\sum_{j=1}^{m} \alpha_j > \frac{12\pi}{k}\frac{k}{2} = 6\pi \quad \text{or} \quad \sum_{j=i_\nu}^{i_\nu+1} \beta_j > \frac{16\pi}{k}\frac{k}{2} = 8\pi$$

which contradicts Lemma 1   □

Lemma 3. *For given boundary nodes* $c_1, \ldots, c_m$ *with*

$$\sum_{i=1}^{m} \alpha_i \leq \frac{1}{2}\sqrt{\epsilon} \quad \text{and} \quad \sum_{i=2}^{m} \beta_i \leq \frac{1}{2}\sqrt{\epsilon} \, ,$$

the disks with center $c_1, \ldots, c_m$ *with radius* $1 - \epsilon$ *are covered by the two disks with center* c_1 *and* c_m *with radius* 1.

Proof. From Eq. 2 we get for all $\ell \leq m$:

$$\left| \sum_{i=2}^{\ell} \gamma_i \right| = \left| \sum_{i=2}^{\ell} \beta_i - \frac{1}{2}(\alpha_1 + \alpha_\ell) - \sum_{i=2}^{\ell-1} \alpha_i \right| \leq \left| \sum_{i=2}^{\ell} \beta_i - \sum_{i=1}^{\ell} \alpha_i \right| \leq \frac{1}{2}\sqrt{\epsilon} \, .$$

Furthermore, we assume $\epsilon < 1$ and use $\tan \frac{1}{2}\alpha_i = |c_i, c_{i+1}|/2(1-\epsilon)$ and $\tan(x) \leq 2x$ for $x \in [0, 1]$.

$$|c_1, c_m| \leq \sum_{i=1}^{m-1} |c_i, c_{i+1}| \leq \sum_{i=1}^{m-1} 2(1-\epsilon)\tan\frac{1}{2}\alpha_i \leq \sum_{i=1}^{m-1} 2(1-\epsilon)\alpha_i \leq \sqrt{\epsilon}$$

Since all sums $\left| \sum_{i=2}^{\ell} \gamma_i \right| \leq \frac{1}{2}\sqrt{\epsilon}$ and the maximum distance $|c_1, c_m| \leq \frac{1}{2}\sqrt{\epsilon}$ we can conclude that $c_1, \ldots, c_m$ fits into a rectangle of length $\frac{1}{2}\sqrt{\epsilon}$ and width $\frac{1}{2}\epsilon$. The rest follows by the following geometric argument.

The worst case placement for the outer nodes c_1 and c_m and some inner node c_i is depicted in Fig. 5. Note that

$$\left(1 - \frac{1}{2}\epsilon\right)^2 + \left(\frac{\sqrt{\epsilon}}{4}\right)^2 = 1 - \epsilon + \frac{1}{16}\epsilon + \frac{1}{4}\epsilon^2 = 1 - \frac{15}{16}\epsilon + \frac{1}{4}\epsilon^2 \leq 1 \, .$$

Therefore the two outermost nodes with radius 1 always cover the disks with radius $1 - \epsilon$.   □

Putting all pieces together, given a start node s it wakes up all neighbor nodes. They report their position and the algorithms chooses $k = \left\lceil \frac{32\pi}{\sqrt{\epsilon}} \right\rceil$ intervals according to Lemma 2. From these k intervals surrounding s we start wake-up calls only from the nodes at the interval borders, i.e. $c_{i_1}, \ldots, c_{i_k}$. These nodes can cover also the area covered by all cover nodes. So, we need $k = O(\epsilon^{-1/2})$ wake-up calls to wake up all nodes in the two-hop $(1 - \epsilon)$-neighborhood. For a grid size of $\frac{1}{\sqrt{2}}(1 - \epsilon)$ this includes all points in the neighborhood of any given

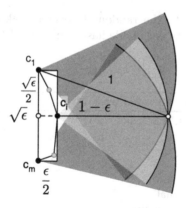

Fig. 5. If the cover nodes are inside a $\sqrt{\epsilon} \times \frac{1}{2}\epsilon$-rectangle, then two disks with radius 1 of the two outermost nodes cover all disks with radius $1 - \epsilon$

point. Therefore, the cell based approach will inform at least the same node set as the adversary.

Another implication from the (constant size) cell based approach is that if we only count cells instead of nodes, there are straight-forward linear upper and lower bounds for the number of wake-up calls. From this observation the competitive ratio of $O(\epsilon^{-1/2})$ follows.   □

5.3 A Position Oblivious Wake Up Algorithm

It seems natural and necessary that the positions of the nodes is used by the wake up algorithms. There are a lot of reasons why the positions might not be known. It is expensive and time-consuming to measure the coordinates and store it on each sensor node. Some sensor nodes might have no persistent memory and cannot store such information. And most important, since the communication range is dependent to the environment it is not clear what the position means in comparison to the unit disk range.

The following oblivious wake-up algorithms come into mind: Flooding, random walk, and epidemic algorithms. While flooding reaches all nodes, it is the worst with respect to energy. Random walks neither reduce the number of wake-up signals nor does it give any delivery guarantees. Epidemic algorithms appear to be the most reasonable solution. The question, however, is how to stop the epidemic wake-up of nodes. We use a push-based epidemic rumor spreading algorithm [12] which will be combined with a simple counter mechanism, which stops the wake-up if k other wake-ups have been received.

However, it is possible to exploit some position information given by the graph structure itself. The random k-covered wake-up of algorithm 3 distinguish between covered and uncovered nodes. A node is covered, if it has received two wake-up signals. The algorithm starts with one node and picks in each round a node which has not been covered twice.

This idea generalizes to the random k-covered wake-up algorithm, where nodes continue to send until each node has been covered $k + 1$ times or has send a wake-up signal.

Algorithm 3: Random k-covered wake-up

Input graph $G = (V, E)$, start node $s \in V$
forall the $u \in V$ **do**
| counter(u) $\leftarrow 0$
end
$W \leftarrow \{s\}$
Node s sends wake-up signal
forall the $u \in V : \{u, s\} \in E$ **do**
| counter(u) $\leftarrow$ counter(u) + 1
end
while $\exists u \in V \setminus W : 0 < counter(u) \leq k$ **do**
| Pick a random node $w \in V \setminus W$ with $0 < counter(u) \leq k$
| $W \leftarrow W \cup \{w\}$
| Node w sends wake-up signal
| **forall the** $u \in V : \{u, w\} \in E$ **do**
| | counter(u) $\leftarrow$ counter(u) + 1
| **end**
end

While we show in Sect. 5.3 that for $k = 1$ the algorithms performs very well, the generalization is necessary since there are situations where the algorithm fails from the start. In Fig. 6 such case is depicted. The node s wakes up four nodes and the near-by node u continues. This node wakes up the same set of nodes as s and so the algorithm stops, since all nodes are covered twice.

With the help of the position information this case could have been clearly avoided. But even without it one could increase k. However, the factor k increases the message complexity. So, a compromise between error rate and message complexity needs to be made.

For the dense case, one can show that the error rate can be reduced to any polynomial if the density is large enough and k is chosen to be logarithmic in n.

Theorem 3. *If n nodes are place randomly in a $m \times m$-grid with $m \leq \sqrt{\frac{n}{c \ln n}}$ for some constant c. Then the Random $O(\log n)$ covered wake-up algorithm computes a CDS with a competitive ratio of $O(\log n)$ high probability, i.e. $1 - n^{-c}$ for some $c \geq 1$.*

Proof. The expected number of nodes in each cell is $\frac{n}{m^2} \geq c \ln n$. Using Chernoff bounds it is possible to prove that this amount is in the range $[\frac{1}{2}\frac{n}{m^2}, 2\frac{n}{m^2}]$ with high probability. The expected number of nodes in the communication range is upper bounded by $5\pi\frac{n}{m^2}$ since the cell length is $\frac{1}{\sqrt{5}}$. Again Chernoff bound can

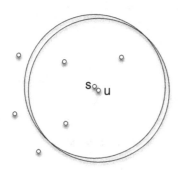

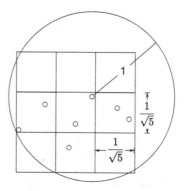

Fig. 6. A counter-example for the random-k-covered-wake-up

Fig. 7. Grid construction in two dimensions.

provide with high probability that the number of nodes reachable in one hop is at most $10\pi\frac{n}{m^2}$ with high probability.

So, the probability that a node of a neighbor cell is activated is at least $\frac{1}{10\pi}$. Therefore, if each node is randomly activated until $10(c+1)\ln n$ wake up calls have been reached, this results in a probability of $1 - n^{-c}$ that each cell starts at least one wake-up call, when a neighbor cell has been activated before.   □

6 Simulations

We have simulated the epidemic random k-covered algorithms to evaluate the efficiency of the covered nodes and the wake-up signals needed from the source node to reach every single node in the area. The grid based flooding algorithm and the competitive algorithm (with $(1-\epsilon)$ unit disk graphs) have high constant factors involved such that they clearly cannot compete.

We randomly deployed varying number of nodes in a square area of $100\,\mathrm{m}$ edge length. The middle node is woken up first and the wake-up communication range is limited to $10\,\mathrm{m}$ based on real-world data. Figures 8 and 9 show how the CDS is constructed in a network with 1,000 randomly deployed nodes using the random 1-covered, resp. 2-covered, wake-up algorithm.

Starting with a source node at a position $(50, 50)$, the algorithm randomly picks the next node to be woken in order to cover the rest of the nodes that are found in the area. The nodes transmitting wake-up signal form a tree from the source node to each covered nodes. Only edges where a new wake-up call is initiated are depicted. In case of k = 1, a node is considered to be covered when it is covered by two wake-up signals or if it transmits one. When the algorithm considers $k > 1$ then a possible intersection in the tree can be formed as it appears in Fig. 9.

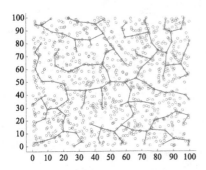

Fig. 8. Random-1-covered-wake-up

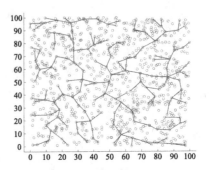

Fig. 9. Random-2-covered-wake-up

The quality of the algorithms are measured according to their coverage, i.e. the ratio of uncovered nodes after the algorithm has terminated, and the complexity, i.e. the number of wake-up calls sent. We have simulated this for the above parameters for increasing density. For this, we increase the number of nodes from 1 to 2,000.

Figure 10 shows that for $k = 1$ the ratio of uncovered nodes is relatively high compared to the set of nodes which can be reached, this percentage is displayed as the result of the flooding algorithm. Increasing k ameliorates this behavior. For high node density all algorithms reach nearly a full coverage. For $k = 1$ a coverage of 95 % happens when 350 nodes are participating, for $k \geq 2$ this already happens for 250 nodes.

Figure 11 indicates that the message complexity grows linearly with k and converges for increasing node density. Surprisingly, the complexity increases from $k = 1$ to $k = 2$ only by around 40 %. So, $k = 2$ appears to be a good compromise between coverage and complexity.

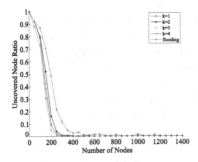

Fig. 10. Uncovered node ratio

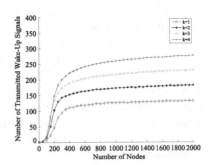

Fig. 11. Wake-up signal transmitted for n nodes

7   Conclusions and Future Work

The improved efficiency of wake-up receivers down to the microwatt range implies a paradigm shift for wireless sensor networks. Now, no busy idling implemented as duty-cycling is necessary until the first sensor information or the first message arrives. However, when the nodes need to be woken up without any prior knowledge we face the wake-up connected dominating set problem, presented here.

For this problem, we provide theoretical and practical solutions. Our algorithms work for random placement and an adversarial setting, where we needed to reduce the power of the adversary, otherwise no efficient algorithms can be found. It turns out that for the random placement the position information is not necessary to find an efficient algorithm with a $O(\log n)$-competitive ratio. We have simulated a simplified variant of this and have seen that it reaches nearly all nodes with small number of wake-up calls.

This raises the hope that duty-cycling might soon be a technique of the past. However, with the available transceiver technology a wake-up call is orders of magnitudes more energy-consuming than standard operation. Taking this into account, it does not make sense to wake up the network from scratch every time a sensor reading appears. At this moment, it is more efficient to put the full network into sleep after some thousand communication cycles. So with the current hardware, a hybrid solution of wake-up calls and duty-cycling is the optimal solution.

Another available technique is the use of IDs for wake-up calls. It is possible to program sensor nodes to be woken up only on a special signal, which are given by a programmable ID. This may help protocols to build up a wake-up infrastructure, where the wake-up signals may trigger different nodes or paths. At this point, it is not clear how this feature can be used in future protocols and what can be achieved with it.

Since, the wake-up transceiver have only become available recently, we are in the process of implementing the given protocols and further research will show, how well these wake-up algorithms behave in real world.

Acknowledgements. This work has been supported by the Ministry for Education and Research (Bundesministerium für Bildung und Forschung, BMBF) within the Research project AURIS (13N11746).

The authors would like to thank SmartExergy WMS GmbH for providing the wake-up receiver's board.

References

1. Bar-Yehuda, R., Goldreich, O., Itai, A.: On the time-complexity of broadcast in radio networks: an exponential gap between determinism randomization. In: Proceedings of the Sixth Annual ACM Symposium on Principles of Distributed Computing, PODC '87, pp. 98–108. ACM, New York (1987)
2. Boukerche, A.: Algorithms and Protocols for Wireless Sensor Networks, vol. 62. Wiley-IEEE, New York (2008)

3. Cheng, X., Huang, X., Li, D., Wu, W., Du, D.-Z.: A polynomial-time approximation scheme for the minimum-connected dominating set in ad hoc wireless networks. Networks **42**(4), 202–208 (2003)
4. Clark, B.N., Colbourn, C.J., Johnson, D.S.: Unit disk graphs. Discrete Math. **86**(1–3), 165–177 (1991)
5. Das, B., Bharghavan, V.: Routing in ad-hoc networks using minimum connected dominating sets. In: IEEE International Conference on Communications, ICC '97 Montreal, Towards the Knowledge Millennium, vol.1, pp. 376–380 (1997)
6. Ding, L., Gao, X., Wu, W., Lee, W., Zhu, X., Du, D.-Z.: Distributed construction of connected dominating sets with minimum routing cost in wireless networks. In: Proceedings of the 2010 IEEE 30th International Conference on Distributed Computing Systems, ICDCS '10, pp. 448–457. IEEE Computer Society, Washington, DC, USA (2010)
7. Du, H., Ye, Q., Zhong, J., Wang, Y., Lee, W., Park, H.: PTAS for minimum connected dominating set with routing cost constraint in wireless sensor networks. In: Wu, W., Daescu, O. (eds.) COCOA 2010, Part I. LNCS, vol. 6508, pp. 252–259. Springer, Heidelberg (2010)
8. Stephan, E.: Online dominating set and variations on restricted graph classes. Technical report, Institute of Theoretical Computer Science, ETH Zürich (2002)
9. Gamm, G.U., Sippel, M., Kostic, M., Reindl, L.M.: Low power wake-up receiver for wireless sensor nodes. In: 2010 Sixth International Conference on Intelligent Sensors, Sensor Networks and Information Processing (ISSNIP), pp. 121–126 (2010)
10. Guha, S., Khuller, S.: Approximation algorithms for connected dominating sets. Algorithmica **20**(4), 374–387 (1998)
11. Gupta, P., Kumar, P.R.: Critical power for asymptotic connectivity. In: Proceedings of the 37th IEEE Conference on Decision and Control, vol. 1, pp. 1106–1110 (1998)
12. Karp, R., Schindelhauer, C., Shenker, S., Vöcking, B.: Randomized rumor spreading. In: IEEE Symposium on Foundations of Computer, Science, pp. 565–574 (2000)
13. Lichtenstein, D.: Planar formulae and their uses. SIAM J. Comput. **11**(2), 329–343 (1982)
14. Rührup, S., Schindelhauer, C.: Competitive time and traffic analysis of position-based routing using a cell structure. In: Proceedings of the 19th IEEE International Parallel and Distributed Processing Symposium, pp. 8 (2005)
15. Wan, P.-J., Alzoubi, K.M., Frieder, O.: Distributed construction of connected dominating set in wireless ad hoc networks. In: Proceedings of the IEEE Twenty-First Annual Joint Conference of the IEEE Computer and Communications Societies, INFOCOM, vol. 3, pp. 1597–1604 (2002)
16. Wu, W., Du, H., Jia, X., Li, Y., Huang, S.C.-H.: Minimum connected dominating sets and maximal independent sets in unit disk graphs. Theor. Comput. Sci. **352**(1), 1–7 (2006)

Reconfiguring Massive Particle Swarms with Limited, Global Control

Aaron Becker[1], Erik D. Demaine[2], Sándor P. Fekete[3](✉), Golnaz Habibi[1], and James McLurkin[1]

[1] Department of Computer Science, Rice University, Houston, TX 77005, USA
aabecker@gmail.com, {gh4,jm23}@rice.edu
[2] Computer Science and Artificial Intelligence Laboratory, MIT,
Cambridge, MA 02139, USA
edemaine@mit.edu
[3] Department of Computer Science, TU Braunschweig,
Mühlenpfordtstr. 23, 38106 Braunschweig, Germany
s.fekete@tu-bs.de

Abstract. We investigate algorithmic control of a large swarm of mobile particles (such as robots, sensors, or building material) that move in a 2D workspace using a global input signal such as gravity or a magnetic field. Upon activation of the field, each particle moves maximally in the same direction, until it hits a stationary obstacle or another stationary particle. In an open workspace, this system model is of limited use because it has only two controllable degrees of freedom—all particles receive the same inputs and move uniformly. We show that adding a maze of obstacles to the environment can make the system drastically more complex but also more useful. The resulting model matches ThinkFun's Tilt puzzle.

If we are given a fixed set of stationary obstacles, we prove that it is NP-hard to decide whether a given initial configuration can be transformed into a desired target configuration. On the positive side, we provide constructive algorithms to design workspaces that efficiently implement arbitrary permutations between different configurations.

Keywords: Robot swarm · Nano-particles · Uniform inputs · Parallel motion planning · Complexity · Array permutations

1 Introduction

Since the first visions of massive sensor swarms, more than ten years of work on sensor networks have yielded considerable progress with respect to hardware miniaturization. The original visions of "Smart Paint" [1] or "Smart Dust" [27] have triggered a considerable amount of theoretical research on swarms of *stationary* processors, e.g., the work in [16–18,29]. Recent developments in the ability to design, produce, and control particles at the nanoscale and the rise of possible applications, e.g., targeted drug delivery, micro and nanoscale construction, and Lab-on-a-Chip, motivate the study of large swarms of *mobile* objects.

P. Flocchini et al. (Eds.): ALGOSENSORS 2013, LNCS 8243, pp. 51–66, 2014.
DOI: 10.1007/978-3-642-45346-5_5, © Springer-Verlag Berlin Heidelberg 2014

But how can we control such a swarm with only limited computational power and a lack of individual control by a central authority? Local, robotics-style motion control by the particles themselves appears hopeless, because (1) the physics of motion at the low Reynold's number nanoscale environment requires overcoming a considerable amount of resistance, and (2) the capacity for storing energy for computation, communication, and motion control shrinks with the third power of object size.

A possible answer lies in applying a global, external force to all particles in the swarm. This resembles the logic puzzle Tilt [38], and dexterity ball-in-a-maze puzzles such as Pigs in Clover and Labyrinth, which involve tilting a board to cause all mobile pieces to roll or slide in a desired direction. Problems of this type are also similar to sliding-block puzzles with fixed obstacles [10, 24–26], except that all particles receive the same control inputs, as in the Tilt puzzle. In the real world, driving ferromagnetic particles with a magnetic resonance imaging (MRI) scanner gives a nano-scale example of this challenge [40]. Becker et al. [7] demonstrate how to apply a magnetic field to simultaneously move cells containing iron particles in a specific direction within a fabricated workspace; see Fig. 1a. Other recent examples include using the global magnetic field from an MRI to guide magneto-tactic bacteria through a vascular network to deliver payloads at specific locations [8], and using electromagnets to steer a magneto-tactic bacterium through a micro-fabricated maze [28]; however, this still involves only individual particles at a time, not the parallel motion of a whole, massive swarm. How can we manipulate the overall swarm with coarse global control, such that individual particles arrive at multiple different destinations in a (known) complex vascular network such as the one in Fig. 1b?

Thus, we study the following basic problem: *Given a map of an environment, such as the vascular network shown in Fig. 1b, along with initial and goal positions for each particle, does there exist a sequence of inputs that will bring each particle to its goal position?*

As it turns out, the deliberate use of existing stationary obstacles leads to a wide range of possible sequences of moves. In the first part of the paper, we show that this may lead to computationally difficult situations. In the second part of the paper (Sect. 5), we develop several positive results. The underlying idea is to construct artificial obstacles (such as walls) that allow arbitrary rearrangements of a given two-dimensional particle swarm. For clearer notation, we will formulate the relevant statements in the language of matrix operations, which is easily translated into plain geometric language.

Our paper is organized as follows. After a formal problem definition in Sect. 2 and a discussion of related work in Sect. 3, we provide our main result on the complexity of the problem in Sect. 4. We then present constructive algorithmic results in Sect. 5, and end with concluding remarks in Sect. 6.

2   Problem Definition

We study the problem on a two-dimensional grid. We assume that particles cannot be individually controlled, but are all simultaneously given a message

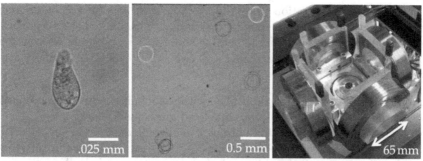

(a) (Left, center) after feeding iron particles to ciliate eukaryon (*Tetrahymena pyriformis*) and magnetizing the particles with a permanent magnet, the cells can be turned by changing the orientation of an external magnetic field. (Right) using two orthogonal Helmholz electromagnets (left), Becker et al. demonstrated steering many living magnetized *T. pyriformis* cells [7]. All cells are steered by the same global field.

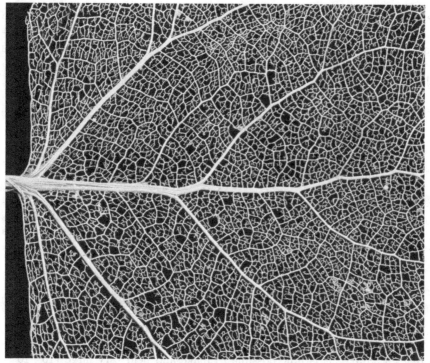

(b) Biological vascular network (cottonwood leaf). Photo: Royce Bair/Flickr/Getty Images. Given such a network along with initial and goal positions of N particles, is it possible to bring each particle to its goal position using a global control signal? Note that this arrangement is *not* a tree, but is a graph structure with loops. MATLAB code for driving n robots through this network available at http://www.mathworks.com/matlabcentral/fileexchange/42892.

Fig. 1. (Top) State of the art in controlling small objects by force fields. (Bottom) A complex vascular network, forming a typical environment for the parallel navigation of small objects.

to travel in a given direction until they collide with an obstacle or another particle. This assumption corresponds to situations with limited state feedback, or for particles that move at unpredictable speeds. More precisely we consider the following scenario, illustrated in Fig. 2, which we call GLOBALCONTROL-MANYPARTICLES:

1. Initially, the planar square grid is filled with some unit-square particles (each occupying a cell of the grid) and some fixed unit-square blocks.
2. All particles are commanded in unison: a valid command is "Go Up" (u), "Go Right" (r), "Go Down" (d), or "Go Left" (l). All particles move in the commanded direction until they hit an obstacle or another particle. A representative command sequence is $\langle u, r, d, l, d, r, u, \ldots \rangle$. We call these global commands *force-field moves*. We assume we can bound the minimum particle speed and can guarantee all particles have moved to their maximum extent.
3. The goal is to get any particle to a specified position.

The algorithmic decision problem GLOBALCONTROL-MANYPARTICLES is to decide whether a given puzzle is solvable. As it turns out, this problem is computationally difficult: we prove NP-hardness in Sect. 4. While this result shows the richness of our model (despite the limited control over the individual parts), it also constitutes a major impediment for constructive algorithmic work.

Fig. 2. In this image, black cells are fixed, white cells are free, solid discs are individual particles, and goal positions are dashed circles. For the simple world at left, it is impossible to maneuver both particles to end at their goals. The world at right has a finite solution: $\langle r, d, l \rangle$.

3   Related Work

Large Robot Populations. Due to the efforts of roboticists, biologists, and chemists (e.g. [9,35,37]), it is now possible to make and field very large (10^3–10^{14}) populations of simple robots. Potential applications for these robots include targeted medical therapy, sensing, and actuation. With large populations come two fundamental challenges: (1) how to perform state estimation for the robots, and (2) how to control these robots.

Traditional approaches often assume independent control signals for each robot, but each additional independent signal requires bandwidth and engineering. These bandwidth requirements grow at $O(n)$. Using independent signals becomes more challenging as the robot size decreases. At the molecular scale, there is a bounded number of modifications that can be made. Especially at the micro- and nano-scales it is not practical to encode autonomy in the robots. Instead, the robots are controlled and interacted with using global control signals.

More recently, robots have been constructed with physical heterogeneity so that they respond differently to a global, broadcast control signal. Examples include *scratch-drive microrobots*, actuated and controlled by a DC voltage signal from a substrate [12]; magnetic structures with different cross-sections that could be independently steered [19]; *MagMite* microrobots with different resonant frequencies and a global magnetic field [20]; and magnetically controlled nanoscale helical screws constructed to stop movement at different cutoff frequencies of a global magnetic field [36]. In our previous work with robots modeled as nonholonomic unicycles, we showed that an inhomogeneity in turning speed is enough to make even an infinite number of robots controllable with regard to position. All these approaches show promise, but they require precise state estimation and heterogeneous robots. In addition, the control law computation required at best a summation over all the robot states $O(n)$ [6] and at worst a matrix inversion $O(n^{2.373})$ [4].

In this paper we take a very different approach. We assume a population of approximately identical planar particles (which could be small robots) and one global control signal that contains the direction all particles should move. In an open environment, this system is not controllable because the particles move uniformly—implementing any control signal translates the entire group identically. However, an obstacle-filled workspace allows us to break symmetry. We showed that if we can command the particles to move one unit distance at a time, some goal configurations have easy solutions [5]. Given a large free space, we have an algorithm showing that a single obstacle is sufficient for position control of N particles (video of position control: http://youtu.be/5p_XIad5-Cw). This result required incremental position control of the group of particles, i.e. the ability to advance them a uniform fixed distance. This is a strong assumption, and one that we relax in this work.

Dexterity Games. The problem we investigate is strongly related to dexterity puzzles—games that typically involve a maze and several balls that should be maneuvered to goal positions. Such games have a long history. *Pigs in Clover*, involving steering four balls through 3 concentric incomplete circles, was invented in 1880 by Charles Martin Crandall. Dexterity games are dynamic and depend on the manual skill of the player. Our problem formulation also applies the same input to every agent, but imposes only kinematic restrictions on agents. This is most similar to the gravity-fed logic maze $Tilt^{TM}$, invented by Vesa Timonen and Timo Jokitalo and distributed by ThinkFun since 2010 [38].

Computational Geometry: Robot Box-Pushing. Many variations of block-pushing puzzles have been explored from a computational complexity viewpoint, with a seminal paper proving NP-hardness by Gordon Wilfong in 1991 [43]. The general case of motion-planning when each command moves particles a single unit in a world composed of even a single robot and both *fixed* and *moveable* squares is in the complexity class PSPACE-COMPLETE [11, 13, 25].

Ricochet Robots [14], Atomix [26], and PushPush [10] have the same constraint that robots when moved must move to their full extent. This constraint reflects physical realities where, due to uncertainties in sensing, control application, and robot models, precise quantified movements in a specified direction is not possible, but the input can be applied for a long period of time and be guaranteed that the robots will move to their fullest extent. In these games the robots move to their full extent with each input, but each robot can be actuated individually. The complexity of the problem with global inputs to all robots has remained an open problem.

Sensorless Manipulation. The algorithms in the second half of our paper do not require feedback, and we have drawn inspiration from work on sensorless manipulation [15]. The basic idea in this work is to explicitly maintain the set of all possible robot configurations and to select a sequence of actions that reduces the size of this set and drives it toward some goal configuration. Carefully selected primitive operations can make this easier. For example, sensorless manipulation strategies often use a sequential composition of primitive operations, "squeezing" a part either virtually with a programmable force field or simply between two flat, parallel plates [23]. Some sensorless manipulation strategies take advantage of limit cycle behavior, for example engineering fixed points and basins of attraction so that parts only exit a feeder when they reach the correct orientation [31, 33]. These two strategies have been applied to a much wider array of mechanisms such as vibratory bowls and tables [21, 41, 42] or assembly lines [2, 23, 39], and have also been extended to situations with stochastic uncertainty [22, 32] and closed-loop feedback [3, 34].

Parallel Algorithms: SIMD. Another related area of research is Single Instruction Multiple Data (SIMD) parallel algorithms [30]. In this model, multiple processors are all fed the same instructions to execute, but they do so on different data. This model has some flexibility, for example allowing command execution selectively only on certain processors and no operations (NOPs) on the remaining processors.

Our model is actually more extreme: the particles all respond in effectively the same way to the same instruction. The only difference is their location, and which obstacles or particles will thus block them. In some sense, our model is essentially Single Instruction, Single Data, Multiple Location.

4   Complexity

We prove that the general problem defined in Sect. 2 is computationally intractable:

Theorem 1. GLOBALCONTROL-MANYPARTICLES *is NP-hard: given an initial configuration of movable particles and fixed obstacles, it is NP-hard to decide whether any particle can be moved to a specified location.*

Proof. We prove hardness by a reduction from 3SAT. Suppose we are given n Boolean variables $x_1, x_2, \ldots, x_n$, and m disjunctive clauses $C_j = U_j \vee V_j \vee W_j$, where each literal U_j, V_j, W_j is of the form x_i or $\neg x_i$. We construct an instance of GLOBALCONTROL-MANYPARTICLES that has a solution if and only if all clauses can be satisfied by a truth assignment to the variables.

Variable gadgets. For each variable x_i that appears in k_i literals, we construct k_i instances of the *variable gadget i* shown in Fig. 3, with a particle initially at the top of the gadget. The gadget consists of a tower of n levels, designed for the overall construction to make n total variable choices. These choices are intended to be made by a move sequence of the form $\langle d, l/r, d, l/r, \ldots, d, l/r, d, l \rangle$, where the ith l/r choice corresponds to setting variable x_i to either true (l) or false (r). Thus variable gadget i ignores all but the ith choice by making all other levels lead to the same destination via both l and r. The ith level branches into two destinations, chosen by either l or r, which correspond to x_i being set true or false, respectively.

In fact, the command sequence may include multiple l and r commands in a row, in which case the last l/r before a vertical u/d command specifies the final decision made at that level, and the others can be ignored. The command sequence may also include a u command, which undoes a d command if done immediately after, or else does nothing; thus we can simply ignore the u command and the immediately preceding d if it exists. We can also ignore duplicate commands (e.g., d, d becomes d) and remove any initial l/r command. After ignoring these superfluous commands, assuming a particle reaches one of the output channels, we obtain a sequence in the canonical form $\langle d, l/r, d, l/r, \ldots, d, l \rangle$ as desired, corresponding uniquely to a truth assignment to the n variables. (If no particle reaches the output port, it is as if the variable is neither true nor false, satisfying no clauses.) Note that all particles arrive at their output ports at exactly the same time.

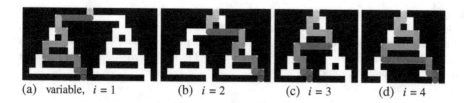

(a) variable, $i = 1$ (b) $i = 2$ (c) $i = 3$ (d) $i = 4$

Fig. 3. Variable gadgets that execute by a sequence of $\langle d, l/r \rangle$ moves. The ith l/r choice sets the variable to true or false by putting the ball in a separate column. This selection move is shown in blue. Each gadget is designed to respond to the ith choice but ignore all others. This lets us make several copies of the same variable by making multiple gadgets with the same i. In the figure $n = 4$, and the input sequence $\langle d, l, d, r, d, l, d, r, d, r, d \rangle$ causes $i = (1, 2, 3, 4)$ to produce (true, false, true, false) (Colour figure online).

Clause gadgets. For each clause, we use the OR gadget shown in Fig. 4a. The OR gadget has three inputs corresponding to the three literals, and input particles are initially at the top of these inputs. For each positive literal of the form x_i, we connect the corresponding input to the left output of an unused instance of variable gadget i. For each negative literal of the form $\neg x_i$, we connect the corresponding input to the right output of an unused instance of a variable gadget i. (In this way, each variable gadget gets used exactly once.)

We connect the variable gadget to the OR gadget in a simple way, as shown in Fig. 5: place the variable gadget above the clause so as to align the vertical output and input channels, and join them into a common channel. To make room for the three variable gadgets, we simply extend the black areas separating the three input channels in the OR gadget. The unused output channel of each variable gadget is connected to a waste receptacle. Any particle reaching that end cannot return to the logic.

If any input channel of the OR gadget has a particle, then it can reach the output port by the move sequence $\langle d, l, d, r \rangle$. Furthermore, because variable gadgets place all particles on their output ports at the same time, if more than one particle reaches the OR gadget, they will move in unison as drawn in Fig. 4a, and only one can make it to the output port; the others will be stuck in the "waste" row, even if extra $\langle l, r, u, d \rangle$ commands are interjected into the intended sequence. Hence, a single particle can reach the output of a clause if and only if that clause (i.e., at least one of its literals) is satisfied by the variable assignment.

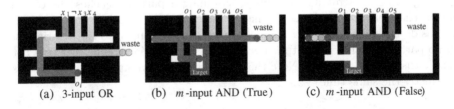

(a) 3-input OR   (b) m-input AND (True)   (c) m-input AND (False)

Fig. 4. Gadgets that use the cycle $\langle d, l, d, r \rangle$. The 3-input OR gadget outputs one particle if at least one particle enters in an input line, and sends any extra particle(s) to a waste receptacle. The m-input AND gadget outputs one particle to the TARGET LOCATION, marked in gray, if at least m inputs are TRUE. Here $m = 5$. Excess particles are sent to a waste receptacle.

Check gadget. As the final stage of the computation, we check that all clauses were simultaneously satisfied by the variable assignment, using the m-input AND gadget shown in Fig. 4b and c. Specifically, we place the clause gadgets along a horizontal line, and connect their vertical output channels to the vertical input channels of the check gadget. Again we can align the channels by extending the black areas that separate the input channels of the AND gadget, as shown in the composite diagram Fig. 5.

The intended solution sequence for the AND gadget is $\langle d, l, d, r \rangle$. The AND gadget is designed with the downward channel exactly m units to the right from

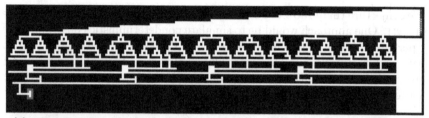

(a) Initial state. The objective is to get one particle to the grey square at lower left.

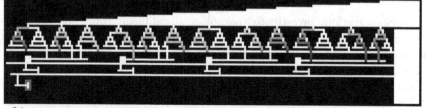

(b) Setting variables to (False, True, False, True) does not satisfy this 3SAT problem.

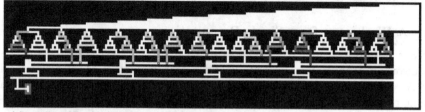

(c) Setting the variables (True, False, False, True) will satisfy this 3SAT problem.

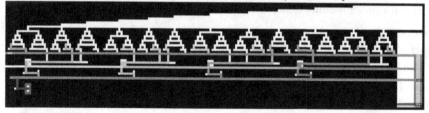

(d) Successful outcome. (True, False, False, True) places a single particle in the goal.

Fig. 5. Combining 12 variable gadgets, three 3-input OR gadgets, and an m-input AND gadget to realize the 3SAT expression $(\neg A \vee \neg C \vee D) \wedge (\neg B \vee \neg C \vee D) \wedge (\neg A \vee B \vee D) \wedge (A \vee \neg B \vee C)$. MATLAB code implementing the examples for each figure in the paper is available online http://www.mathworks.com/matlabcentral/fileexchange/42892.

the left wall, and $> 2\,m$ units from the right wall, so for any particle to reach the downward channel (and ultimately, the target location), at least m particles must be presented as input. Because each input channel will present at most one particle (as argued in a clause), a particle can reach the final destination if and only if all m clauses output a particle, which is possible exactly when all clauses are satisfied by the variable assignment.

This completes the reduction and the NP-hardness proof.

We conjecture that GLOBALCONTROL-MANYPARTICLES is in fact PSPACE-COMPLETE. One approach would be to simulate nondeterministic constraint logic [24], perhaps using a unique move sequence of the form $\langle d, l/r, d, l/r, \ldots \rangle$ to identify and "activate" a component. One challenge is that all gadgets must properly reset to their initial state, without permanently trapping any particles. We leave this for future work.

5   Matrix Permutations

The previous sections investigated pathologically difficult configurations. This section investigates a complementary problem. Given the same particle and world constraints as before, what types of control are possible and economical if we are free to design the environment?

First, we describe an arrangement of obstacles that implement an arbitrary matrix permutation in four commands. Then we provide efficient algorithms for sorting matrices, and finish with potential applications.

5.1   A Workspace for a Single Permutation

For our purposes, a *matrix* is a 2D array of particles (each possibly a different color). For an $a_r \times a_c$ matrix A and a $b_r \times b_c$ matrix B, of equal total size $N = a_r \cdot a_c = b_r \cdot b_c$, a *matrix permutation* assigns each element in A a unique position in B. Figures 6 and 7 show example constructions that execute matrix permutations of total size $N = 25$ and 100, respectively. For simplicity of exposition, we assume henceforth that all matrices are $n \times n$ squares.

Theorem 2. *Any matrix permutation can be executed by a set of obstacles that transforms matrix A into matrix B in just four moves. For N particles, the arrangement requires $(3N + 1)^2$ space, $4N + 1$ obstacles, and $12N/\text{speed}$ time.*

Proof. Refer to Figs. 6 and 7 for examples. The move sequence is $\langle u, r, d, l \rangle$.
Move 1: We place n obstacles, one for each column, spaced n units apart, such that moving u spreads the particle array into a staggered vertical line. Each particle now has its own row. **Move 2:** We place N obstacles to stop each particle during the move r. Each particle has its own row and can be stopped at any column by its obstacle. We leave an empty column between each obstacle to prevent collisions during the next move. **Move 3:** Moving d arranges the particles into their desired rows. These rows are spread in a staggered horizontal line. **Move 4:** Moving l stacks the staggered rows into the desired permutation, and returns the array to the initial position.

By reapplying the same permutation enough times, we can return to the original configuration. The permutations shown in Fig. 6 return to the original image in 2 cycles, while Fig. 7 requires 740 cycles. For a two-color image, we can always construct a permutation that resets in 2 cycles. We construct an *involution*, a function that is its own inverse, using cycles of length two that transpose two particles. This technique does not extend to images with more than two colors.

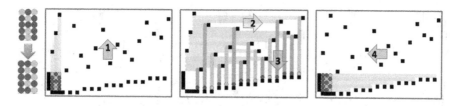

Fig. 6. In this image for $N = 15$, black cells are obstacles, white cells are free, and colored discs are individual particles. The world has been designed to permute the particles between 'A' into 'B' every four steps: $\langle u, r, d, l \rangle$. See video at http://youtu.be/3tJdRrNShXM.

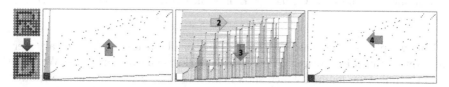

Fig. 7. In this larger example with $N = 100$, the different control sections are easier to see than in Fig. 6. (1) The staggered obstacles on the left spread the matrix vertically, (2) the scattered obstacles on the upper right permute each element, and (3) the staggered obstacles along the bottom reform each row, which are collected by (4). The cycle resets every 740 iterations. See http://youtu.be/eExZO0HrWRQ for an animation of this gadget.

5.2   A Workspace for Arbitrary Permutations

There are various ways in which we can exploit Theorem 2 in order to generate larger sets of (or even all) possible permutations. As it turns out, there is a tradeoff between the number of introduced obstacles and the number of moves required for realizing a permutation.

We start with obstacle sets that require only few moves.

Theorem 3. *For any set of k fixed, but arbitrary, permutations of $n \times n$ pixels, we can construct a set of $O(kN)$ obstacles, such that we can switch from a start arrangement into any of the k permutations using at most $O(\log k)$ force-field moves.*

Proof. See Fig. 8. Build a binary tree of depth $\log k$ for choosing between the permutations by a sequence of $\langle r, d, (r/l), d, (r/l), \ldots, d, (r/l), d, l, u \rangle$ with $\log k$ (r/l) decisions between the initial prefix $\langle r, d \rangle$ and final suffix $\langle d, l, u \rangle$. This gets the pixels to the set of obstacles for performing the appropriate permutation.

Corollary 1. *For any $\varepsilon > 0$, we can construct a set of $(N!)^\varepsilon$ obstacles such that any permutation of $n \times n = N$ pixels can be achieved by at most $O(N \log N)$ force-field moves.*

Proof. Follows from Theorem 3 by $k = (N!)^\varepsilon / N$.

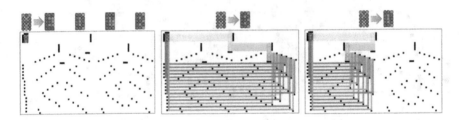

Fig. 8. For any set of k fixed, but arbitrary permutations of $n \times n$ pixels, we can construct a set of $O(kN)$ obstacles, such that we can switch from a start arrangement into any of the k permutations using at most $O(\log k)$ force-field moves. Here $k = 4$ and transforms 'A' into 'B', C', 'D', or 'E' in eight moves: $\langle r, d, (r/l), d, (r/l), d, l, u \rangle$.

Now we proceed to more economical sets of obstacles, with arbitrary permutations realized by clockwise and counterclockwise move sequences. We make use of the following lemma, which shows that two base permutations are enough to generate any desired rearrangement.

Lemma 1. *Any permutation of N objects can be generated by the two base permutations $p = (1,2)$ and $q = (1, 2, \cdots N)$. Moreover, any permutation can be generated by a sequence of length at most N^2 that consists of p and q.*

Proof. Similar to BUBBLE SORT, we use two nested loops of N. Each move consists of performing q once, and p when appropriate.

This allows us to establish the following result.

Theorem 4. *We can construct a set of $O(N)$ obstacles such that any $n \times n$ arrangement of N pixels can be rearranged into any other $n \times n$ arrangement π of the same pixels, using at most $O(N^2)$ force-field moves.*

Proof. See Fig. 9. Use Theorem 2 to build two sets of obstacles, one each for p and q, such that p is realized by the sequence $\langle u, r, d, l \rangle$ (clockwise) and q is realized by $\langle r, u, l, d \rangle$ (counterclockwise). Then we use the appropriate sequence for generating π in $O(N^2)$ moves.

Using a larger set of generating base permutations allows us to reduce the number of necessary moves. Again, we make use of a simple base set for generating arbitrary permutations.

Lemma 2. *Any permutation of N objects can be generated by the N base permutations $p_1 = (1,2), p_2 = (1,3), \ldots, p_{N-1} = (1, (N-1))$ and $q = (1, 2 \cdots N)$. Moreover, any permutation can be generated by a sequence of length at most N that consists of the p_i and q.*

Proof. Straightforward, analogous to Theorem 4: in each step i, apply q once, and swap element $\pi(i)$ into position i.

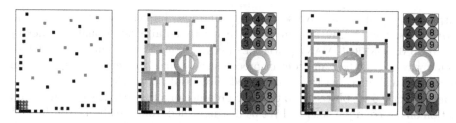

Fig. 9. Repeated application of two base permutations can generate any permutation, when used in a manner similar to BUBBLE SORT. The obstacles above generate the base permutation $p = (1, 2)$ in the clockwise direction $\langle u, r, d, l \rangle$ and $q = (1, 2, \cdots N)$ in the counter-clockwise direction $\langle r, u, l, d \rangle$.

Theorem 5. *We can construct a set of $O(N^2)$ obstacles such that any $n \times n$ arrangement of N pixels can be rearranged into any other $n \times n$ arrangement π of the same pixels, using at most $O(N \log N)$ force-field moves.*

Proof. Use Theorem 2 to build N sets of obstacles, one each for $p_1, \ldots, p_{N-1}, q$. Furthermore, use Lemma 2 for generating all permutations with at most N different of these base permutation, and Theorem 3 for switching between these $k = N$ permutations. Then we can get π with at most N cycles, each consisting of at most $O(\log N)$ force-field moves.

This is the best possible with respect to the number of moves, in the following sense:

Theorem 6. *Suppose we have a set of obstacles such that any permutation of an $n \times n$ arrangement of pixels can be achieved by at most M force-field moves. Then M is at least $\Omega(N \log N)$.*

Proof. Each permutation must be achieved by a sequence of force-field moves. Because each decision for a force-field move $\langle u, d, l, r \rangle$ partitions the remaining set of possible permutations into at most four different subsets, we need at least $\Omega(\log(N!)) = \Omega(N \log N)$ such moves.

6 Conclusions

In this paper we analyzed the complexity of steering many particles with uniform inputs in a 2D environment with obstacles. We are motivated by practical challenges in steering magnetically-actuated robots through vascular networks. Many examples of natural, locally 2D vascular networks exist, such as the leaf example in Fig. 1b, and the endothelial networks on the surface of organs.

Clearly, there are many exciting new challenges that lie ahead. The next step is to extend the complexity analysis to PSPACE-COMPLETE. We are also exploring using particles and obstacles to construct logic gates. These results let us implement AND and OR gates. Using *dual-rail logic*, where the signal and its inverse are explicitly represented by the presence of a particle along either the signal or the inverse rail, we could also implement NOT, NAND, and NOR gates. Generating fan-out gates seems to require additional complexity in GLOBALCONTROL-MANYPARTICLES because conservation rules are violated. Some way of encoding an order of precedence so that a reversible operation on particle a will affect particle b is needed. Potential approaches use either 2×1 particles, or 0.5×1 obstacles so that the presence of a first particle can enable an action on a second particle, and yet be distinguished from the absence of the first particle and the presence of the second. With uniform 1×1 obstacles and particles, these cases are indistinguishable. Finally, many exciting applications require platforms that can navigate in three dimensions. This poses a large number of additional challenges, both for the theory and the physical implementation.

Acknowledgments. We acknowledge the helpful discussion and motivating experimental efforts with *T. pyriformis* cells by Yan Ou and Agung Julius at RPI and Paul Kim and MinJun Kim at Drexel University. This work was supported by the National Science Foundation under CPS-1035716.

References

1. Abelson, H., Allen, D., Coore, D., Hanson, C., Homsy, G., Thomas, J., Knight, F., Nagpal, R., Rauch, E., Sussman, G.J., Weiss, R.: Amorphous computing. Commun. ACM **43**(5), 74–82 (2000)
2. Akella, S., Huang, W.H., Lynch, K.M., Mason, M.T.: Parts feeding on a conveyor with a one joint robot. Algorithmica **26**(3), 313–344 (2000)
3. Akella, S., Mason, M.T.: Using partial sensor information to orient parts. Int. J. Robot. Res. **18**(10), 963–997 (1999)
4. Becker, A., Bretl, T.: Approximate steering of a unicycle under bounded model perturbation using ensemble control. IEEE Trans. Robot. **28**(3), 580–591 (2012)
5. Becker, A., Habibi, G., Werfel, J., Rubenstein, M., McLurkin, J.: Massive uniform manipulation: controlling large populations of simple robots with a common input signal. In: IEEE International Conference on Intelligent Robots and Systems, pp. 520–527, November 2013
6. Becker, A., Onyuksel, C., Bretl, T.: Feedback control of many differential-drive robots with uniform control inputs. In: IEEE International Conference on Intelligent Robots and Systems, October 2012
7. Becker, A., Ou, Y., Kim, P., Kim, M., Julius, A.: Feedback control of many magnetized tetrahymena pyriformis cells by exploiting phase inhomogeneity. In: IEEE International Conference on Intelligent Robots and Systems, pp. 3317–3323, November 2013
8. Chanu, A., Felfoul, O., Beaudoin, G., Martel, S.: Adapting the clinical mri software environment for real-time navigation of an endovascular untethered ferromagnetic bead for future endovascular interventions. Magn. Reson. Med. **59**(6), 1287–1297 (2008)
9. Chiang, P.-T., Mielke, J., Godoy, J., Guerrero, J.M., Alemany, L.B., Villagómez, C.J., Saywell, A., Grill, L., Tour, J.M.: Toward a light-driven motorized nanocar: Synthesis and initial imaging of single molecules. ACS Nano **6**(1), 592–597 (2011)

10. Demaine, E.D., Demaine, M.L., O'Rourke, J.: PushPush and Push-1 are NP-hard in 2D. In: Proceedings of the 12th Annual Canadian Conference on Computational Geometry (CCCG), pp. 211–219, August 2000
11. Demaine, E.D., Hearn, R.A.: Playing games with algorithms: algorithmic combinatorial game theory. In: Albert, M.H., Nowakowski, R.J. (eds.) Games of No Chance 3, vol. 56, pp. 3–56. Mathematical Sciences Research Institute Publications, Cambridge University Press, Cambridge (2009)
12. Donald, B.R., Levey, C.G., Paprotny, I., Rus, D.: Planning and control for microassembly of structures composed of stress-engineered mems microrobots. Int. J. Robot. Res. **32**(2), 218–246 (2013)
13. Dor, D., Zwick, U.: Sokoban and other motion planning problems. Comput. Geom. **13**(4), 215–228 (1999)
14. Engels, B., Kamphans, T.: On the complexity of Randolph's robot game. Technical report, Rheinische Friedrich-Wilhelms-Universität Bonn Institut für Informatik I, University of Cologne, Germany (2005)
15. Erdmann, M., Mason, M.: An exploration of sensorless manipulation. IEEE J. Robot. Autom. **4**(4), 369–379 (1988)
16. Fekete S.P., Kröller, A.: Geometry-based reasoning for a large sensor network. In: Proceedings of the 22nd Annual ACM Symposium on Computational Geometry, pp. 475–476 (2006). http://www.computational-geometry.org/SoCG-videos/socg06video/
17. Fekete, S.P., Kröller, A.: Topology and routing in sensor networks. In: Kutyłowski, M., Cichoń, J., Kubiak, P. (eds.) ALGOSENSORS 2007. LNCS, vol. 4837, pp. 6–15. Springer, Heidelberg (2008)
18. Fekete, S.P., Kröller, A., Pfisterer, D., Fischer, S., Buschmann, C.: Neighborhood-based topology recognition in sensor networks. In: Nikoletseas, S.E., Rolim, J. (eds.) ALGOSENSORS 2004. LNCS, vol. 3121, pp. 123–136. Springer, Heidelberg (2004)
19. Floyd, S., Diller, E., Pawashe, C., Sitti, M.: Control methodologies for a heterogeneous group of untethered magnetic micro-robots. Int. J. Robot. Res. **30**(13), 1553–1565 (2011)
20. Frutiger, D., Kratochvil, B., Vollmers, K., Nelson, B.J.: Magmites - wireless resonant magnetic microrobots. In: IEEE International Conference on Robotics and Automation, Pasadena, CA, May 2008
21. Goemans, O.C., Goldberg, K., van der Stappen, A.F.: Blades: a new class of geometric primitives for feeding 3D parts on vibratory tracks. In: International Conference on Robotics and Automation, pp. 1730–1736, May 2006
22. Goldberg, K., Mirtich, B.V., Zhuang, Y., Craig, J., Carlisle, B.R., Canny, J.: Part pose statistics: estimators and experiments. IEEE Trans. Robot. Autom. **15**(5), 849–857 (1999)
23. Goldberg, K.Y.: Orienting polygonal parts without sensors. Algorithmica **10**(2), 201–225 (1993)
24. Hearn R.A., Demaine, E.D.: PSPACE-completeness of sliding-block puzzles and other problems through the nondeterministic constraint logic model of computation. arXiv:cs/0205005, cs.CC/0205005 (2002)
25. Hoffmann, M.: Motion planning amidst movable square blocks: Push-* is NP-hard. In: Canadian Conference on Computational Geometry, pp. 205–210, June 2000
26. Holzer, M., Schwoon, S.: Assembling molecules in atomix is hard. Theoret. Comput. Sci. **313**(3), 447–462 (2004)
27. Kahn, J., Katz, R., Pister, K.: Emerging challenges: mobile networking for smart dust. J. Comm. Net. **2**, 188–196 (2000)

28. Khalil, I.S.M., Pichel, M.P., Reefman, B.A., Sukas, O.S., Abelmann, L., Misra, S.: Control of magnetotactic bacterium in a micro-fabricated maze. In: IEEE International Conference on Robotics and Automation, Karlsruhe, Germany, pp. 5488–5493, May 2013

29. Kröller, A., Fekete, S.P., Pfisterer, D., Fischer, S.: Deterministic boundary recognition and topology extraction for large sensor networks. In: Proceedings of the 17th ACM-SIAM Symposium Discrete Algorithms, pp. 1000–1009 (2006)

30. Leighton, F.T.: Introduction to Parallel Algorithms and Architectures: Arrays, Trees, Hypercubes. Morgan Kaufmann, San Mateo (1991)

31. Lynch, K.M., Northrop, M., Pan, P.: Stable limit sets in a dynamic parts feeder. IEEE Trans. Robot. Autom. **18**(4), 608–615 (2002)

32. Moll, M., Erdmann, M.: Manipulation of pose distributions. Int. J. Robot. Res. **21**(3), 277–292 (2002)

33. Murphey, T.D., Bernheisel, J., Choi, D., Lynch, K.M.: An example of parts handling and self-assembly using stable limit sets. In: International Conference on Robots and System, pp. 1624–1629, August 2005

34. Murphey, T.D., Burdick, J.W.: Feedback control methods for distributed manipulation systems that involve mechanical contacts. Int. J. Robot. Res. **23**(7–8), 763–781 (2004)

35. Ou, Y., Kim, D.H., Kim, P., Kim, M.J., Julius, A.A.: Motion control of magnetized tetrahymena pyriformis cells by magnetic field with model predictive control. Int. J. Rob. Res. **32**(1), 129–139 (2013)

36. Peyer, K.E., Zhang, L., Nelson, B.J.: Bio-inspired magnetic swimming microrobots for biomedical applications. Nanoscale **5**, 1259–1272 (2013)

37. Rubenstein, M., Ahler, C., Nagpal, R.: Kilobot: a low cost scalable robot system for collective behaviors. In: IEEE International Conference on Robotics and Automation, pp. 3293–3298, May 2012

38. ThinkFun. Tilt: Gravity fed logic maze. http://www.thinkfun.com/tilt

39. van der Stappen, A.F., Berretty, R.-P., Goldberg, K., Overmars, M.: Geometry and part feeding. In: Hager, G.D., Christensen, H.I., Bunke, H., Klein, R. (eds.) Sensor Based Intelligent Robots. LNCS, vol. 2238, pp. 259–281. Springer, Heidelberg (2002)

40. Vartholomeos, P., Akhavan-Sharif, M., Dupont, P.E.: Motion planning for multiple millimeter-scale magnetic capsules in a fluid environment. In: IEEE International Conference on Robotics and Automation, pp. 1927–1932, May 2012

41. Vose, T.H., Umbanhowar, P., Lynch, K.M.: Friction-induced velocity fields for point parts sliding on a rigid oscillated plate. In: Robotics: Science and Systems, Zurich, Switzerland, June 2008

42. Vose, T.H., Umbanhowar, P., Lynch, K.M.: Sliding manipulation of rigid bodies on a controlled 6-DoF plate. Int. J. Robot. Res. **31**(7), 819–838 (2012)

43. Wilfong, G.: Motion planning in the presence of movable obstacles. Ann. Math. Artif. Intell. **3**(1), 131–150 (1991)

Polygon-Constrained Motion Planning Problems

Davide Bilò[1], Yann Disser[2], Luciano Gualà[3], Matúš Mihal'ák[4],
Guido Proietti[5,6]([✉]), and Peter Widmayer[4]

[1] Dipartimento di Scienze Umanistiche e Sociali, Università di Sassari, Sassari, Italy
[2] Institut für Mathematik, Technische Universität Berlin, Berlin, Germany
[3] Dipartimento di Ingegneria dell'Impresa, Università di Roma Tor Vergata,
Rome, Italy
[4] Institut für Theoretische Informatik, ETH, Zürich, Switzerland
[5] Dipartimento di Ingegneria e Scienze dell'Informazione e Matematica,
Università dell'Aquila, Coppito L'Aquila, Italy
[6] Istituto di Analisi dei Sistemi ed Informatica, CNR, Roma, Italy
guido.proietti@univaq.it

Abstract. We consider the following class of polygon-constrained motion planning problems: Given a set of k centrally controlled mobile agents (say *pebbles*) initially sitting on the vertices of an n-vertex simple polygon P, we study how to plan their vertex-to-vertex motion in order to reach with a minimum (either *maximum* or *total*) movement (either in terms of *number of hops* or *Euclidean distance*) a final placement enjoying a given requirement. In particular, we focus on final configurations aiming at establishing some sort of *visual connectivity* among the pebbles, which in turn allows for wireless and optical intercommunication. Therefore, after analyzing the notable (and computationally tractable) case of gathering the pebbles at a *single* vertex (i.e., the so-called *rendezvous*), we face the problems induced by the requirement that pebbles have eventually to be placed at: (i) a set of vertices that form a *connected* subgraph of the *visibility graph* induced by P, say $G(P)$ (*connectivity*), and (ii) a set of vertices that form a *clique* of $G(P)$ (*clique-connectivity*). We will show that these two problems are actually hard to approximate, even for the seemingly simpler case in which the hop distance is considered.

1 Introduction

In many practical applications a number of centrally controlled devices need to be moved from an initial positioning towards a final configuration so that a desired task can be completed. In particular, in settings like robotics and sensor networking, the devices generally happen to have a limited transmission and reception capability, and thus to establish some kind of reciprocal communication they need

This work was partially supported by the Research Grant PRIN 2010 "ARS TechnoMedia", funded by the Italian Ministry of Education, University, and Research. Part of this work was developed while Guido Proietti was visiting ETH.

P. Flocchini et al. (Eds.): ALGOSENSORS 2013, LNCS 8243, pp. 67–82, 2014.
DOI: 10.1007/978-3-642-45346-5_6, © Springer-Verlag Berlin Heidelberg 2014

to build an obstacle-free *ad-hoc network*. However, by any respects, movements are expensive, and so this repositioning procedure should be accomplished in such a way that some distance-related objective function is minimized.

In this paper, we assume the underlying environment is a *simple polygon*, say P, and the moving devices (*pebbles*, in the sequel) are initially placed on vertices of P. In our setting, pebbles can only move within the polygon through a *vertex-to-vertex polygonal path*, and so they will reach a final position which coincides with a polygon vertex. This restriction about the initial, intermediate, and ending position of the pebbles is motivated by the fact that vertices are a notable position in a polygon, for which several well-studied classes of computational geometry problems (e.g., art-gallery guarding, facility location, etc.) have been considered. Moreover, from a more practical point of view, we point out that recently there has been a growing attention towards *limited-sensing* robotic devices, which are built in such a way that they are able to only detect very minimal information about the surrounding environment. In particular, the so-called *combinatorial robots* [12] are only able to move to visible corners of the (planar) region they are embedded in, i.e., the vertices of a polygon. Therefore, we study a set of *motion planning* (i.e., centrally managed) problems that arise by the combination of three different final positioning goals and a pair of movement optimization functions, which will be computed with respect to two different distance concepts. More precisely, we first focus our study on a scenario where we want the pebbles to be moved to a *single vertex* (RV, which stands for *rendez-vous*) of P. In fact, gathering at a single vertex will enable pebbles to exchange information in a setting where long-range communication is not allowed. Then, we turn our attention to the more general case in which pebbles have to form a *connected subgraph* (CON) of the *visibility graph* of P. Recall that such a graph has a node for each polygon vertex, and an edge for each pair of polygon vertices which can be joined by a straight line contained in the interior or the boundary of polygon P. Thus, quite naturally, we focus on the visibility graph of P, since intervisibility between polygon vertices turns out to enable wireless or optical connection among devices. Finally, in order to consider the plausible case in which a mutual direct connection among pebbles is needed, we analyze the problem in which they have to form a *clique* (CLIQUE) in the visibility graph. For all these problems, we consider both the minimization of the *overall movement* (SUM) and the *maximum movement* (MAX) of the pebbles. To this respect, these functions will be measured both in terms of the classic *Euclidean distance* (ED) covered by the pebbles, and with regard to the *hop distance* (HD) measure, i.e., that in which the distance between two vertices in P is given by the minimum number of edges in any vertex-to-vertex polygonal path in P connecting the two vertices. This latter type of distance is important in many practical cases since it resorts to the number of turns that a device must take all along the way.

Related Work. Although movement problems were deeply investigated in a distributed setting (see [11] for a survey), quite surprisingly the centralized counterpart has received attention from the scientific community only very recently.

The first paper which defines and studies these problems in this latter setting is [6]. In their work, the authors study the problem of moving the pebbles on a graph G of n vertices so that their final positions form any of the following configurations: connected component, path (directed or undirected) between two specified nodes, independent set, and matching. Regarding connectivity problems, the authors show that both variants are hard and that the approximation ratio of CON-MAX is between 2 and $O(1 + \sqrt{k/\text{Opt}})$, where k is the number of pebbles and Opt denotes the measure of an optimal solution. This result has been improved in [3], where the authors show that CON-MAX can be approximated within a constant factor, more precisely 136. In [6] it is also shown that CON-SUM is not approximable within $O(n^{1-\epsilon})$ (for any positive ϵ), while it admits an approximation algorithm with ratio of $O(\min\{n\log n, k\})$ (where k is the number of pebbles). Moreover, they also provide an exact polynomial-time algorithm for CON-MAX on trees.

More recently, in [7], a variant of the classical facility location problem has been studied. This variant, called *mobile facility location*, can be modelled as a motion planning problem and is approximable within $(3+\epsilon)$ (for any positive ϵ) if we seek to minimize the total movement [1]. On the other hand, a variant where the maximum movement has to be minimized admits a tight 2-approximation [1,6].

Finally, for CON and CLIQUE, in [4] the authors present a set of improved (in)approximability results both for general and special classes of graphs, and moreover they also study the problem of moving pebbles to an independent set.

Our Problems and Results. More formally, our problems can be stated as follows. Let P be a simple polygon delimited by the set of vertices $V(P) = \langle v_1, \ldots, v_n \rangle$, in this order. Let $A = \{p_1, \ldots, p_k\}$ be a set of pebbles. Each pebble initially sits on a polygon vertex (multiple pebbles can occupy the same position). Thus, by $S = (s_1, \ldots, s_k)$ we denote the initial configuration of the pebbles. Given a *target* vertex $v_i \in P$, we denote by $d(s_i, v_i)$ the length of a shortest path in P starting at s_i and ending at v_i. Such a shortest path is actually a vertex-to-vertex polygonal path, which is in compliance with our setting. Let $U = (u_1, \ldots, u_k)$, with $u_i \in V$, denote the final configuration of the pebbles, and let $|d(S, U)| = \sum_{i=1}^{k} d(s_i, u_i)$, and $\|d(S, U)\| = \max_{i=1,\ldots,k}\{d(s_i, u_i)\}$. With a small abuse of notation, when in the final configuration all the pebbles sit on a same vertex u, we denote these quantities by $|d(S, u)|$ and $\|d(S, u)\|$. Finally, let $G(P)$ be the visibility graph of P. We study the following problems:

1. *Rendez-vous:* The questions we address are:
   (i) RV-MAX: find $u^* = \arg\min_{u \in V(P)}\{\|d(S, u)\|\}$;
   (ii) RV-SUM: find $u^* = \arg\min_{u \in V(P)}\{|d(S, u)|\}$.
2. *Connectivity:* Let $\mathcal{C}$ denote the set of subsets of vertices of P which induce a connected subgraph in $G(P)$. Then, the questions we address are:
   (i) CON-MAX: find $U^* = \arg\min_{U \in \mathcal{C}}\{\|d(S, U)\|\}$;
   (ii) CON-SUM: find $U^* = \arg\min_{U \in \mathcal{C}}\{|d(S, U)|\}$.
3. *Clique:* Let $\mathcal{K}$ denote the set of subsets of vertices of P which induce a clique in $G(P)$. Then, the questions we address are:

    (i) CLIQUE-MAX: find $U^* = \arg\min_{U \in \mathcal{K}}\{\|\|d(S, U)\|\|\}$;

    (ii) CLIQUE-SUM: find $U^* = \arg\min_{U \in \mathcal{K}}\{|d(S, U)|\}$.

Besides the above problems, we also define the corresponding ones associated with the hop distance in P between v_i and v_j, say $h(v_i, v_j)$. An example of solutions for our problems w.r.t. both distance models is given in Fig. 1, while the results we present in the paper are summarized in Table 1, where by p and m we denote the size of the set of vertices of P initially occupied by pebbles and of the set of edges of $G(P)$, respectively.

Table 1. New (in bold) and old (with the reference therein) results for the various motion problems, where ρ denotes the best approximation ratio for the corresponding problem. All the inapproximability results hold under the assumption that P $\neq$ NP.

	MAX	SUM
RV	HD: solvable in $O(pm)$ time	HD: solvable in $O(pm + k)$ time
	ED: solvable in $O(n \log n)$ time	ED: solvable in $O(pn + k)$ time
CON	HD: NP-hard; $\rho \geq 2$; $\rho \leq 136$ [3]	HD: NP-hard; $\rho \geq n^{1-\epsilon}$; $\rho \leq 1 + O(nk/\text{Opt})$
	ED: polyAPX-hard	ED: polyAPX-hard
CLIQUE	HD: NP-hard; $\rho \geq 3/2$; $\rho \leq 1 + 1/\text{Opt}$ [4]	HD: NP-hard; $\rho \leq 2$ [4]
	ED: open	ED: open

2 Rendez-vous

As far as the hop distance is concerned, RV-MAX and RV-SUM have a naïve $O(pm)$ and $O(pm + k)$ time solution, respectively, whose improvement is a challenging open problem. Indeed, let $V(S)$ be the set of vertices of P initially occupied by the pebbles. Observe that a shortest hop-distance path is just a shortest path in the visibility graph $G(P)$ of P. Then, first of all we compute $G(P)$ in $O(n + m)$ time [9]. After, and only for RV-SUM, in $O(n + k)$ time we associate with each vertex v of P the multiplicity of pebbles initially sitting on it, say $\mu(v)$. Then, in $O(pm)$ time we find the p breadth-first search trees of $G(P)$ rooted at the vertices of $V(S)$. From these trees, it is easy to see that we are able in $O(pn)$ time to solve both problems, by computing for RV-MAX and for RV-SUM respectively

$$x^* = \arg\min_{x \in V(P)} \{\max\{h(x, v) | v \in V(S)\}\},$$

$$x^* = \arg\min_{x \in V(P)} \left\{ \sum_{v \in V(S)} h(x, v) \cdot \mu(v) \right\}.$$

Concerning the Euclidean distance, once again RV-MAX and RV-SUM have a trivial $O(pn)$ and $O(pn + k)$ time solution, respectively, which work as follows. First, for RV-SUM only, in $O(n + k)$ time we associate with each vertex v of P the multiplicity $\mu(v)$. Then, for each vertex $v \in V(S)$ we can find its distance

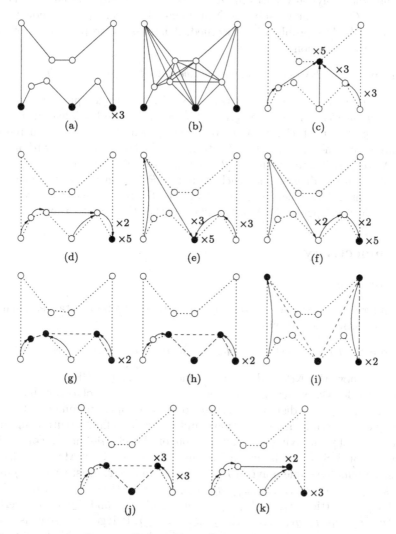

Fig. 1. An example of our studied problems for both HD and ED. Polygon P and its visibility graph $G(P)$ are shown in (a) and (b), respectively. Black vertices are occupied by pebbles, whose movement is depicted with directed paths. Optimal solutions for RV-MAX and RV-SUM w.r.t. ED are shown in (c) and (d), respectively, while (e) and (f) show optimal solutions for the corresponding problems w.r.t. HD, respectively. Optimal solutions for CON-MAX and CON-SUM w.r.t. ED are shown in (g) and (h), respectively, while an optimal solution for the corresponding problems w.r.t. HD is shown in (i). Finally, in (j) it is shown an optimal solution for CLIQUE-MAX w.r.t. to both ED and HD and CLIQUE-SUM w.r.t. ED, while an optimal solution for CLIQUE-SUM w.r.t. HD is shown in (k). Notice that dashed lines in (g–k) show the subgraph of $G(P)$ induced by the final position of the pebbles.

to all the other polygon vertices in $O(n)$ time [5,8]. Finally, similarly to the hop distance, in $O(pn)$ time we solve both problems. However, we now show that as far as the RV-MAX problem is concerned, it is possible to provide an efficient $O(n \log n)$ time solution:

Theorem 1. *The* RV-MAX *problem can be solved in* $O(n \log n)$ *time.*

Proof. Observe that in $O(n \log n)$ time (see [2]) we can compute the so-called *furthest-site geodesic Voronoi diagram* of $V(S)$ w.r.t. the Euclidean distance in P, i.e., a partition of P into a set of regions such that each region remains associated with the farthest point (in terms of Euclidean distance within P) in $V(S)$. Moreover, it can be shown [2] that the size of such a diagram is $O(n)$, and that given the diagram, for each vertex of P we can find in $O(1)$ time the farthest point in $V(S)$, i.e., the farthest pebble. Finally, we select the vertex for which the farthest pebble is closest, and we gather the pebbles there.   □

3   Connectivity

3.1   Con-Max

Concerning CON-MAX, let us start by focusing on the hop distance. Then, we are able to prove the following.

Theorem 2. *The* CON-MAX *problem w.r.t. the hop distance is* NP-*hard.*

Proof. We show the NP-hardness by reduction from the NP-complete 3-SAT problem. In 3-SAT, we are given a set $X = \{x_1, \ldots, x_\eta\}$ of η variables, a set $Y = \{c_1, \ldots, c_m\}$ of m disjunctive clauses over X, each containing exactly three literals (i.e., a variable or its negation), and we want to find a truth assignment $\tau : X \to \{0, 1\}$ satisfying the conjunction of the clauses in Y. For a given instance $\mathcal{I}$ of 3-SAT, we build an instance $\mathcal{I}'$ for the CON-MAX problem as follows: we build a simple polygon P, illustrated in Fig. 2, consisting of 2η *literal* vertices $V_L = \{x_1, \bar{x}_1, \ldots, x_\eta, \bar{x}_\eta\}$, η *assignment* vertices $V_A = \{a_1, \ldots, a_\eta\}$, $2\eta + 2m$ *gate* vertices $V_G = \{g_1, g'_1, \ldots, g_{\eta+m}, g'_{\eta+m}\}$, and $5m$ *clause* vertices $V_C = \{c_{11}, c_{12}, c_{13}, p_1, q_1, \ldots, c_{m1}, c_{m2}, c_{m3}, p_m, q_m\}$. Polygon P is so constructed such that, among the others, the following visibility constraints hold:

- literal vertices see each other reciprocally;
- each assignment vertex a_i can see only $g_i, g'_i, x_i, \bar{x}_i$;
- each clause vertex c_{ij} can see only $g_{\eta+i}, g'_{\eta+i}, p_i, q_i$, the other two clause vertices in its clause, and the literal vertex corresponding to the jth literal of its clause;
- each clause vertex p_i can see only $c_{i1}, c_{i2}, c_{i3}, q_i$.

Then, we put a pebble in each assignment vertex, and a pebble in each p_i, $i = 1, \ldots, m$, so the number of pebbles is $k = \eta + m$.

We now show that the 3-SAT instance $\mathcal{I}$ has a satisfying truth assignment iff there exists a solution for $\mathcal{I}'$ having maximum hop distance of 1. One direction

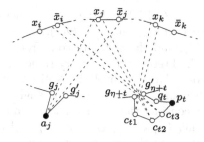

Fig. 2. The polygon P used for proving the NP-hardness of Con-Max problem w.r.t. the hop distance. Pebbles sit initially on black vertices.

is immediate. Given a satisfying assignment τ, we indeed define the following movement: each pebble in an assignment vertex moves to the appropriate literal verified by τ, while each pebble in a clause moves to any clause vertex seeing a verified literal. In this way, the pebbles originally sitting on assignment vertices will form a clique in the visibility graph $G(P)$, while the remaining pebbles are connected to (i.e., see) exactly an occupied literal vertex.

Concerning the other direction, suppose that there is a solution for $\mathcal{I}'$ having value at most 1. We show that such a solution can be transformed in polynomial time into a satisfying assignment for $\mathcal{I}$. First of all, notice that by construction of P, each assignment pebble must be moved to either an associated gate vertex or to an associated literal vertex to guarantee mutual visibility among assignment pebbles. Then, observe that in a single hop a pebble in a clause can see a literal vertex only if it moves to either of the three clause vertices that can see the respective literal vertices. Thus, to guarantee connectedness among assignment and clause pebbles, it is required that at least one of these three literals is occupied by an assignment pebble. Hence, the satisfying assignment for $\mathcal{I}$ is given exactly by the placement onto these literal vertices of the assignment pebbles that will guarantee the connectedness with the clause pebbles. Notice that some of the assignment pebbles may need not move to any literal vertex (i.e., the corresponding variable is not instrumental to guarantee the satisfiability of $\mathcal{I}$), and so they could simply move to an associated gate vertex in order to be connected to the assignment pebbles which moved towards the literal vertices. For these pebbles, we arbitrarily assign a value to the associated variables.

It remains to prove that P can be constructed in polynomial time. It suffices to show that P can be embedded on integer grid points with polynomial area and using a polynomial number of algebraic operations, similarly to the approach used in [10]. Let $r = \eta + m$. W.l.o.g., assume that $m = \Theta(\eta)$, and so $r = \Theta(\eta) = \Theta(m)$. Consider a circle C with radius $\Theta(r^2)$ centered at a point o. We position the literal vertices $x_1, \bar{x}_1, \ldots, x_\eta, \bar{x}_\eta$ on the upper side of C, regularly spaced. Let the angle (in radians) from o to any two contiguous vertices be $\Theta(1/r)$, and so the angle $\angle x_1 o \bar{x}_\eta$ can be less than a fraction of π (i.e., all the vertices stay on the upper side). Observe that in this way, the distance between two contiguous vertices is $\Theta(r)$. Now, position the assignment spikes on the lower-left side of C, so that the angle $\angle g_j a_j g'_j$ is $\Theta(1/r)$ and the distance between a_j and g_j, g'_j is $\Theta(r)$ (i.e., the distance between g_j and g'_j is $\Theta(1)$, and so we can actually put these vertices on the grid), and the visibility cone from a_j towards the upper side of C includes only the literal vertices $x_j, \bar{x}_j$ (indeed, the projection of the cone on the upper side of C has length $\Theta(r)$). Notice that each spike has area $\Theta(r)$. Moreover, again we can guarantee that the angle $\angle g_1 o g'_n$ is less than a fraction of π. Let us now consider the set of vertices in a clause, along with the associated gate vertices. Let $\{g_{n+t}, g'_{n+t}, c_{t1}, c_{t2}, c_{t3}, p_t, q_t\}$ be this set of vertices for the tth clause. We will embed these points on an $r \times r$ grid drawn at the lower-right side of C. Let o_t be the center of such a grid. We draw a circle C_t centered at o_t of radius $\Theta(r)$, and we append it to C by the gate vertices g_{n+t}, g'_{n+t}. We let the angle $\angle g_{n+t} o_t g'_{n+t}$ be $\Theta(1/r)$. Observe now that it is not hard to see that the angle $\angle g_{n+t} x g'_{n+t}$ is less than $\angle g_{n+t} o_t g'_{n+t}$, for any point x on the semi-circumference of C_t opposite to the gate vertices, since any such point is farther from g_{n+t}, g'_{n+t} than o_t. So, the visibility cone from any such point towards the upper side of C has an angle $O(1/r)$ and then a projection on C of length $O(r)$. Thus, it includes a portion of C which is in the order of the distance between two contiguous literal vertices. Then, we place c_{t1}, c_{t2}, c_{t3} on the projection through the midpoint of g_{n+t}, g'_{n+t} of the respective associated literal vertex. Finally, we suitably deform C_t so as to put p_t, q_t in such a way that p_t can only see c_{t1}, c_{t2}, c_{t3} and q_t. Again, the angle $\angle g_{n+1} o g'_{n+m}$ is less than a fraction of π. It can now be verified that this construction gives the desired polygon, and its area is $\Theta(r^4)$. $\qquad\square$

Thus, since the problem is hard already when the optimal solution costs 1, we immediately have the following:

Corollary 1. *For any $\epsilon > 0$, the* CON-MAX *problem w.r.t. the hop distance cannot be approximated within $2 - \epsilon$, unless* P = NP.

Moreover, the following implication is also easy to prove:

Corollary 2. *Deciding whether* CON-MAX *admits a solution with at most h hops is* NP-*complete, for any $h \geq 1$.*

Proof. Case $h = 1$ follows directly from Theorem 2. For $h > 1$, it suffices to suitably modify the polygon P in Fig. 2 in such a way that the pebbles need to move for $h - 1$ steps in order to see the literal and the clause vertices. $\qquad\square$

Concerning the approximability, we recall that in [3] the authors provide a 136-approximation for the very same problem on general unweighted graphs, which can therefore be applied to visibility graphs as well.

The above NP-hardness proof can be modified in order to show that the general CON-MAX problem with Euclidean distances is NP-hard as well.

Theorem 3. *The* CON-MAX *problem is* NP-*hard.*

Proof. We show the NP-hardness again by reduction from 3-SAT. For a given instance $\mathcal{I}$ of 3-SAT, we build an instance $\mathcal{I}'$ for CON-MAX as follows: we build a simple polygon P with 2η *literal* vertices $V_L = \{x_1, \bar{x}_1, \ldots, x_\eta, \bar{x}_\eta\}$, η *assignment* vertices $V_A = \{a_1, \ldots, a_\eta\}$, $2m$ *gate* vertices $V_G = \{g_1, g'_1, \ldots, g_m, g'_m\}$, and $3m$ *clause* vertices $V_C = \{c_{11}, c_{12}, c_{13}, \ldots, c_{m1}, c_{m2}, c_{m3}\}$. Polygon P is so constructed such that, among the others, the following visibility constraints hold (see Fig. 3):

- literal and assignment vertices see each other reciprocally;
- each clause vertex c_{ij} can see only $g_{\eta+i}, g'_{\eta+i}$, the other two clause vertices in its clause, and the literal vertex corresponding to the jth literal of its clause.

Then, we put a pebble in each assignment vertex, and a pebble in each clause vertex, so the number of pebbles is $k = \eta + 3m$. Let us see how polygon P is actually constructed in polynomial time and with polynomial area. Let $r = \eta + m$. W.l.o.g., assume that $m = \Theta(\eta)$, and so $r = \Theta(\eta) = \Theta(m)$. Consider a circle C with radius $\Theta(r^3)$ centered at a point o. We position each triple of vertices $x_i, a_i, \bar{x}_i$ on the upper side of C, regularly spaced at a distance $\Theta(r)$. Then, we let the angle (in radians) from o to any two contiguous triplets be $\Theta(1/r)$ (i.e., the distance between two contiguous triplets is $\Theta(r^2)$). In this way, the angle $\angle x_1 o \bar{x}_\eta$ can be less than a fraction of π, and then assume that all these vertices lie in the $[\pi/4, 3\pi/4]$ sector. Let us now consider the set of vertices in a clause, along with the associated gate vertices. Let $\{g_t, g'_t, c_{t1}, c_{t2}, c_{t3}\}$ be this set of vertices for the tth clause. We will embed these points on an $r^2 \times r^2$ grid drawn at the lower side of C, in the $[5\pi/4, 7\pi/4]$ sector. Let o_t be the center of such a grid. We draw a circle C_t centered at o_t of radius $\Theta(r^2)$, and we append it to C by the gate vertices g_t, g'_t. We let the angle $\angle g_t o_t g'_t$ be $\Theta(1/r^2)$ (i.e., the distance between g_t and g'_t is $\Theta(1)$, and so we can actually put these vertices on the grid). Then, the visibility cone from any point on the semi-circumference of C_t opposite to the gate vertices towards the upper side of C has an angle of $O(1/r^2)$ and then a projection on C of length $O(r)$. Thus, it sees a portion of C including a single literal vertex. Then, we place c_{t1}, c_{t2}, c_{t3} on the projection through the midpoint of g_t, g'_t of the respective associated literal vertex. Notice that by construction, these vertices will lie in the lower side of C_t, and so they will be at $\Theta(r^2)$ distance from the respective gate vertices. It can now be verified that this construction gives the desired polygon, and its area is $\Theta(r^6)$.

Then, it is not hard to see that the 3-SAT instance $\mathcal{I}$ has a satisfying truth assignment iff there exists a solution for $\mathcal{I}'$ having maximum distance $\Theta(r)$. One direction is immediate. Given a satisfying assignment τ, we indeed define the following movement: each pebble in an assignment vertex moves to the appropriate literal verified by τ, while each pebble in a clause stands still. In this way, the pebbles originally sitting on assignment vertices will form a clique in the visibility graph $G(P)$, while for each clause there is at least a pebble connected to a literal vertex satisfying the clause, and so the other pebbles in the clause will remain connected to it. Notice that the maximum movement is $\Theta(r)$.

Concerning the other direction, suppose that there is a solution for $\mathcal{I}'$ having value $\Theta(r)$. We show that such a solution can be transformed in polynomial

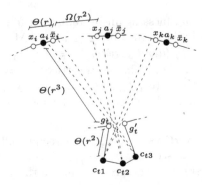

Fig. 3. The polygon P used for proving the NP-hardness of CON-MAX, along with a specification of distances among vertices. Pebbles sit initially on black vertices.

time into a satisfying assignment for $\mathcal{I}$. First of all, notice that by construction of P, no assignment pebble can move beyond the adjacent literal vertices, and similarly no pebble in a clause can move to the associated gate vertices. Then, in order to guarantee connectedness, we have that each assignment vertex must move to a literal vertex, and moreover there must be at least a pebble in each clause seeing a literal vertex occupied by a pebble. Thus, this corresponds to a satisfying assignment for $\mathcal{I}$. □

The above result has a very strong implication:

Corollary 3. CON-MAX *is not approximable within any polynomial, unless* P = NP.

Proof. Observe that the construction of the polygon P given in Theorem 3 can be inflated as follows: for any integer $k > 2$, we let the circle C have radius $\Theta(r^{k+1})$, we let the distance between two contiguous triplets be $\Theta(r^k)$, we embed each clause on an $r^k \times r^k$ grid, and we finally let the angle $\angle g_t o_t g'_t$ be $\Theta(1/r^k)$. It can now be verified that this construction gives a polygon of area $\Theta(r^{2(k+1)})$ for which an optimal solution of cost $\Theta(r)$ exists iff there is a satisfying assignment, while any approximate solution will require a pebble to be moved by $\Omega(r^k)$. Hence, since $r = \Theta(n)$, the claim follows. □

3.2   Con-Sum

Concerning CON-SUM, the reduction shown in Theorem 2 can be modified to prove the following two results:

Theorem 4. *The* CON-SUM *problem w.r.t. the hop distance is* NP*-hard.*

Proof. We use the same construction as in Theorem 2. The claim is that the instance $\mathcal{I}$ of 3-SAT is satisfiable iff there is a solution for the instance $\mathcal{I}'$ of CON-SUM of cost at most $\eta + m$. Given a truth assignment, the existence of a solution of cost $\eta + m$ is immediate, since we have shown how to move every pebble at most by one to obtain connectedness. Now assume that we have a solution U with total movement of $h := ||d(S, U)|| \leq \eta + m$. First of all, we show that $h < \eta + m$ is unfeasible, and so it must be $h = \eta + m$.

For the sake of contradiction, assume that $h < \eta + m$ hops are enough. This means that there is at least a pebble that does not move. But then observe that the distance in the visibility graph $G(P)$ of P between any two initial positions of pebbles is at least 3, and so to be visually connected to other pebbles, a pebble that stands still asks for (at least) another pebble being moved by at least 2 hops. Moreover observe that each vertex of P guards at most a single vertex on which pebbles initially sit, and so no pebble which moves for at least 2 hops can be visually connected to more than one pebble which remained still. From this, we have that to guarantee connectedness, it must be $h \geq \eta + m$, a contradiction.

Then, let $h = \eta + m$. If each pebble has moved, we are done, since this implies that each clause pebble is connected in $G(P)$ to a literal pebble, and therefore we can compute (in polynomial time) a truth assignment for $\mathcal{I}$ by using the same arguments used in Theorem 2. Otherwise, assume this is not the case, and so there is at least one pebble that remained still. We will show that U can be modified into another solution U' such that (i) U' still has total movement h, and (ii) every pebble moves exactly one step in U'. The modification of the solution U is quite simple. Let H be the (connected) subgraph of the visibility graph induced by the final positions of the pebbles in U. Moreover, let p be a pebble sitting in the node v and which did not move in U. Consider a node v' which is adjacent to v in H onto which a pebble p' sits. In order to reach v', as explained before pebble p' moved by $t \geq 2$ hops. Moreover, observe that by construction the set of vertices which are visible from v is a subset of the set of vertices which are visible from v'. Then, we modify U as follows. We move p from v to v', and we move p' backwards by one step w.r.t. its path towards v'. In this way, the movement of p' is now $t - 1 \geq 1$, all the vertices which where guarded by p' are now guarded by p, and p and p' are connected. So the new pebble configuration is still connected and the total movement remains h. By proceeding in this way, we will arrive to the aimed configuration U'. $\square$

Corollary 4. *For any $0 < \epsilon < 1$, the* CON-SUM *problem w.r.t. the hop distance cannot be approximated within $n^{1-\epsilon}$, unless* P=NP.

Proof. We adapt the reduction of Theorem 2 as follows: we modify the gadgets of the assignment vertices and of the clauses by adding $2N$ vertices and $N + 1$ pebbles for each gadget, where $N = (\eta + m)^{2/\epsilon - 1}$ (see Fig. 4).

Then, it can be shown that (see also the proof of Theorem 4):

(i) if there exists a satisfying truth assignment for $\mathcal{I}$, then there exists a solution for $\mathcal{I}'$ having total movement of $\eta + m$;

(ii) if there exists a solution for $\mathcal{I}'$ with total movement less than or equal to N, then there exists a satisfying truth assignment for $\mathcal{I}$. Indeed, as the total movement is less than or equal to N, a pebble placed on vertex a_j or vertex p_t has been moved by at most 1 (otherwise, all the other N pebbles placed in the same gadget would have been moved by at least 1).

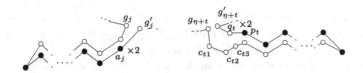

Fig. 4. The assignment and clause gadgets for proving the inapproximability of CON-SUM w.r.t. the hop distance. Pebbles sit initially on black vertices. Vertex a_j and vertex p_t contain two pebbles each.

Since 3-SAT is NP-complete and $n = \Theta\big((\eta + m)^{2/\epsilon}\big)$, the approximation ratio of any polynomial time algorithm for the CON-SUM problem must be at least

$$\frac{N}{\eta + m} = \frac{(\eta + m)^{2/\epsilon - 1}}{\eta + m} = (\eta + m)^{2/\epsilon - 2} = \Omega(n^{\epsilon/2})^{2/\epsilon - 2} = \Omega(n^{1-\epsilon}).$$

□

On the positive side, we have the following:

Theorem 5. *The* CON-SUM *problem w.r.t. the hop distance can be solved optimally up to an additive term of* $O(nk)$.

Proof. It suffices to observe that any solution which will bring all the pebbles to sit on a same vertex cannot require more than n additional hops for each of the k pebbles w.r.t. an optimal solution.   □

On the other hand, when we consider the Euclidean distance, CON-SUM becomes much harder, as the following two results show:

Theorem 6. *The* CON-SUM *problem is* NP-*hard.*

Proof. The NP-hardness follows again by reduction from 3-SAT, by slightly modifying the construction given in Theorem 3. More precisely, we let the circle C have radius $\Theta(r^4)$, we let the distance between two contiguous triplets be $\Theta(r^3)$, we embed each clause on an $r^3 \times r^3$ grid, and we finally let the angle $\angle g_t o_t g'_t$ be $\Theta(1/r^3)$. It can now be verified that this construction gives a polygon of area $\Theta(r^8)$, for which it can be shown that there exists a satisfying assignment for 3-SAT iff there exists a solution for CON-SUM of costs $\Theta(r^2)$.   □

Corollary 5. CON-SUM *is not approximable within any polynomial, unless* P=NP.

Proof. Observe that the construction of the polygon P given in Theorem 6 can be inflated as follows: for any integer $k > 3$, we let the circle C have radius $\Theta(r^{k+1})$, we let the distance between two contiguous triplets be $\Theta(r^k)$, we embed each clause on an $r^k \times r^k$ grid, and we finally let the angle $\angle g_t o_t g'_t$ be $\Theta(1/r^k)$. It can now be verified that this construction gives a polygon of area $\Theta(r^{2(k+1)})$ for which an optimal solution costs $\Theta(r^2)$ (and can be found in polynomial time iff P=NP), while any approximate solution will require a pebble to me moved by $\Omega(r^k)$. Hence, since $r = \Theta(n)$, the claim follows.   □

4   Clique-Connectivity

As far as the clique-connectivity problems are concerned, we are able to provide results only for the hop distance case, while the Euclidean case remains open.

4.1   Clique-Max

Concerning CLIQUE-MAX, it is easy to see that the problem can be solved optimally up to an additive term of 1, by just guessing a vertex belonging to an optimal solution onto which all the pebbles are moved (see [4]). In spite of that, the problem is hard, as proven in the following:

Theorem 7. *The* CLIQUE-MAX *problem w.r.t. the hop distance is* NP-*hard.*

Proof. We suitably modify the reduction of Theorem 2. So, the reduction is still from the 3-SAT problem. For a given instance $\mathcal{I}$ of 3-SAT, we build an instance $\mathcal{I}'$ for the CLIQUE-MAX problem as follows: we build a simple polygon P consisting of 3η *literal* vertices $V_L = \{x_1, \check{x}_1, \bar{x}_1, \ldots, x_\eta, \check{x}_\eta, \bar{x}_\eta\}$, $5m$ *clause* vertices $V_C = \{c_{11}, c_{12}, c_{13}, p_1, q_1, \ldots, c_{m1}, c_{m2}, c_{m3}, p_m, q_m\}$, $2m$ *gate* vertices $V_G = \{g_1, g'_1, \ldots, g_m, g'_m\}$, 3 *obstacle* vertices y_1, y_2, y_3, and finally five *auxiliary* vertices $z_1, z_2, z_3, z_4, \bar{p}$. Polygon P is so constructed that, among the others, the following visibility constraints hold (see Fig. 5):

- every literal vertex x_i sees all the other literal vertices but $\bar{x}_i$, and vice versa;
- each clause vertex c_{ij} can see only $g_{\eta+i}, g'_{\eta+i}, p_i, q_i$, the other two clause vertices in its clause, and the literal vertex corresponding to the jth literal of its clause;
- each clause vertex p_i can see only $c_{i1}, c_{i2}, c_{i3}, q_i$;
- gate vertices cannot see auxiliary vertices due to the obstacle made up by y_1, y_2, y_3;
- $\bar{p}$ can see only z_3 and z_4, z_3 and z_4 see each other and can see only $z_1, z_2, \bar{p}$, while z_1 and z_2 can see all the literal vertices but not the gate vertices.

Then, we put a pebble in each clause vertex p_i, $i = 1, \ldots, m$, and a pebble in $\bar{p}$.

We now show that the 3-SAT instance $\mathcal{I}$ has a positive answer iff there exists a solution for $\mathcal{I}'$ having maximum hop distance of 2. One direction is simple. Given a satisfying assignment τ, we indeed define the following movement: each pebble in a clause moves first (with a single hop) to any clause vertex seeing a verified literal, and then it reaches the corresponding literal vertex with an additional hop. Moreover, we move the pebble in $\bar{p}$ to z_1. In this way, the assignment vertex will form a clique in the visibility graph $G(P)$, and each pebble makes 2 hops.

Concerning the other direction, suppose that there is a solution for $\mathcal{I}'$ having value at most 2. We show that such a solution can be transformed in polynomial time into a satisfying assignment for $\mathcal{I}$. First of all, notice that the pebble in $\bar{p}$ must be either in z_1 or z_2. Moreover, by construction, since the final positions of the pebbles induce a clique, it must be the case that every pebble is on a literal

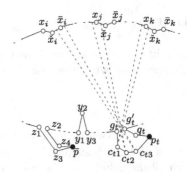

Fig. 5. The polygon P used for proving the NP-hardness of CLIQUE-MAX w.r.t. the hop distance. Pebbles sit initially on black vertices.

vertex. Indeed, the only vertices that can be reached by 2 hops from p_t and which are visible by z_1 or z_2 are the literal vertices associated with the tth clause. Moreover, we cannot have two pebbles on x_i and $\bar{x}_i$, because these two vertices cannot see each other. Hence, the final positions of the clause pebbles define a truth assignment for the formula. Notice that it can be the case that there is no pebble in x_j nor in $\bar{x}_j$ (i.e., the corresponding variable is not instrumental to guarantee the satisfiability of $\mathcal{I}$). In this case we assign an arbitrary value to x_j.

It remains to show that P can be constructed in polynomial time. Actually, the construction is similar to that used in Theorem 2, and so we leave it to the reader. We just point out that the angle $\angle \bar{x}_i x_i \check{x}_i$ must be $\Theta(1/r^2)$, in order to hide only $\bar{x}_i$ to x_i (i.e., the distance between $\check{x}_i$ and the ray passing through $x_i, \bar{x}_i$ will be $\Theta(1)$). $\qquad\square$

Since the problem is hard already when the optimal solution costs 2, we have:

Corollary 6. *For any $\epsilon > 0$, CLIQUE-MAX w.r.t. the hop distance cannot be approximated within $3/2 - \epsilon$, unless P = NP.*

Moreover, the following implication is also easy to prove:

Corollary 7. *Deciding whether CLIQUE-MAX admits a solution with at most h hops is NP-complete, for any $h \geq 2$.*

Proof. Case $h = 2$ follows directly from Theorem 7. For $h > 2$, it suffices to suitably modify the polygon P used in Theorem 7 in such a way that the pebbles need to move for $h - 1$ steps in order to see the literal and the clause vertices. $\square$

4.2   Clique-Sum

Concerning CLIQUE-SUM, once again we restrict ourselves to the hop distance case. First of all, notice that in this case the problem is 2-approximable [4]. However, it turns out that a slight modification of the reduction used for CLIQUE-MAX yields the following:

Theorem 8. *The* CLIQUE-SUM *problem w.r.t. the hop distance is* NP-*hard.*

Proof. We use the same construction as in Theorem 7. W.l.o.g. we assume that in the instance of 3-SAT the mth clause c_m contains only the variable x_η (either negate or not), and that x_η occurs only in c_m. We claim that $\mathcal{I}$ is satisfiable iff $\mathcal{I}'$ admits a solution of total movement at most $2(m+1)$. One direction is immediate, since we have proved that if $\mathcal{I}$ has a satisfying truth assignment then we can move the pebbles towards a clique with maximum movement of 2. Now, assume that we have a solution of total movement of at most $2m + 2$. We will show that every pebble must move by at least 2 hops to guarantee the clique constraint (and so actually at least $2m + 2$ hops are needed). This immediately implies the claim, since this means that each pebble moves exactly 2 steps, and so we can compute (in polynomial time) a truth assignment for $\mathcal{I}$ by using the same arguments used in Theorem 7. First of all, observe that the hop distance between any two p_t and $p_{t'}$ is at least 4 (i.e., it is 4 if c_t and $c_{t'}$ share a literal, otherwise is 5). Moreover, the hop distance between any p_t and $\bar{p}$ is 5. Finally, for our assumption about instance $\mathcal{I}$, the hop distance between p_t and p_m is 5, for every $t \neq m$. Let h be the movement of a pebble p. In order to move all the pebbles in a feasible configuration, we have that two pebbles have been moved by at least $4 - h$ hops, and the remaining $m - 2$ pebbles have been moved by at least $3 - h$ hops. Summing up over all the pebbles, the total movement is at least $3m + 2 - (m - 1)h$, which is less than or equal to $2m + 2$ only when $h \geq \lceil m/(m-1) \rceil$, i.e., $h \geq 2$.   □

5   Discussion and Open Problems

Motion planning in a constrained environment is susceptible of a deep investigation in several respects. Here we have limited our attention to planar vertex-to-vertex motion in a simple polygon and with the objective of achieving very basic configurations, but it is easy to imagine more challenging scenarios. For instance, notice that relaxing the assumption that pebbles have to start, turn, and stop at vertices only will make the planning task substantially more difficult. On the other hand, a simplifying yet very interesting setting is that in which the constraining polygon is orthogonal.

As far as the problems in our setting are concerned, we point out that it remains open to understand the computational properties of CLIQUE (both MAX and SUM) w.r.t. the Euclidean distance. Moreover, establishing whether CLIQUE-MAX for the hop distance is hard already when an optimal solution costs 1 is very intriguing: indeed, such a case retains a strong connection with the CLIQUE DOMINATING SET (CDS) problem (i.e., deciding whether a graph has a dominating clique). For general graphs, it is known that this problem is NP-complete, while it is unknown whether CDS is NP-complete for visibility graphs. Notice that if CDS was NP-complete for visibility graphs, we would have the NP-hardness of CLIQUE-MAX already restricted to instances where an optimal solution costs 1 (indeed, it suffices to consider instances with a pebble in each vertex). Conversely, if we prove that CLIQUE-MAX is polynomially solvable for $h = 1$, then this implies that CDS for visibility graphs is also decidable in polynomial time. Finally, we feel that an improvement of the 136-approximation algorithm for CON-MAX w.r.t. the hop distance might be possible, by exploiting the special nature of visibility graphs.

Acknowledgements. The authors wish to thank an anonymous referee for her/his insightful comments, which helped us in improving the paper.

References

1. Ahmadian, S., Friggstad, Z., Swamy, C.: Local-search based approximation algorithms for mobile facility location problems. arXiv preprint:1301.4478 (2013)
2. Aronov, B., Fortune, S., Wilfong, G.T.: The furthest-site geodesic Voronoi diagram. Discrete Comput. Geom. **9**, 217–255 (1993)
3. Berman, P., Demaine, E.D., Zadimoghaddam, M.: $O(1)$-approximations for maximum movement problems. In: Goldberg, L.A., Jansen, K., Ravi, R., Rolim, J. (eds.) RANDOM 2011 and APPROX 2011. LNCS, vol. 6845, pp. 62–74. Springer, Heidelberg (2011)
4. Bilò, D., Gualà, L., Leucci, S., Proietti, G.: Exact and approximate algorithms for movement problems on (special classes of) graphs. In: Moscibroda, T., Rescigno, A.A. (eds.) SIROCCO 2013. LNCS, vol. 8179, pp. 322–333. Springer, Heidelberg (2013)
5. Chazelle, B.: Triangulating a simple polygon in linear time. Discrete Comput. Geom. **6**, 485–524 (1991)
6. Demaine, E.D., Hajiaghayi, M., Mahini, H., Sayedi-Roshkhar, A.S., Oveisgharan, S., Zadimoghaddam, M.: Minimizing movement. In: Proceedings of the 18th ACM-SIAM Symposium on Discrete Algorithms (SODA'07), pp. 258–267 (2007)
7. Friggstad, Z., Salavatipour, M.R.: Minimizing movement in mobile facility location problems. ACM Trans. Algorithms 7(3), article **28**, 1–22 (2011)
8. Guibas, L.J., Hershberger, J., Leven, D., Sharir, M., Tarjan, R.E.: Linear-time algorithms for visibility and shortest path problems inside triangulated simple polygons. Algorithmica **2**, 209–233 (1987)
9. Hershberger, J.: An optimal visibility graph algorithm for triangulated simple polygons. Algorithmica 4(1), 141–155 (1989)
10. Lin, Y.-W., Skiena, S.S.: Complexity aspects of visibility graphs. Int. J. Comput. Geom. Appl. 5(3), 289–312 (1995)
11. Prencipe, G., Santoro, N.: Distributed algorithms for autonomous mobile robots. In: Navarro, G., Bertossi, L., Kohayakawa, Y. (eds.) TCS 2006. IFIP, vol. 206, pp. 47–62. Springer, Heidelberg (2006)
12. Suri, S., Vicari, E., Widmayer, P.: Simple robots with minimal sensing: From local visibility to global geometry. Int. J. Robot. Res. 27(9), 1055–1067 (2008)

Fast Localized Sensor Self-Deployment
for Focused Coverage

László Blázovics[1] and Tamás Lukovszki[2](✉)

[1] Department of Automation and Applied Informatics,
Budapest University of Technology and Economics, Budapest, Hungary
`laszlo.blazovics@aut.bme.hu`
[2] Faculty of Informatics, Eötvös Lóránd University, Budapest, Hungary
`lukovszki@inf.elte.hu`

Abstract. We consider the focused coverage self-deployment problem in mobile sensor networks, where an area with maximum radius around a Point of Interest (POI) must be covered without sensing holes. Li et al. [9,10] described several algorithms solving this problem. They showed that their algorithms terminate in finite time. We present a modified version of the Greedy-Rotation-Greedy (GRG) algorithm by Li et al., which drive sensors along the equilateral triangle tessellation (TT) graph to surround a POI. We prove that our modified GRG (mGRG) algorithm is collision free and always ends up in a hole-free network around the POI with maximum radius in $O(D)$ steps, where D is the sum of the initial distances of the sensors from the POI. This significantly improves the previous bound on the coverage time. The theoretical results are also validated by simulations.

Keywords: Self-deployment · Mobile sensor network · Localized algorithms

1 Introduction

Mobile sensor networks (MSN) are distributed collections of nodes, where each node has sensing, computation, communication and locomotion capabilities.

Partially supported by the EU EIT project SmartUC and the grant EITKIC-12-1-2012-0001 of the National Development Agency. Performed in cooperation with the EIT ICT Labs Budapest Associate Partner Group. Partially supported by the European Union and the European Social Fund through project FuturICT.hu (grant no.: TAMOP-4.2.2.C-11/1/KONV-2012-0013) organized by VIKING Zrt. Balatonfüred. Partially supported by the Hungarian Government, managed by the National Development Agency, and financed by the Research and Technology Innovation Fund (grant no.: KMR_12-1-2012-0441). This work is connected to the scientific program of the "Development of quality-oriented and cooperative R+D+I strategy and functional model at BUTE" project.

P. Flocchini et al. (Eds.): ALGOSENSORS 2013, LNCS 8243, pp. 83–94, 2014.
DOI: 10.1007/978-3-642-45346-5_7, © Springer-Verlag Berlin Heidelberg 2014

By assuming a large scale sensor network with unpredictable sensor failure, limited sensing and communication range, decentralized or localized sensor self-deployment methods are more beneficial and scale invariant than centralized solutions. In this context *localized* means that each sensor makes independent decisions using neighborhood information only.

There are situations where sensors should cover a dedicated area around a Point of Interest (POI). These scenarios are typical in such cases like area discovery for survivors around the epicenter of disaster. In these cases the area close to the POI has higher priority and it is more important to be covered than the distant one. This type of coverage is called focused coverage or F-coverage [10].

In this article we present a localized, synchronous algorithm for the sensor self-deployment problem with optimal F-coverage.

1.1 Focused Coverage (F-Coverage)

We follow the terminology of Li et al. [10]. The *coverage region* of a sensor network is the region which is enclosed by the outer boundary of the network. If the coverage is not complete there will be still *sensing* (or coverage) *holes*. Sensing holes are closed areas inside the coverage region which are not covered by the sensing range of the sensors.

The *coverage radius* (or *radius of an F-coverage*) is the radius of the maximal hole-free disc enclosed by sensors and centered as POI. The optimal F-coverage has maximized coverage radius. If the number of sensors is unlimited and the sensing radius of the sensors approaching zero then the maximum hole-free disc has a circular shape. Since the sensing radius of the sensors is finite, we consider a discrete variant of coverage radius measured by *layer distance*. Layer distance, also called convex layers in computational geometry represents the number of successive complete convex polygons adjacently surrounding POI. More precisely, we consider a discrete set of convex polygons $P_i, (i = 1, 2, ...)$ composed of sensors, centered at POI, and having a diameter of $i \cdot d$ for some constant d. Then the coverage radius is the maximum value k, such that P_k is completely in the coverage region.

1.2 The Equilateral Triangle Tessellation

The equilateral triangle tessellation is a tiling of the plane in equilateral triangles with no overlaps and no gaps. The equilateral triangle tessellation (TT) maximizes the coverage area of a given number of sensors without coverage gap when sensor separation is equal to $r_s\sqrt{3}$, where r_s is the sensing radius of the sensors [1,11]. If the communication radius r_c of the sensors is at least $r_s\sqrt{3}$, the deployment of the sensors corresponding to a TT layout guarantees the connectivity of the network. The convex polygons defining the layers and the layer distance of the F-coverage are hexagons centered at the POI.

1.3 Problem Statement

We are given n mobile sensors with communication radius r_c and sensing radius is r_s of each, $r_c \geq r_s\sqrt{3}$. We assume that the n mobile sensors are initially placed at the vertices of the TT, such that each sensor is placed in a different vertex. This is an unrealistic assumption if the sensors are dropped from a plane. In that case the sensors can perform the Snap and Spread algorithm by Bartolini et al. [2] to achieve the above condition. The sensors may be disconnected at the beginning. All sensors have a common coordinate system and they all know the location of the POI. Without loss of generality, the POI at the origin of the coordinate system. Furthermore, the sensors only have information about their 1- and 2-hop neighbors. The sensors are able to move only on the edges of the TT graph (see Fig. 2). They all move synchronously with uniform speed, s.t. they travel an edge of the TT in one time unit.

The sensors operate corresponding to the *Look-Compute-Move* model. In one cycle, a sensor takes a snapshot of the current configuration (Look), makes a decision to stay idle or to move to one of its adjacent nodes (Compute), and in the latter case makes an instantaneous move to this neighbor (Move).

The motion ends when the sensors uniformly surround the POI by forming hole-free network with maximized coverage radius. From now on we will use the terms node and sensor interchangeably.

1.4 Our Contribution

We present a modified version of the GRG/CV algorithm of Li et al. [10]. We prove that our modified GRG (mGRG) algorithm guarantees, that after $O(D)$ steps each node reaches its final layer, where D is the sum of initial hop distances of the nodes from the POI in the TT. We validate our theoretical results also with simulations.

An important difference between the requirements of the GRG of Li et al. [10] and our mGRG algorithm is that the GRG in [10] uses the knowledge about the 1-hop neighborhood of the sensors, while our mGRG algorithm needs the knowledge of the 2-hop neighborhood. We give examples, that show that the knowledge about the 2-hop neighborhood of the sensors is necessary to avoid collision situations and make the deployment process faster.

This paper is organized as follows. Section 2 gives an overview of related work. In Sect. 3 we introduce our mGRG algorithm and mathematical notations. We prove the convergence of the mGRG algorithm in Sect. 4 and present an $O(D)$ upper bound on the surrounding time. Section 5 presents our experimental results. Finally, Sect. 6 summarizes the work.

2 Related Work

In the field of mobile sensor networks sensor self-deployment problem has been an important research topic that deals with autonomous coverage formation.

In the article of Gage et al. [6] three type of formation was introduced. However it was a military oriented article, from the perspective of the F-coverage only the *blanket* formation is relevant. In this formation the nodes form static connected group in order to maximizes the detection rate of targets appearing within the coverage area.

The most common sensor self-deployment method is the vector or virtual-force-based approach. The algorithms which rely on this approach use potential fields, generated around the sensors which moves the neighbors by attract or repulse them (depending on the distance). The first work which used this approach was published by Howard et al. [7].

Large amount of research deals with sensor deployment algorithms for coverage formation over a Region of Interest (ROI). An excellent summary can be found in the works of Nayak et al. [12] and Brass et al. [3].

Cortes et al. [5] proposed Voronoi diagram based sensor self-deployment method for the coverage of the ROI. The main idea of self-deployment with Voronoi diagrams is to move sensors to minimize their local uncovered areas (equivalently speaking, to maximize their sensing-effective areas) by aligning their sensing range with their Voronoi regions as much as possible.

Li at al. [8–10] introduced the F-coverage problem. They solved the problem in a discrete case on an equilateral triangle tessellation. Collision of sensors during the deployment was allowed, i.e. more than one sensors can occupy the same triangle vertex at the same time. They presented a proof of the convergence of their solution within finite time. The convergence time, energy consumption and number of collisions has been evaluated by simulations.

In the work of Yang et al. [13] a distributed load-balancing sensor self-deployment algorithm was presented which partitions the plane into a 2D mesh, and treats nodes as load. By this algorithm, nodes in each cell form a cluster covering the cell and are managed by an elected cluster head. This approach also requires dense network coverage and inter-agent communication.

Bartolini et al. [2] have presented a localized algorithm on a hexagonal grid map in which the entities simultaneously use the *snap* and the *spread* activities in order to cover the given area. The nodes are dispersing from their initial position while occupying the free hexagons. On each occupied hexagon only the occupier allowed to stay, which forwards the others towards the borderline of the covered area.

Cord-Landwehr et al. [4] studied the problem of gathering mobile robots with an extent at a fixed position as dense as possible to form a disk of minimum radius around the gathering point. The authors present an algorithm for the continuous case and the discrete case, where the robots are moving on a grid. They prove an $O(nR)$ upper bound for the gathering time, where n is the number of robots and R is the distance of the farthest robot from the gathering point. They empirically studied the continuous case, where in they report a few deadlock situations in the simulations.

3   The Modified GRG

Before the introduction of our modified GRG algorithm, we briefly review the collision avoidance version GRG/CV [10] of the GRG algorithm.

3.1   The GRG/CV Algorithm

The Greedy-Rotation-Greedy (GRG) algorithm and its version with collision avoidance (GRG/CV) are designed for the asynchronous model. The self-deployment decision of the sensors only uses the information about the 1-hop neighborhood. In order to keep the description simple, we declare the POI unoccupyable. Therefore, the POI related rules were not used. The sensors try to move toward the POI along the TT edges and decrease the hop distance to the POI step by step. This movement is called greedy advance movement. If the greedy advance movement is blocked, the sensors use another type of movement, called rotation movement, i.e. they try to move on the same layer. The rotation is restricted to a particular, say the counterclockwise, direction so as to avoid unnecessary collision among rotating nodes. The key is that a sensor should not move away from the POI once it moves closer to it. A sensor stops rotating when it reaches a vertex where greedy advance can resume, or when it returns to the vertex where it started rotating or the rotational movement is blocked. In the case that a greedy advance movement and a rotation movement target the same vertex, a competition rule is applied, which gives higher priority to the greedy advance movement.

Although the GRG/CV is an asynchronous algorithm, in most cases it works also in a synchronous environment. However, as illustrated on Fig. 1(a), there are situations where a sensor is unable occupy an empty vertex. The sensor u on the second layer is unable to move to the empty vertex, because it only knows its 1-hop neighborhood. Thus, u does not see the empty vertex. Therefore, u rotates counterclockwise. In the same time step v moves to the empty place, and the previous position of v becomes empty. After one step (See Fig. 1(b)) the u is in front of the empty place, however due to the *Safety Rule* in [10] it is not allowed to occupy the empty place[1]. The empty space moves in the first layer in clockwise direction and u on the second layer in counterclockwise direction. After three steps the same situation appears as in Fig. 1(a), just rotated around the POI by an angle of $2\pi/3$ counterclockwise. After making a full circle around the target, u will stop without occupying the empty place.

It is not solved in [10].

3.2   The mGRG Algorithm

The main ideas of our concept to make the surrounding of the POI faster are the following. First, each hexagonal layer is assigned a heading direction, such

---

[1] The Safety Rule in [10] describes that a node must not greedily advance unless it knows the movement is definitely safe. It says that a node u does not choose inward vertex neighbor x as greedy next hop if the neighbor of x on the layer of x in clockwise direction is not a one-hop neighbor of u.

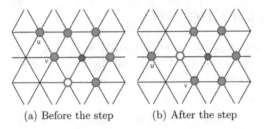

| (a) Before the step | (b) After the step |

Fig. 1. Endless loop with the modified GRG/CV

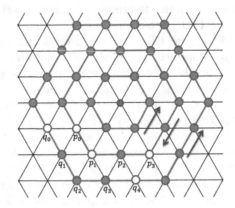

Fig. 2. The hexagonal layers (trajectories) and their heading direction in the equilateral triangle tessellation

that any two neighboring layers have opposite heading direction. For example, odd layers have counterclockwise and even layers clockwise heading direction. When a sensor on a certain layer performs a rotation step, it moves around the POI in the given heading direction. Second, if a greedy advance movement and a rotation movement target the same vertex, the rotational movement gets higher priority. This principle will ensure that each sensor can keep moving in each time step, since rotation will be always possible.

The sensors move straight towards the POI until they reach the innermost hexagonal layer. This is the *primary trajectory* (T_1).

Similarly to the base GRG, if a node is unable to the get closer to the POI – because of another node is in front of it, or it has reached the innermost layer – it should rotate on the current layer.

If the node is able to move to an inner layer it should check whether an other node is trying to get to the same place. An example can be seen on Fig. 2.

Now we define the priority rule more precisely. Consider a vertex x on the layer T_i The vertex x has at most four neighboring vertices from which it can be occupied in the next step if x is a corner vertex, and at most three, otherwise. One such neighboring vertex is on the same layer T_i and at most three vertices on the next higher layer T_{i+1}. Regarding x the highest priority is assigned to the neighbor vertex on T_i. The heading direction T_{i+1} defines an order on the neighbor vertices on T_{i+1}. The first one in this order gets the second highest priority, the second the next highest, etc. For example, in Fig. 2 vertex p_1 can be occupied from vertices q_1, q_2, q_3, p_2. The priority order from highest to lowest is p_2, q_3, q_2, q_1. A sensor u obtains the same priority than the vertex currently occupied by u. A sensor u can occupy a vertex x, if no other sensor resides on a vertex with higher priority regarding x. Note that each sensor is aware of the sensors that can occupy the same vertex, since they are in the 2-hop neighborhood of each other. Thus, each sensor can decide locally, whether it has the highest priority among them.

If a sensor u is equally far from two vertices closer to the POI than u (like q_1 from p_0 and p_1 in Fig. 2 and the heading direction of u is counterclockwise (clockwise), then the vertex left (right) from the direction of the POI is prefered. If there is another sensor with higher priority regarding this vertex, then u choose the other. If this vertex also can be occupied by a higher priority sensor, then u must rotate.

These rules imply that each sensor either moving towards the target or rotating around it, it never stays on same place in the next time step. They also imply that a node must know the 2-hop neighborhood in order to avoid collision and to detect an occupyable vertex in the next inner layer.

4   Analysis

In this section we prove that by using the mGRG algorithm a group of mobile sensors will always enclose a given POI with maximum coverage radius.

We assume that at the beginning each sensor resides on different vertices of the TT and tries to move on the edges towards of the POI.

We say that two sensor u and v are in *conflict* if they may target the same vertex x of the TT in one step.

We prove that the sensors always can move, never stuck in deadlock situation and we give a convergence guarantee of the surrounding process.

4.1   Upper Bound on the Coverage Time

First we show that each sensors can move either into the direction of the POI or rotate on the same layer around the POI. Therefore, the sum of the distances between the nodes and the POI never increases during the process.

Lemma 1. *Each sensor v, which is not on the innermost layer T_1, can move towards the POI if v is not in conflict with another sensor u with higher priority. Otherwise, v can move on its current layer around the POI in the corresponding heading direction. The distance between the POI and the sensor never increases during the coverage process.*

Proof. First we consider a sensor v which is already on the innermost layer T_1. We show that v can move on that layer and its distance never increases. Since all sensors on T_1 (if any) move in the same direction around the POI synchronously, their distance to each other remains the same, and thus, they do not cause a collision. Another sensor u is only allowed to move to T_1, if it does not cause a collision.

Now we consider a sensor v which not in T_1. If v is not in conflict with any another sensor u regarding a neighboring vertex on the next inner layer, then it moves towards the POI and its distance strictly decreases. Otherwise, by similar argument than above, v can rotate on its current layer and its distance from the POI does not change.   □

Now we are able to prove a guarantee of the convergence of the coverage process.

Theorem 1. *Until the inner layers have not been occupied with sensors (i.e. an inner orbit T_{in} contains an unoccupied vertex), the sum of hop-distances of the sensors from the POI decreases by at least 1 within $3(i+1)+1$ steps, where i is the index of the innermost layer with an unoccupied vertex T_i.*

Proof. Our rules guarantee that for each sensor the distance from the POI never increases. If a sensor is not prohibited by other sensors, it is moving towards the POI, until it reaches the innermost layer. If a sensor does not decrease its hop-distance to the POI by one unit in a time step, then it is either on the innermost layer or it is in conflict with another sensor.

Let T_i be the innermost layer which contains an unoccupied vertex. Consider a sensor v on the layer T_j, such that $j > i$ and j is smallest among them. If no such sensor exists, then we are done, all inner layers are filled. Otherwise, if v is not at a corner vertex of the hexagonal layer T_j, then v can only be prohibited to move in the direction of the POI by sensors on T_i. If v is at a corner vertex of T_j, it also can be prohibited to move in the direction of the POI by another node v' on a neighboring vertex of the same layer. Then we substitute v by v'. The sensor v decreases its distance until it reaches layer T_{i+1} where it starts the rotation. The sensor v and the unoccupied vertex on T_i rotate in opposite direction. Within $3(i+1)+1$ steps either v can move into the unoccupied vertex on T_i or another sensor filled it before v. Thus, within $3(i+1)+1$ steps at least one sensor decreased its hop-distance to the POI at least by one.   □

Theorem 2. *After $O(D)$ time steps all inner layers are filled, where D is the sum of initial hop-distances of the sensors to the POI.*

Proof. For $i \geq 1$, the number of vertices of layer T_i is $6i$. Let i^* be the smallest index, such that the number of sensors n is less than or equal to $\sum_{i=1}^{i^*} 6i$. For $1 \leq i < i^*$, let $m_i = 6i$ and let $m_{i^*} = n - \sum_{1 \leq i < i^*} 6i$. We show that after $O(D)$ time steps all vertices of layers T_i, $1 \leq i < i^*$, become occupied and layer T_{i^*} contains m_{i^*} sensors.

For $1 \leq i \leq i^*$ and $1 \leq j \leq m_i$, let $t_{i,j}$ be the time, when all layers T_ℓ, $\ell < i$, are already filled and the number of sensors on layer T_i increases from $j-1$ to j.

Let $v_{i,j}$ be the sensor moving to layer T_i at time $t_{i,j}$. T_i will be the final layer of $v_{i,j}$, since all layers closer to the POI are already filled. Let $V_{i,j}$ be the set of sensors that have already reached their final layer at time $t_{i,j}$. To simplify the description let $t_{1,0}$ be the starting time and $t_{i+1,0} := t_{i,m_i}$, for $1 \leq i < i^*$. Consider the time $dt_{i,j} := t_{i,j} - t_{i,j-1}, 1 \leq i \leq i^*, 1 \leq j \leq m_i$. Let $u_{i,j} \in V \backslash V_{i,j-1}$ be a sensor at time $t_{i,j-1}$ which is closest to the POI and not prohibited to move in the direction of the POI by nodes in $V \backslash V_{i,j-1}$. The sensor $u_{i,j}$ can move in the direction of the POI in each time step, until it reaches the layer T_{i+1}. (In case $u_{i,j}$ could not decrease its hop-distance to the POI, then it is in conflict with another sensor w on the same layer. Then we simply replace $u_{i,j}$ by w. Note that at time $t_{i,j-1}$, w was also on the same layer as $u_{i,j}$, since $u_{i,j}$ was closest to the POI.) Then after $O(i)$ rotation steps on layer T_{i+1} it can occupy an unoccupied vertex in T_i, if T_i has not been filled before this time step. Thus, we obtain that $dt_{i,j} \leq d_{i,j-1}(u_{i,j}, o) + O(i)$, where $d_{i,j-1}(u_{i,j}, o)$ is the distance of $u_{i,j}$ to the POI at time $t_{i,j-1}$, which is not greater than the initial distance $d(u_{i,j}, o)$ of $u_{i,j}$ and the POI. Furthermore, $d(u_{i,j}, o) \geq i$. Therefore, $dt_{i,j} \leq c \cdot d(u_{i,j}, o)$, for some constant $c > 1$. Thus we can conclude that the time t until all the sensors reach their final layer is:

$$t = \sum_{i=1}^{i^*} \sum_{j=1}^{m_i} dt_{i,j} \leq \sum_{u \in V} c \cdot d(u, o) = O(D).$$

□

5   Evaluation

In order to evaluate our solution we have implemented both the GRG/CV and the mGRG algorithms in our custom synchronous simulation environment. We performed simulations where the nodes were placed uniformly at random on the vertices of a TT graph. In all scenarios the POI was placed on one of the centre vertex of the TT graph. The sensors know their 2-hop neighborhood in the TT. Due to the synchronous environment, the speed and the taken distance were the same for each sensor in each time step. We made two group of simulations during the evaluation process.

In the first group we measured the performance of the GRG/CV and the mGRG when the *dropping area* was a fixed 30×30 square area and network size was varying from 30 to 315 nodes. With each parameter we performed 20-20 simulations. In the second group we kept the number of the nodes fixed (90), while we varied the size of the dropping area from 25×25 to 45×45. We performed 40-40 simulations with each parameter.

5.1   GRG/CV vs. mGRG

Below we will introduce the results of the two simulation groups. The results are visualised on Fig. 3. Figure 3(a) and (b) show, that the mGRG always required less time steps than GRG/CV for finalising the coverage.

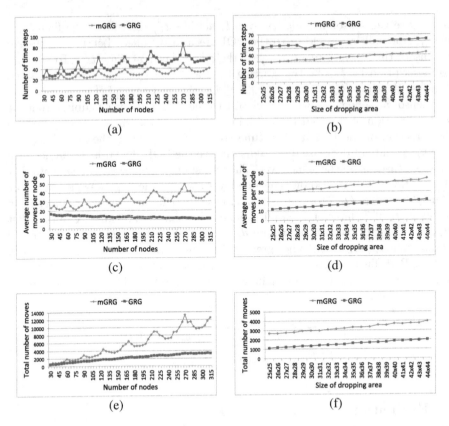

Fig. 3. Simulation results for mGRG (blue) GRG/CV (red), (a)–(c) var. size network fixed size dropping area, (d)-(f) fixed size network, var. size dropping area, (a)(d) Coverage time (b)(e) average moves per node (c)(f) overall moves of nodes (Colour figure online)

5.2 Fixed Sized Dropping Area and Varied Network Size

Figure 3(a) shows the coverage time of the algorithms. As it was noticed in simulations of Li et al. [10] for the GRG/CV algorithm the curves are tendentially increasing however they contain similar intervals in which they descend. Our mGRG algorithm shows a similar behavior. This is because both algorithms do not converge until there is no node which are able to move closer to the POI. However if the outermost layer has more free vertices than moving nodes, the occupation of these vertices required less steps than those situations when all vertices must be occupied. The latter case can be observed in the peaks of Fig. 3(a) where the number of nodes enables to fully fill all the layers. In both situations mGRG always performs better than GRG/CV. It can be observed that difference will be more significant with increasing number of sensor nodes.

Figure 3(c) shows the average number of steps taken by the sensor nodes. Because the nodes in mGRG never stop moving, each node make the same amount of steps which is equal to the convergence time. In contrast, the nodes in GRG/CV often stop moving in order to allow neighbors with higher priority to continue their motion and to avoid collisions. The moving of the nodes is prohibited more frequently as the density of the nodes increases. That is why the average number of steps in GRG/CV decreases and that is the main reason why the GRG/CV is slower than mGRG.

Figure 3(e) shows the total number of moves taken during the coverage process. GRG/CV performs less moves than the mGRG it is less sensitive to number of nodes. However we should note this is because that nodes in GRG/CV often stop and then start moving again. The nodes in mGRG keep moving even they reached their final layer.

5.3 Fixed Sized Network and Varied Sized Dropping Area

Figure 3(b) shows that the mGRG performs the converage faster than the GRG/CV. In addition it is less dependent on the startup constellation. It can be also observed that these curves are monotonically increasing. This is because the number of the entities is fixed and the coverage time only depends on the size of the dropping area.

Figure 3(d) shows, that both the GRG/CV and the mGRG require more steps as the dropping area getting larger. It can be also observed that in that case nodes in the GRG/CV take less steps. This is mainly caused by the stopped nodes as we have already described.

Similar to the results of the first group the overall number of time steps of GRG/CV was less in various field sizes too, as it can be seen of Fig. 3(f). As it was already described this is because the nodes in mGRG are always moving even when they have reached their final layer.

5.4 Simulation Summary

In the simulations our mGRG algorithm were always faster but it required more moving steps than the GRG/CV. We note in certain situations like aerial application, the difference between standing or hovering and moving is not significant from the perspective of energy consumption. In those scenarios the coverage time is more relevant.

6 Conclusions

We have presented a new algorithm mGRG to solve the focused coverage problem in self-deploying mobile sensor networks. Our algorithm is a modified version of the GRG/CV algorithm by Li et al. [10]. We have proved that our algorithm always guarantees that the sensor nodes enclose the POI without sensing holes in $O(D)$ time step, where D is the sum of distances of the nodes from the POI in the initial configuration. This significantly improves the previous bound on the coverage time. The theoretical results are also validated by simulations. The simulations show that our mGRG algorithms results in a faster coverage than the GRG/CV.

References

1. Bai, X., Kumar, S., Xuan, D., Yun, Z., Lai, T.H.: Deploying wireless sensors to achieve both coverage and connectivity. In: MobiHoc, pp. 131–142 (2006)
2. Bartolini, N., Calamoneri, T., Fusco, E.G., Massini, A., Silvestri, S.: Snap and spread: a self-deployment algorithm for mobile sensor networks. In: Nikoletseas, S.E., Chlebus, B.S., Johnson, D.B., Krishnamachari, B. (eds.) DCOSS 2008. LNCS, vol. 5067, pp. 451–456. Springer, Heidelberg (2008)
3. Brass, P.: Bounds on coverage and target detection capabilities for models of networks of mobile sensors. ACM Trans. Sen. Netw. **3**(2), 9 (2007)
4. Cord-Landwehr, A., et al.: Collisionless gathering of robots with an extent. In: Černá, I., Gyimóthy, T., Hromkovič, J., Jefferey, K., Královič, R., Vukolić, M., Wolf, S. (eds.) SOFSEM 2011. LNCS, vol. 6543, pp. 178–189. Springer, Heidelberg (2011)
5. Cortes, J., Martinez, S., Karatas, T., Bullo, F.: Coverage control for mobile sensing networks. IEEE Trans. Robot. Autom. **20**(2), 243–255 (2004)
6. Gage, D.W.: Command control for many-robot systems. Naval Command Control and Ocean Surveillance Center RDT and E Div San Diego CA (1992).
7. Howard, A., Matarić, M.J., Sukhatme, G.S.: An incremental self-deployment algorithm for mobile sensor networks. Auton. Robots **13**(2), 113–126 (2002)
8. Li, X., Frey, H., Santoro, N., Stojmenovic, I.: Localized sensor self-deployment for guaranteed coverage radius maximization. In: IEEE ICC '09. pp. 1–5 (2009)
9. Li, X., Frey, H., Santoro, N., Stojmenovic, I.: Focused-coverage by mobile sensor networks. In: 6th IEEE International Conference on Mobile Adhoc and Sensor Systems (MASS), pp. 466–475 (2009)
10. Li, X., Frey, H., Santoro, N., Stojmenovic, I.: Strictly localized sensor self-deployment for optimal focused coverage. IEEE Trans. Mob. Comput. **10**(11), 1520–1533 (2011)
11. Ma, M., Yang, Y.: Adaptive triangular deployment algorithm for unattended mobile sensor networks. IEEE Trans. Comput. **56**(7), 946–958 (2007)
12. Nayak, A., Stojmenovic, I.: Wireless Sensor and Actuator Networks: Algorithms and Protocols for Scalable Coordination and Data Communication. Wiley-Interscience, New York (2010)
13. Yang, S., Li, M., Wu, J.: Scan-based movement-assisted sensor deployment methods in wireless sensor networks. IEEE Trans. Parallel Distrib. Syst. **18**(8), 1108–1121 (2007)

Minimal Solvers for Unsynchronized TDOA Sensor Network Calibration

Simon Burgess[1]([✉]), Yubin Kuang[1], Johannes Wendeberg[2],
Kalle Åström[1], and Christian Schindelhauer[2]

[1] Centre for Mathematical Sciences, Lund University, Lund, Sweden
{simonb,yubin,kalle}@maths.lth.se
[2] Department of Computer Science, University of Freiburg, Freiburg, Germany
{wendeber,schindel}@informatik.uni-freiburg.de

Abstract. We present two novel approaches for the problem of self-calibration of network nodes using only TDOA when both receivers and transmitters are unsynchronized. We consider the previously unsolved minimum problem of far field localization in three dimensions, which is to locate four receivers by the signals of nine unknown transmitters, for which we assume that they originate from far away. The first approach uses that the time differences between four receivers characterize an ellipsoid. The second approach uses linear algebra techniques on the matrix of unsynchronized TDOA measurements. This approach is easily extended to more than four receivers and nine transmitters. In extensive experiments, the algorithms are shown to be robust to moderate Gaussian measurement noise and the far field assumption is reasonable if the distance between transmitters and receivers is at least four times the distance between the receivers. In an indoor experiment using sound we reconstruct the microphone positions up to a mean error of 5 cm.

1 Introduction

In this paper we study the problem of node localization using only Unsynchronized Time difference Of Arrival (UTOA) measurements between nodes, where either receivers or transmitters are far away from the other group. The problem arises naturally in microphone arrays for audio sensing. Is it possible to calculate both multiple microphone positions as well as the timings and directions of the sound sources, if the microphones are unsynchronized, i.e. do not use the same clock, just from sounds emanating from far away at unknown locations and times? An example application could be to locate several cell phones just by environmental sounds, where cell phone positions and sound directions are to be recovered without synchronizing the phones first.

1.1 Related Work

Although time of arrival (TOA) and time difference of arrival (TDOA) problems have been studied extensively in the literature in the form of localization of e.g. a

P. Flocchini et al. (Eds.): ALGOSENSORS 2013, LNCS 8243, pp. 95–110, 2014.
DOI: 10.1007/978-3-642-45346-5_8, © Springer-Verlag Berlin Heidelberg 2014

sound source using a calibrated array, see e.g. [4,6–8], the problem of calibration of a sensor array from only measurements, i.e. the node localization problem, has received less attention.

In [21] and refined in [12] a far field approximation was utilized to solve the TOA and TDOA case, with the minimal number of four receivers and six unsynchronized far field transmitters in 3D. Under the assumption that signals and receivers are distributed in the unit disk, the distance between receivers can be approximated by evaluation of the range of time differences [3,16,18] or by statistical analysis of their distribution [10,20], although these approaches depend on the availability of a large number of signals. Calibration of TOA networks using only measurements has been studied in [14,19], where solutions to the minimal cases of three transmitters and three receivers in the plane, or six transmitters and four receivers in 3D are given. Calibration of TDOA networks is studied in [17] and further improved upon in [13], where the non-minimal case of eight transmitters and five receivers is solved. In [2,23] a TDOA setup is used for indoor navigation based on non-linear optimization, but the methods can get stuck in local minima and are dependent on initialization.

The problem of node localization using only UTOA measurements from unsynchronized receivers and transmitters in a far field setting has been considered in [5], however the approach requires at least five receivers, which is more than the minimum case. Minimal algorithms are of importance in RANSAC schemes [9] to weed out outliers in noisy data which is a common problem in TOA/TDOA/UTOA applications. The problem has been addressed in a different manner estimating ellipse coefficients in [22], but no analysis of degenerate cases has been done and the algorithm is only described for the planar case.

In this paper we expand on previous work and propose two novel algorithms for parameter estimation of a receiver array, the Ellipsoid method in 3D and the Matrix Factorization method for UTOA measurements, that both consider the minimum case of four receivers and nine transmitters in three dimensions. We compare the methods on simulated and real data where we demonstrate their numerical stability. The methods are also evaluated on overdetermined cases using more than four receivers and nine transmitters.

2   Problem Setting

In the following treatment, we make no difference between real and virtual transmitters. Assume that the transmitters are stationary at position $\mathbf{b}_j \in \mathbb{R}^3$, $j = 1, \ldots, k$ and that the receivers are at positions $\mathbf{r}_i \in \mathbb{R}^3$, $i = 1, \ldots, m$. By measuring how long time the signals take to reach the receiver and knowing the speed of the signals, distances $\delta_{ij} = \|\mathbf{r}_i - \mathbf{b}_j\|$ can be measured, $\|\cdot\|$ denoting the Euclidean norm. These are TOA measurements.

When neither receivers or transmitters are synchronized, for instance external sound sources recorded on different computers, the measurements will be of the form $\delta_{ij} = \|\mathbf{r}_i - \mathbf{b}_j\| + f_i + \tilde{g}_j$ where $f_i, \tilde{g}_j$ are unknown offsets for receivers and transmitters respectively. We denote measurements of this kind Unsynchronized

Time difference Of Arrival (UTOA) measurements. Furthermore, if the transmitters are so far from the receivers that a transmitter can be considered to have a common direction to the receivers, the measurements can be approximated by

$$\delta_{ij} = \|\mathbf{r}_i - \mathbf{b}_j\| + f_i + \tilde{g}_j \approx \|\mathbf{r}_1 - \mathbf{b}_j\| + (\mathbf{r}_i - \mathbf{r}_1)^T \mathbf{n}_j + f_i + \tilde{g}_j = \mathbf{r}_i^T \mathbf{n}_j + \bar{g}_j + f_i + \tilde{g}_j \quad (1)$$

where $\bar{g}_j = \|\mathbf{r}_1 - \mathbf{b}_j\| - \mathbf{r}_1^T \mathbf{n}_j$ and $\mathbf{n}_j$ is the direction of unit length from transmitter j to the receivers. By setting $g_j = \bar{g}_j + \tilde{g}_j$ we get the far field approximation

$$\delta_{ij} \approx \mathbf{r}_i^T \mathbf{n}_j + f_i + g_j.$$

When the approximation is good, we will call δ_{ij} Far Field UTOA (FFUTOA) measurements.

2.1 The FFUTOA Calibration Problem

We assume that (i) the speed of signals v is known, and thus all time measurements are transformed to distances by multiplication by v and (ii) receivers can distinguish which TOA signal comes from which sender. This can be done in practice by e.g. separating the signals temporally or by frequency.

Problem 1. Given mk FFUTOA measurements $\delta_{ij} \in \mathbb{R}, i = 1, \ldots, m, j = 1, \ldots, k,$ taken from m receivers and k transmitters, estimate receiver positions $\mathbf{r}_i \in \mathbb{R}^3$, directions $\mathbf{n}_j \in \mathbb{R}^3$ from transmitter j to receivers, receiver and transmitter offsets $f_i \in \mathbb{R}, g_j \in \mathbb{R}$ such that

$$\delta_{ij} = \mathbf{r}_i^T \mathbf{n}_j + f_i + g_j, \quad \text{and} \quad \|\mathbf{n}_j\| = 1. \quad (2)$$

Note that the problem is symmetric in receivers and transmitters, i.e. if each receiver instead could be viewed as having a common direction to all transmitters, the same problem can be solved for transmitter positions and receiver directions. We denote $\mathbf{f} = [f_1, \ldots, f_m]^T$, $\mathbf{g} = [g_1, \ldots, g_k]$, $\mathbf{r} = [\mathbf{r}_1, \ldots, \mathbf{r}_m]$ and $\mathbf{n} = [\mathbf{n}_1, \ldots, \mathbf{n}_k]$.

The problem of determining full transmitter positions $\mathbf{b}_j$ instead of directions $\mathbf{n}_j$, see (1), seems harder than using the far field approximation as in Problem 1. The measurements are now bilinearly dependent on $\mathbf{r}_i$ and $\mathbf{n}_j$. Algorithms that explicitly consider the far field assumption are also required, as the problem of determining general positions of transmitters when the far field approximation is in effect, is an ill-conditioned problem.

We denote the problem as minimal if the number of solutions for generic distance measurements δ_{ij} is finite and positive, disregarding solutions that are the same up to gauge freedom.

2.2 Gauge Freedom

The unknown parameters $(\mathbf{r}, \mathbf{n}, \mathbf{f}, \mathbf{g})$ have certain degrees of freedom that does not change the measurements, called gauge freedom. Any translation $\mathbf{t}$, rotation matrix $\mathbf{R}$ and offset change K can be applied to the solution according to

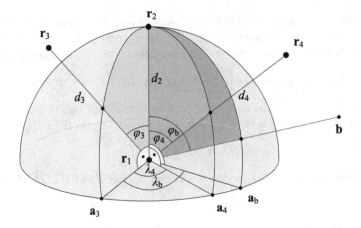

Fig. 1. Scheme of the Ellipsoid method. Three distances d_2, d_3, d_4, and three angles φ_3, φ_4, and θ define a tetrahedron of four receivers $\mathbf{r}_1$, $\mathbf{r}_2$, $\mathbf{r}_3$, $\mathbf{r}_4$. Transmitter $\mathbf{b}$ is assumed to be far away from the receivers. Its signal arrives from the angles φ_b, λ_b.

$$\mathbf{r}_{i,\text{trans}} = \mathbf{r}_i + \mathbf{t}, \; g_{j,\text{trans}} = g_j - \mathbf{t}^T \mathbf{n}_j$$
$$\mathbf{r}_{i,\text{rot}} = \mathbf{R}\mathbf{r}_i, \; \mathbf{n}_{j,\text{rot}} = \mathbf{R}\mathbf{n}_j$$
$$f_{i,\text{offs}} = f_i + K, \; g_{j,\text{offs}} = g_j - K$$

without changing the measurements δ_{ij}. Thus, we can only hope to reconstruct the unknowns up to these seven degrees of freedom.

3   The Ellipsoid Method in Three-Dimensional Space

We propose the Ellipsoid TDOA method which solves the FFUTOA calibration problem for four receivers using at least nine transmitters. The time differences of signals from distant emitters form an ellipsoid which characterizes the distances and angles between four receivers. An elegant representation can be derived from the knowledge that an ellipsoid corresponds to a covariance matrix. Once this covariance matrix is known, one can extract the parameters that generate the ellipsoid from the matrix, i.e. the configuration of four receivers.

3.1   Definition of the Covariance Ellipsoid

A rigid tetrahedron of four receivers is defined by three distances $d_2 = \|\mathbf{r}_1 - \mathbf{r}_2\|$, $d_3 = \|\mathbf{r}_1 - \mathbf{r}_3\|$, $d_4 = \|\mathbf{r}_1 - \mathbf{r}_4\|$, two height angles $\varphi_3 = \angle_{\mathbf{r}_2\mathbf{r}_1\mathbf{r}_3}$, $\varphi_4 = \angle_{\mathbf{r}_2\mathbf{r}_1\mathbf{r}_4}$, and the azimuth angle $\lambda_4 = \angle_{\mathbf{a}_3\mathbf{r}_1\mathbf{a}_4}$, see Fig. 1. Furthermore we define $\theta = \angle_{\mathbf{r}_3\mathbf{r}_1\mathbf{r}_4}$.

A signal arrives from the angles $\varphi_b = \angle_{\mathbf{r}_2\mathbf{r}_1\mathbf{b}}$ and $\lambda_b = \angle_{\mathbf{a}_3\mathbf{r}_1\mathbf{a}_b}$, uniquely determining the direction. The signal angles with respect to two receivers are $\gamma_2 = \angle_{\mathbf{r}_2\mathbf{r}_1\mathbf{b}}$, $\gamma_3 = \angle_{\mathbf{r}_3\mathbf{r}_1\mathbf{b}}$, and $\gamma_4 = \angle_{\mathbf{r}_4\mathbf{r}_1\mathbf{b}}$. Omitting the signal index, these angles are defined by the UTOA measures according to the cosine law as

$$x = \delta_1 - \delta_2 = d_2 \cos(\gamma_2) \,, \quad y = \delta_1 - \delta_3 = d_3 \cos(\gamma_3) \,, \tag{3}$$
$$\text{and} \quad z = \delta_1 - \delta_4 = d_4 \cos(\gamma_4) \,.$$

The auxiliary points a_3, a_4, and a_b are projections of r_3, r_4, and b respectively, onto the plane orthogonal to $r_1 - r_2$ through r_1.

In the following we derive the covariance matrix for time differences in the Eq. (3) assuming uniform signal source positions. This matrix characterizes a covariance ellipsoid, [15], describing the ellipsoid which the time differences reside on. If this matrix is known, the distances and angles between the receivers can be directly read from the matrix. We state the following definition.

Definition 1. *The Σ-ellipsoid for covariance matrix Σ is the ellipsoid with center μ where for all points x holds*

$$d_{Mah}(x, \mu, \Sigma) = \sqrt{(x - \mu)^T \Sigma^{-1} (x - \mu)} = 1 .$$

The metric $d_{\text{Mah}}(x, \mu, \Sigma)$ is the Mahalanobis distance. For Σ-ellipsoids the following holds.

Lemma 1. *The covariance of points uniformly distributed over a Σ-ellipsoid in $\mathbb{R}^3$ is $\hat{\Sigma} = \frac{1}{3}\Sigma$. In the two-dimensional case the covariance is $\hat{\Sigma} = \frac{1}{2}\Sigma$.*

Lemma 1 can be verified by integration over all points of the Σ-ellipsoid and calculating the covariance. Given the definition of the covariance ellipsoid we propose the following theorem.

Theorem 1. *The time differences (x, y, z) of distant signals arriving at four receivers r_1, r_2, r_3, r_4 in space $\mathbb{R}^3$ form a $3\hat{\Sigma}$-ellipsoid with covariance matrix*

$$\hat{\Sigma} = \frac{1}{3} \begin{pmatrix} d_2^2 & d_2 d_3 \cos(\varphi_3) & d_2 d_4 \cos(\varphi_4) \\ d_2 d_3 \cos(\varphi_3) & d_3^2 & d_3 d_4 \cos(\theta) \\ d_2 d_4 \cos(\varphi_4) & d_3 d_4 \cos(\theta) & d_4^2 \end{pmatrix} .$$

Proof. The proof is directly based on the definition of a covariance ellipsoid. The first thing to show is that the matrix $\hat{\Sigma}$ is actually a covariance matrix, therefore is positive semi-definite. For simplicity we assume that the receivers are synchronized, therefore the mean μ is zero. In case they are not, synchronize the receivers by regression as described in the next Sect. 3.2.

Now, consider the continuous distribution of synchronized time differences over uniformly distributed directions of origin. Such a uniform distribution of signal origins $\hat{b}$ in space $\mathbb{R}^3$ can be created by points

$$\hat{b} = R \cdot \left(r \cos(\lambda), \ r \sin(\lambda), \ \ell \right)^T ,$$

where $\lambda \in [0, 2\pi]$ and $\ell \in [-1, 1]$ are uniformly independently distributed random variables, and $r = \sqrt{1 - \ell^2}$.

The density function of the distribution is $h(\lambda, \ell) = g(\lambda)f(\ell) = \frac{1}{4\pi}$. Without loss of generality, the tetrahedron is aligned such that r_1 is the origin, r_2 is parallel to the $\hat{z}$-axis, and r_3 resides on the $\hat{x}/\hat{z}$-plane. Assuming that the sphere is large, i.e. the signals $\hat{b}$ originate from far away, the angles of the signals are

$$\lambda_b = \lambda \quad \text{and} \quad \cos(\varphi_b) = \ell . \tag{4}$$

By using spherical trigonometry and the Eq. (3) we calculate the time differences $\hat{\mathbf{x}} = [x, y, z]^T$ with respect to the tetrahedron angles as follows

$$
\begin{aligned}
x &= d_2 \cos(\gamma_2) = d_2\Big(\cos(\varphi_b)\Big) \\
y &= d_3 \cos(\gamma_3) = d_3\Big(\cos(\varphi_3)\cos(\varphi_b) + \sin(\varphi_3)\sin(\varphi_b)\cos(\lambda_b)\Big) \qquad (5) \\
z &= d_4 \cos(\gamma_4) = d_4\Big(\cos(\varphi_4)\cos(\varphi_b) + \sin(\varphi_4)\sin(\varphi_b)\cos(\lambda_b - \lambda_4)\Big) .
\end{aligned}
$$

Note that the angles γ_2, γ_3, γ_4, are not uniformly distributed in the three-dimensional case, in contrast to the planar case. We express θ as

$$
\cos(\theta) = \sin(\varphi_3)\sin(\varphi_4)\cos(\lambda_4) + \cos(\varphi_3)\cos(\varphi_4) . \qquad (6)
$$

Using the uniform distribution of signals (4) and the time differences $\hat{\mathbf{x}}$ from Eq. (5) that follow, we show by integration that the time differences characterize a covariance matrix as stated in Theorem 1:

$$
\hat{\boldsymbol{\Sigma}} = \int_0^{2\pi}\int_{-1}^1 \hat{\mathbf{x}}\hat{\mathbf{x}}^T\, h(\gamma, \ell)\, d\ell\, d\lambda = h(\gamma, \ell)\int_0^{2\pi}\int_{-1}^1 \begin{pmatrix} x^2 & xy & xz \\ xy & y^2 & yz \\ xz & yz & z^2 \end{pmatrix} d\ell\, d\lambda
$$

$$
\overset{(4)-(6)}{=} \frac{1}{3}\begin{pmatrix} d_2^2 & d_2 d_3 \cos(\varphi_3) & d_2 d_4 \cos(\varphi_4) \\ d_2 d_3 \cos(\varphi_3) & d_3^2 & d_3 d_4 \cos(\theta) \\ d_2 d_4 \cos(\varphi_4) & d_3 d_4 \cos(\theta) & d_4^2 \end{pmatrix} . \qquad (7)
$$

Due to the quadratic form is $\hat{\boldsymbol{\Sigma}}$ positive semidefinite. Furthermore, the matrix is definite, which follows from the fact that the time differences are bounded.

The next step is to verify that the time differences are actually characterized by the matrix. The distribution of signal directions (λ_b, φ_b) is irrelevant for this step, and for application of the algorithm. However, as the points $\mathbf{b}$ in Eq. (4) cover the complete sphere, all signal directions are considered. Calculating the Mahalanobis distance by inserting $\hat{\mathbf{x}}$ and $\hat{\boldsymbol{\Sigma}}$ yields

$$
d_{\text{Mah}}\big(\hat{\mathbf{x}}, \vec{0}, 3\hat{\boldsymbol{\Sigma}}\big) = \sqrt{\hat{\mathbf{x}}^T\big(3\hat{\boldsymbol{\Sigma}}\big)^{-1}\hat{\mathbf{x}}} = 1 ,
$$

revealing that all time difference points have constant Mahalanobis distance from the origin, therefore reside on an ellipsoid, which is according to Lemma 1 the $3\hat{\boldsymbol{\Sigma}}$-ellipsoid.   □

3.2   Transformation of the Covariance Matrix

We now describe the transformation of parameters from a regression polynomial to the parameters of the covariance matrix. Under the assumption of a zero-mean ellipsoid, i.e. the receivers are synchronized, an ellipsoid is described by a polynomial equation

$$
ax^2 + by^2 + cz^2 + dxy + exz + fyz = 1 . \qquad (8)
$$

Regression of at least $m \geq 6$ signals in the equation system

$$\underbrace{\begin{pmatrix} x_1^2 & y_1^2 & z_1^2 & x_1y_1 & x_1z_1 & y_1z_1 \\ \vdots & \vdots & \vdots & \vdots & \vdots & \vdots \\ x_m^2 & y_m^2 & z_m^2 & x_my_m & x_mz_m & y_mz_m \end{pmatrix}}_{Q} \underbrace{(a,b,c,d,e,f)^T}_{u} = \vec{1}$$

and solving a least squares scheme for $u = (Q^T Q)^{-1} (Q^T \vec{1})$ yields ellipsoid parameters a to f.

An ellipsoid in space $\mathbb{R}^3$ can be represented by the matrix form $x^T \Sigma^{-1} x = 1$, where $x = [x, y, z]^T$ is a vector and Σ is a symmetric positive definite matrix

$$\Sigma = \begin{pmatrix} \sigma_1^2 & \omega_1 & \omega_2 \\ \omega_1 & \sigma_2^2 & \omega_3 \\ \omega_2 & \omega_3 & \sigma_3^2 \end{pmatrix} .$$

Substitution and conversion of the parameter set yields the parameters of the covariance matrix

$$\begin{aligned}
\sigma_1^2 &= (f^2 - 4bc) / Z & \omega_1 &= (2cd - ef) / Z \\
\sigma_2^2 &= (e^2 - 4ac) / Z & \omega_2 &= (2be - df) / Z \\
\sigma_3^2 &= (d^2 - 4ab) / Z & \omega_3 &= (2af - de) / Z
\end{aligned} \qquad (9)$$

where $Z = be^2 + cd^2 + af^2 - 4abc - def$.

In case the receivers are not synchronized, the ellipsoid is shifted to zero-mean by converting the general ellipsoid polynomial equation to a translation-invariant form. In three dimensions the general form is

$$ax^2 + by^2 + cz^2 + dxy + exz + fyz + gx + hy + jz = 1 , \qquad (10)$$

for which the parameters a to j are calculated by regression of at least nine signals. The parameters are converted to the following translation-invariant form

$$\begin{aligned}
& \hat{a}(x - \hat{u})^2 + \hat{b}(y - \hat{v})^2 + \hat{c}(z - \hat{w})^2 \\
& + \hat{d}(x - \hat{u})(y - \hat{v}) + \hat{e}(x - \hat{u})(z - \hat{w}) + \hat{f}(y - \hat{v})(z - \hat{w}) = 1.
\end{aligned} \qquad (11)$$

Calculation of $\hat{a}$ to $\hat{f}$ and $\hat{u}, \hat{v}, \hat{w}$ from the coefficients of Eq. (10) can be done in a computer algebra software by expansion of Eq. (11) and substitution of the constant term. The coefficients $\hat{a}$ to $\hat{f}$ are converted for the covariance matrix using Eq. (9). The coefficient vector $(\hat{u}, \hat{v}, \hat{w})^T$ equals the center point of the ellipse and the synchronization offset of the receivers.

According to Theorem 1, the distances and angles in the tetrahedron of receivers are now directly characterized by the coefficients of the covariance matrix. The distances and angles are calculated by

$$\begin{aligned}
d_2 &= \sqrt{3}\,\sigma_1 , & \cos(\varphi_3) &= \frac{\omega_1}{\sigma_1\sigma_2} , \\
d_3 &= \sqrt{3}\,\sigma_2 , & \cos(\varphi_4) &= \frac{\omega_2}{\sigma_1\sigma_3} , \\
d_4 &= \sqrt{3}\,\sigma_3 , & \cos(\theta) &= \frac{\omega_3}{\sigma_2\sigma_3} .
\end{aligned}$$

3.3 Degenerate Cases

When measurements δ_{ij} are corrupted by noise, or the far field assumption is violated, the solution of parameters in (10) might not yield an ellipsoid, but another type of quadric surface. For four receivers and nine transmitters, this constitutes a case of given measurements where there is no exact real solution to (2), as such time differences must lie on an ellipsoid by Theorem 1. Instead of using the regression scheme, one can obtain an approximation based on Theorem 1 by covariance estimation of the given time differences (x, y, z), denoted $\mathbf{\Sigma}^*$. Using $\hat{\mathbf{\Sigma}} = \frac{1}{3}\mathbf{\Sigma}^*$, distance and angle parameters can be estimated as in Sect. 3.2.

Other degenerate cases are when the ellipsoid is collapsed to a ellipse surface, or when transformed time differences in (11) lie on two intersecting quadric surfaces, thus giving infinite number of solutions.

4 Matrix Factorization Method

The matrix factorization method uses linear techniques to solve Problem 1 for receiver positions, transmitter directions and offsets. At least four receivers and nine transmitters are needed. Without loss of generality we assume that the solution is partially normalized for gauge freedom as the first receiver $\mathbf{r}_1 = \mathbf{0}$ and $f_1 = 0$, see Sect. 2.2.

Using the FFUTOA measurements δ_{ij}, collected in the matrix $\tilde{\mathbf{D}} = [\delta_{ij}]_{m \times k}$ we immediately obtain the unknowns g_j since $\delta_{1j} = \mathbf{r}_1^T \mathbf{n}_j + f_1 + g_j = g_j$, since $\mathbf{r}_1 = \mathbf{0}$ and $f_1 = 0$. We then subtract the first row containing g_j from all other rows of $\tilde{\mathbf{D}}$ and remove the first row of zeros to obtain a new matrix that fulfill

$$\mathbf{D}_2 = \begin{bmatrix} \mathbf{r}^T & \mathbf{f} \end{bmatrix} \begin{bmatrix} \mathbf{n} \\ \vec{1} \end{bmatrix} \tag{12}$$

where $\vec{1}$ is a vector of ones. $\mathbf{D}_2$ is a product of two matrices of rank ≤ 4 and is thus itself of rank ≤ 4. This is used in [5]. Here we further reduce the rank of the factorization by subtracting the first column of $\mathbf{D}_2$ from all the other columns and remove the first row of columns. Both steps manipulating $\tilde{\mathbf{D}}$ can be done using the compaction matrices $\mathbf{C}_m$ of size $(m-1) \times m$ and $\mathbf{C}_k$ of size $k \times (k-1)$:

$$\mathbf{C}_m = \begin{bmatrix} -1 & 1 & 0 & \ldots & 0 \\ -1 & 0 & 1 & \ldots & 0 \\ \vdots & \vdots & \vdots & \ddots & \vdots \\ -1 & 0 & 0 & \ldots & 1 \end{bmatrix}, \quad \mathbf{C}_k = \begin{bmatrix} -1 & -1 & \ldots & -1 \\ 1 & 0 & \ldots & 0 \\ 0 & 1 & \ldots & 0 \\ \vdots & \vdots & \ddots & \vdots \\ 0 & 0 & \ldots & 1 \end{bmatrix}. \tag{13}$$

Then we have

$$\mathbf{D} = \mathbf{C}_m \tilde{\mathbf{D}} \mathbf{C}_k = \begin{bmatrix} \tilde{\mathbf{r}}^T & \mathbf{f} \end{bmatrix} \begin{bmatrix} \tilde{\mathbf{n}} \\ \vec{0} \end{bmatrix} = \tilde{\mathbf{r}}^T \tilde{\mathbf{n}}, \tag{14}$$

where $\tilde{\mathbf{r}}$ equals $\mathbf{r}$ with the first receiver removed as $\mathbf{r}_1 = \mathbf{0}$, and $\tilde{\mathbf{n}}$ is a $3 \times (k-1)$ matrix with the jth column $\tilde{\mathbf{n}}_j = \mathbf{n}_{j+1} - \mathbf{n}_1$. Now we have a rank-3 factorization, thus requiring at least four receivers and four transmitters. After applying SVD to $\mathbf{D} = \mathbf{USV}^T$ we obtain the rank-3 factorization such that $\mathbf{D} = \bar{\mathbf{r}}^T \bar{\mathbf{n}}$ where $\bar{\mathbf{r}} = \mathbf{U}_3 \mathbf{S}_3$ and $\bar{\mathbf{n}} = \mathbf{V}_3^T$. $\mathbf{U}_3$, $\mathbf{S}_3$ and $\mathbf{V}_3$ are the truncated parts of the SVD corresponding to the three largest singular values. This factorization of $\mathbf{D}$ is unique up to an unknown transformation $\mathbf{H}$ i.e. $\mathbf{D} = \bar{\mathbf{r}}^T \mathbf{H}^{-1} \mathbf{H} \bar{\mathbf{n}}$. We will find $\tilde{\mathbf{n}}_j = \mathbf{H} \bar{\mathbf{n}}$ i.e. $\mathbf{n}_{j+1} - \mathbf{n}_1 = \mathbf{H} \bar{\mathbf{n}}_j$ by using the constraints that

$$\tilde{\mathbf{n}}_j^T \tilde{\mathbf{n}}_j = (\mathbf{n}_{j+1} - \mathbf{n}_1)^T (\mathbf{n}_{j+1} - \mathbf{n}_1) = 2 - 2\mathbf{n}_{j+1}^T \mathbf{n}_1 = 2 - 2(\mathbf{H}\bar{\mathbf{n}}_j + \mathbf{n}_1)^T \mathbf{n}_1$$
$$= -2\bar{\mathbf{n}}_j^T \mathbf{H}^T \mathbf{n}_1 = \bar{\mathbf{n}}_j^T \mathbf{H}^T \mathbf{H} \bar{\mathbf{n}}_j . \tag{15}$$

We apply a change of variables with a 3×3 symmetric $\mathbf{C} = \mathbf{H}^T \mathbf{H}$ and a 3×1 vector $\mathbf{v} = \mathbf{H}^T \mathbf{n}_1$. From (15), we have the following equation for transmitter j:

$$\bar{\mathbf{n}}_j^T \mathbf{C} \bar{\mathbf{n}}_j + 2\bar{\mathbf{n}}_j^T \mathbf{v} = 0. \tag{16}$$

These equations are linear in the elements of $\mathbf{C}$ and $\mathbf{v}$ which have in total 9 variables. In general, with 8 such equations (thus 9 transmitters), we can solve this homogeneous linear equation system uniquely up to scale.

We can extract the solutions for $\mathbf{C}$ and $\mathbf{v}$ from the solution to the linear equation which is valid up to an unknown scaling factor and sign. We can determine the sign by using that $\mathbf{C}$ is positive definite and compute $\mathbf{H}$ by applying Cholesky factorization $\mathbf{C} = \mathbf{H}^T \mathbf{H}$. As $\mathbf{H}^T \mathbf{H} = \mathbf{H}^T \mathbf{R}^T \mathbf{R} \mathbf{H}$ for a rotation/mirroring matrix $\mathbf{R}$, this will give $\mathbf{H}$ uniquely up to $\mathbf{R}$. But as $\mathbf{R}$ corresponds to rotating/mirroring the coordinate system, $\mathbf{R}$ is a gauge freedom according to Sect. 2.2 and can be set to the identity matrix.

We can find the scale by using the constraint $\|\mathbf{n}_1\| = \|\mathbf{H}^{-T} \mathbf{v}\| = 1$. Note that fixing the scale in this way will also guarantee that $\mathbf{n}_j^T \mathbf{n}_j = (\mathbf{H}^T \bar{\mathbf{n}}_j + \mathbf{n}_1)^T (\mathbf{H}^T \bar{\mathbf{n}}_j + \mathbf{n}_1) = \underbrace{\bar{\mathbf{n}}_j^T \mathbf{H}^T \mathbf{H} \bar{\mathbf{n}}_j + 2\bar{\mathbf{n}}_j^T \mathbf{H}^T \mathbf{n}_1}_{=0 \text{ by } (15)} + \mathbf{n}_1^T \mathbf{n}_1 = \mathbf{n}_1^T \mathbf{n}_1 = 1$. Summarizing these steps yields Algorithm 1.

Algorithm 1. *Input:* FFUTOA measurement matrix $\tilde{\mathbf{D}}$ of size $(m = 4) \times (k = 9)$. *Output:* Receiver positions $\mathbf{r}$, transmitter directions $\mathbf{n}$, receiver offsets $\mathbf{f}$ and transmitter offsets $\mathbf{g}$. *Conditions:* (i) $\mathbf{D}$ must have rank 3, (ii) the linear equations (16) must only have a null space of dimension one, (iii) $\mathbf{C}$ must be positive definite.

1. Set $g_j := \tilde{\mathbf{D}}_{1j}$ and $\mathbf{D} := \mathbf{C}_m \tilde{\mathbf{D}} \mathbf{C}_k$ where $\mathbf{C}_l, \mathbf{C}_m$ is the compaction matrices in (13)
2. Calculate the SVD $\mathbf{D} = \mathbf{USV}^T$ and set $\bar{\mathbf{r}}$ to first three columns of $\mathbf{US}$ and $\bar{\mathbf{n}}$ to first three rows of $\mathbf{V}^T$
3. For the unknowns in the symmetric matrix $\mathbf{C}$ and vector $\mathbf{v}$, get the solution space for the equations $\bar{\mathbf{n}}_j^T \mathbf{C} \bar{\mathbf{n}}_j + 2\bar{\mathbf{n}}_j^T \mathbf{v} = 0$ where $\bar{\mathbf{n}}_j$ is the jth column of $\bar{\mathbf{n}}$
4. Set the sign of the solution $\mathbf{C}, \mathbf{v}$ such that $C_{11} > 0$
5. Calculate the Cholesky decomposition $\mathbf{C} = \mathbf{H}^T \mathbf{H}$
6. Lock the scale of the solutions $\mathbf{H}, \mathbf{v}$ so that $\|\mathbf{H}^{-T} \mathbf{v}\| = 1$
7. Set $\mathbf{n}_1 := \mathbf{H}^{-T} \mathbf{v}$, $\mathbf{n}_{j+1} := \mathbf{H} \bar{\mathbf{n}}_j + \mathbf{n}_1$ and $\mathbf{r} := \mathbf{H}^{-T} \hat{\mathbf{r}}$

4.1 Degenerate Cases

Theorem 2. *Degenerate cases for the minimal algorithm are when (i) The transformed measurement matrix* $\mathbf{D}$ *has* $Rank(\mathbf{D}) \leq 2$ *or (ii) The difference of the transmitter directions* $\mathbf{n}_j - \mathbf{n}_1$ *lie on the intersection of two or more quadric surfaces with constant term 0.*

Case (i) happens iff receivers or transmitter directions lie in a plane. For (ii), the case when the transmitter directions $\mathbf{n}_j$ *lie on the intersection of two or more a quadric surfaces is a special case.*

Proof. The only time the algorithm fails is when the prerequisites are not fulfilled. This happens iff (i) $Rank(\mathbf{D}) \leq 2$ or (ii) the linear equations (16) have a null space of dimension two or more.

For case (i), step (4) will extract data from the SVD that are not uniquely determined from the measurements, but has several degrees of freedom. This will result in a reconstruction of $\mathbf{r}, \mathbf{n}$ that fulfills the measurements, but is not unique, as there are an infinite number of solutions.

$Rank(\mathbf{D}) \leq 2$ iff either receivers $\mathbf{r}$ or difference of transmitter directions $\mathbf{n}_j - \mathbf{n}_1$ in (14) are embedded in a lower dimensional subspace than assumed. Remembering that receiver positions can be translated as in Sect. (2.2), this is equivalent to receivers or transmitter directions being embedded in a plane.

For case (ii), there are at least two non linearly dependent solutions to (16). The solutions can be seen as constants for a quadric surface with radius 0 that $\bar{\mathbf{n}}_j$ should lie on, i.e.

$$\bar{\mathbf{n}}_j^T \mathbf{C}_1 \bar{\mathbf{n}}_j + \bar{\mathbf{n}}_j^T \mathbf{D}_1 = 0, \quad \bar{\mathbf{n}}_j^T \mathbf{C}_2 \bar{\mathbf{n}}_j + \bar{\mathbf{n}}_j^T \mathbf{D}_2 = 0,$$

where $[\mathbf{C}_1 \ \mathbf{D}_1] \neq \lambda[\mathbf{C}_2 \ \mathbf{D}_2]$ for all $\lambda \in \mathbb{R} \setminus \{0\}$ and $\mathbf{C}_i$ symmetric. As $\mathbf{n}_{j+1} - \mathbf{n}_1 = \mathbf{H}\bar{\mathbf{n}}_j$, this is equivalent to

$$(\mathbf{n}_j - \mathbf{n}_1)^T \mathbf{H}^{-T} \mathbf{C}_1 \mathbf{H}^{-1} (\mathbf{n}_j - \mathbf{n}_1) + (\mathbf{n}_j - \mathbf{n}_1)^T \mathbf{H}^{-T} \mathbf{D}_1 = 0,$$
$$(\mathbf{n}_j - \mathbf{n}_1)^T \mathbf{H}^{-T} \mathbf{C}_2 \mathbf{H}^{-1} (\mathbf{n}_j - \mathbf{n}_1) + (\mathbf{n}_j - \mathbf{n}_1)^T \mathbf{H}^{-T} \mathbf{D}_2 = 0, \quad (17)$$

which is equivalent of the difference of the receiver directions $\mathbf{n}_j - \mathbf{n}_1$ lying on two or more quadric surfaces with constant term 0. As a special case, if the transmitter directions $\mathbf{n}_j$ lie on two or more different quadric surfaces, then the differences $\mathbf{n}_j - \mathbf{n}_1$ will fulfill (17). □

Note that the degenerate cases characterized in (i) is inherent to the problem, not the algorithm. There are fewer degrees of freedom to estimate than assumed, and thus there is not a unique solution. If both receivers and transmitter directions lie in the same plane, a similar algorithm for 2D based on the same factorization steps and equations can readily be constructed.

A special case is when $\mathbf{C}$ is not positive definite. Then there exists no real factorization $\mathbf{C} = \mathbf{H}^T \mathbf{H}$. There exists complex factorizations, e.g. obtained using eigenvalue decomposition $\mathbf{C} = \mathbf{Q}^T \mathbf{D} \mathbf{Q} = \mathbf{Q}^T \sqrt{\mathbf{D}}^T \sqrt{\mathbf{D}} \mathbf{Q} = \mathbf{H}^T \mathbf{H}$, which results in complex solutions. These cases equate exactly to the cases where the ellipsoid method does not get an ellipsoid from solving (10), as these are the cases where there are no exact real solutions to the given measurements.

5 Extension to Overdetermined Cases and Noise

Both algorithms solve a minimal case, meaning that there are only a finite positive number of solutions to (2) given arbitrary measurements in general enough position. This can be seen from the fact that the matrix factorization algorithm does not lose any solutions from the solution space by any particular choice in any of the steps. Thus there one solution discounting gauge freedom. Another way of seeing it is by counting degrees of freedom. When using $m = 4$ receivers and $k = 9$ transmitters, the number of measurements $mk = 36$ equals to the number of unknowns, $4m + 3k - 7 = 36$ accounting for gauge freedom.

When having more than four receivers, more than nine transmitters and the measurements d_{ij} are not true FFUTOA measurements, due to noise or that the far field assumption does not hold, both methods can be extended in a straightforward manner.

For the ellipsoid method, two modifications are made. (i) When having more than nine receivers, the least squares solution to (11) can be calculated. (ii) When having more than four receivers, subproblems using only four receivers at a time are solved. With overlap of receivers used in the different subproblems, all distances between receivers can be calculated and multidimensional scaling [1] can be used to get the full coordinates of all receivers.

For the matrix factorization method, the three following modifications are made. (i) In step 4, the best rank 3 approximation can still be obtained by SVD, although $\mathbf{D}$ is not necessarily rank 3. (ii) The system of equations in step 4 will in general only have the trivial solution, but is approximated to rank 8 by SVD to still attain the expected one dimensional solution set. (iii) $\|\mathbf{n}_j\|$ is only approximately 1, so $\mathbf{n}_j$ is normalized to be of length 1.

From here on, the extended methods will be used. Note that when only minimal number of measurements are available, the extended methods are equivalent to the minimal ones.

6 Experimental Validation

To be able to evaluate the quality of a solution, receivers positions $\mathbf{r}_i$ are compared to ground truth receiver positions $\mathbf{r}_{i,gt}$. Receivers are rotated, mirrored and translated so that $\sum_i \|\mathbf{R}(\mathbf{r}_i - \mathbf{t}) - \mathbf{r}_{i,gt}\|^2$ is minimized, where $\mathbf{R}$ and $\mathbf{t}$ is a rotation/mirroring and translation respectively. Finding $\mathbf{R}$ and $\mathbf{t}$ is done by using [11]. For all experiments, relative errors are then defined as $\|\mathbf{r} - \mathbf{r}_{gt}\|_{Fro}/\|\mathbf{r}_{gt}\|_{Fro}$ where $\|\cdot\|_{Fro}$ is the Frobenius norm. All algorithms were implemented and run on a standard desktop computer in Scilab.

6.1 Simulations

For all simulations, offsets f_i and g_j are drawn from i.i.d. uniform distributions over $[0, 10]$. To evaluate the assumption that transmitters have a common direction to the receivers, transmitter positions $\mathbf{b}_j$ were uniformly distributed over a

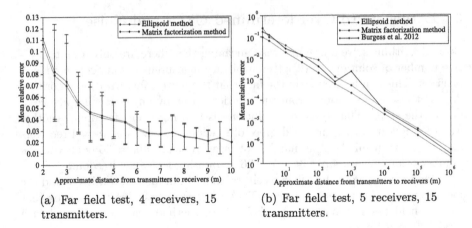

(a) Far field test, 4 receivers, 15 transmitters.

(b) Far field test, 5 receivers, 15 transmitters.

Fig. 2. Mean relative error of reconstructed receiver positions for 100 runs, plotted against the approximate distance from receivers to transmitters. (a) Bars are ±1 standard deviation for the different transmitter distances.

sphere of radius d. To be able to control how much further away transmitters were from receivers than the inter distance between receivers, four receivers were placed at a tetrahedron around the origin with side length 1 m. As signal sources are often easily obtained in applications, 15 transmitters were used. UTOA Measurements were constructed as $\delta_{ij} = \|\mathbf{r}_i - \mathbf{b}_j\| + f_i + g_j$. The mean relative error for 100 runs each plotted against the transmitter distance d to the origin can be seen for different radii d in Fig. 2a for the minimal four receivers and in Fig. 2b for five receivers. The extra receiver was uniformly distributed in the cube of which the tetrahedron of the four first receivers were inscribed to.

Figure 2a shows that using only four receivers, both algorithms can get under 5 % relative error with having transmitter approximately four times further away than the inter distance between receivers. For the experiment in Fig. 2b, we compare the results to the method in [5] as we now have the five receivers for the method to be applicable. The results indicate the ellipsoid method being slightly worse on short distances and the matrix factorization method being generally more accurate. Mean execution time was 8.0 ms, 2.1 ms and 30 ms for the ellipsoid method, matrix factorization method and the method in [5] each.

To test the robustness of the methods, white Gaussian noise was added to the measurements. The same setup as for the far field experiments was used, with transmitter distance of 10^7 from the receivers. In Fig. 3 relative error of reconstructed receiver positions are plotted against the standard deviation of the noise. The results indicate the ellipsoid method being slightly better with higher noise level when using the minimum four receivers, and the matrix factorization outperforming both the ellipsoid and the method in [5] using five receivers.

The numerical performance of the minimal algorithms were evaluated by generating problems where the measurement matrix $\tilde{\mathbf{D}}$ are FFUTOA (2). Receivers are drawn from i.i.d. uniform distributions in a cube of unit volume centered

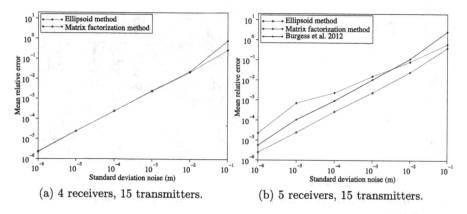

(a) 4 receivers, 15 transmitters.   (b) 5 receivers, 15 transmitters.

Fig. 3. Measurements with additive Gaussian white noise. The standard deviation is plotted against the mean relative error of reconstructed receiver positions for 100 runs.

around the origin. Nine transmitter directions and four receivers were simulated. The error distribution for 1000 such experiments can be seen in Fig. 4a. Mean execution time for the ellipsoid method and the matrix factorization method was 3.2 and 1.9 ms respectively.

6.2   Real Data

The same data as in [5] was used, where the measurements d_{ij} were obtained from an experimental setup using eight SHURE SV100 microphones as receivers and random distinct manually made sounds as transmitters. The microphones were connected to a M-Audio Fast Track Ultra 8R audio interface. The 19 sound sources were approximately 30 m away from the receivers. Microphones were set in the corners of a cuboid of roughly $100 \times 105 \times 60\,cm^3$. A picture of the experiment setup can be seen in Fig. 4b. The microphone offsets were created by adding uniformly i.i.d. silences between 0-1 s long to the beginning of each sound track, effectively starting the recordings at different unknown times. The beginning of each sound were matched by a heuristic cross correlation algorithm to create TDOA measurements.

As we have more than five microphones, the algorithms were also compared using the method in [5]. The mean reconstruction error on the microphone positions were 15 cm, 5 cm and 14 cm for the ellipsoid method, matrix factorization method and the method in [5] respectively. Most of the error are in the floor-to-roof direction. This can be explained by the sounds all being made close to ground level and thus the transmitter directions will be close to being in a plane, giving poor resolution in floor-to-roof direction.

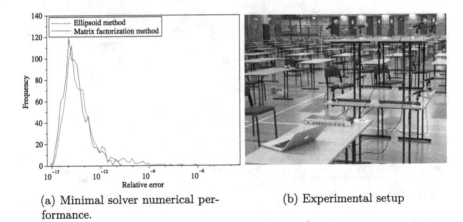

(a) Minimal solver numerical per-   (b) Experimental setup
formance.

Fig. 4. (a) Numerical performance of minimal solver in 1000 simulated experiments. (b) Setup for indoor experiment using microphones and distinct manually made sounds.

7   Conclusions

We have presented two methods for solving the previously unsolved problem of sensor network calibration using only a minimal number of unsynchronized TDOA measurements in a far field setting. The assumption of far field signals is important, as the problem of trying to determine exact positions for transmitters is ill conditioned when the far field assumption is close to true.

Simulated experiments support the feasibility of the methods, and show that the minimal algorithms are numerically stable and fast, making them suitable in RANSAC schemes to weed out outliers. They also handle additive Gaussian noise well. The far field assumption gives good results as long as transmitter-receiver distances are four times larger than inter-receiver distances.

A comparison between the two methods, running on the minimal case of four receivers and nine transmitters, indicates the matrix factorization method being slightly faster and having better worst case precision than the ellipsoid method. The ellipsoid method however has a more plausible way of handling the case when the measurements are such that no exact real solutions exist, as per Sects. 3.3 and 4.1. The ellipsoid method estimates the covariance of the time differences for parameter estimation, whereas the matrix factorization finds a complex solution.

When having more than the minimum amount of four receivers and nine transmitters, the matrix factorization is easily extended to handle more than the minimal number of receivers and transmitters, and usually exhibit better average case performance than both the ellipsoid method and the method in [5], applicable when five or more receivers are available. The ellipsoid method is easily extended to handle more than the minimal nine transmitters, but not easily extended to handle more than the minimal four receivers. Although none of the methods are formally optimal in any sense, they are in closed form, fast, and can serve as initializations for further nonlinear optimization if need be. Both methods are significantly faster than the method in [5], and the matrix factorization method performs better in reconstructing the receiver positions.

In a real world experiment in an indoor environment, both methods perform well and the matrix factorization method reconstructs microphone positions with an average of 5 cm error from the previous 14 cm in [5].

Future work of interest is developing a method for calibration of UTOA networks not using the far field assumption, thus being able to solve problems when the far field assumption is far from true.

Acknowledgements. The research leading to these results has received funding from the German Research Foundation (Deutsche Forschungsgemeinschaft, DFG) within the Research Training Group 1103 (Embedded Microsystems), the strategic research projects ELLIIT and eSSENCE, and Swedish Foundation for Strategic Research projects ENGROSS and VINST (grants no. RIT08-0075 and RIT08-0043).

References

1. Birchfield, S.T., Subramanya, A.: Microphone array position calibration by basis-point classical multidimensional scaling. IEEE Trans. Actions Speech Audio Process. **13**(5), 1025–1034 (2005)
2. Biswas, R., Thrun, S.: A passive approach to sensor network localization. In: IROS (2004)
3. Biswas, R., Thrun, S.: A distributed approach to passive localization for sensor networks. In: Proceedings of the National Conference on Artificial Intelligence, vol. 20, p. 1248. AAAI Press, Menlo Park, CA, ; MIT Press, London, Cambridge 1999 (2005)
4. Brandstein, M., Adcock, J., Silverman, H.: A closed-form location estimator for use with room environment microphone arrays. EEE Trans. Speech Audio Process. **5**(1), 45–50 (1997)
5. Burgess, S., Kuang, Y., Åström, K.: Node localization in unsynchronized time of arrival sensor networks. In: Proceedings of 21st International Conference on Pattern Recognition (ICPR 2012), pp. 2042–2046. International Association for Pattern Recognition (IAPR) & IEEE (2012)
6. Cirillo, A., Parisi, R., Uncini, A.: Sound mapping in reverberant rooms by a robust direct method. In: IEEE International Conference on Acoustics, Speech and Signal Processing, ICASSP 2008, pp. 285–288, March 31–April 4, 2008
7. Cobos, M., Marti, A., Lopez, J.: A modified srp-phat functional for robust real-time sound source localization with scalable spatial sampling. IEEE Sig. Process. Lett. **18**(1), 71–74 (2011)
8. Do, H., Silverman, H., Yu, Y.: A real-time srp-phat source location implementation using stochastic region contraction(src) on a large-aperture microphone array. In: IEEE International Conference on Acoustics Speech on, Signal Processing, vol. 1, pp. I-121-I-124, April 2007
9. Fischler, M.A., Bolles, R.C.: Random sample consensus: a paradigm for model fitting with applications to image analysis and automated cartography. Commun. ACM **24**(6), 381–395 (1981)

10. Janson, T., Schindelhauer, C., Wendeberg, J.: Self-localization application for iphone using only ambient sound signals. In: Proceedings of the 2010 International Conference on Indoor Positioning and Indoor Navigation (IPIN), pp. 259–268, November 2010
11. Arun, K.S., Huang, T.S., Blostein, S.D.: Least-squares fitting of two 3-d point sets. IEEE Trans. Patter, Anal. Mach. Intell. **9**(5), 698–700 (1987)
12. Kuang, Y., Ask, E., Burgess, S., Åström, K.: Understanding toa and tdoa network calibration using far field approximation as initial estimate. In: ICPRAM (2012)
13. Kuang, Y., ÅAström, K.: Stratified sensor network self-calibration from tdoa measurements. In: EUSIPCO (2013)
14. Kuang, Y., Burgess, S., Torstensson, A., Åström, K.: A complete characterization and solution to the microphone position self-calibration problem. In: Proceedings of ICASSP (2013)
15. Nawri, N.: Berechnung von kovarianzellipsen. http://imkbemu.physik. uni-karlsruhe.de/~eisatlas/covariance_ellipses.pdf (1996)
16. Pertila, P., Mieskolainen, M., Hamalainen, M.: Passive self-localization of microphones using ambient sounds. In: 2012 Proceedings of the 20th European Signal Processing Conference (EUSIPCO), pp. 1314–1318. IEEE (2012)
17. Pollefeys, M., Nister, D.: Direct computation of sound and microphone locations from time-difference-of-arrival data. In: IEEE International Conference on Acoustics, Speech and Signal Processing, pp. 2445–2448. IEEE (2008)
18. Schindelhauer, C., Lotker, Z., Wendeberg, J.: Network synchronization and localization based on stolen signals. In: Kosowski, A., Yamashita, M. (eds.) SIROCCO 2011. LNCS, vol. 6796, pp. 294–305. Springer, Heidelberg (2011)
19. Stewénius, H.: Gröbner Basis Methods for Minimal Problems in Computer Vision. Ph.D. thesis, Lund University (2005)
20. Sun, Z., Purohit, A., Chen, K., Pan, S., Pering, T., Zhang, P.: Pandaa: physical arrangement detection of networked devices through ambient-sound awareness. In: Proceedings of the 13th International Conference on Ubiquitous Computing (UbiComp), pp. 425–434. ACM (2011)
21. Thrun, S.: Affine structure from sound. In: Proceedings of Conference on Neural Information Processing Systems (NIPS). MIT Press, Cambridge (2005)
22. Wendeberg, J., Janson, T., Schindelhauer, C.: Self-localization based on ambient signals. Theor. Comput. Sci. **453**, 98–109 (2011)
23. Wendeberg, J., Höflinger, F., Schindelhauer, C., Reindl, L.: Calibration-free tdoa self-localization. J. Location Based Services **5**(1), 1–24 (2013)

Data Delivery by Energy-Constrained Mobile Agents

Jérémie Chalopin[1], Shantanu Das[1], Matúš Mihaľák[2]([✉]),
Paolo Penna[2], and Peter Widmayer[2]

[1] LIF, Aix-Marseille University & CNRS, Marseille, France
[2] Institute of Theoretical Computer Science, ETH Zurich, Zürich, Switzerland
`matus.mihalak@inf.ethz.ch`

Abstract. We consider mobile agents of limited energy, which have to collaboratively deliver data from specified sources of a network to a central repository. Every move consumes energy that is proportional to the travelled distance. Thus, every agent is limited in the total distance it can travel. We ask whether there is a schedule of agents' movements that accomplishes the delivery. We provide hardness results, as well as exact, approximation, and resource-augmented algorithms for several variants of the problem. Among others, we show that the decision problem is NP-hard already for a single source, and we present a 2-approximation algorithm for the problem of finding the minimum energy that can be assigned to each agent such that the agents can deliver the data.

Keywords: Mobile agents and robots · Data aggregation and delivery · Power-awareness · Algorithms

1 Introduction

Recent progress in designing and producing small, simple, and cheap mobile micro-robots raised new algorithmic challenges in deploying these robots in various tasks. In this paper, we study the question of whether and how such simple robots can collaboratively transfer information from specified sources to a single repository. Due to their simplistic construction, the robots have only very limited capabilities, for example, little memory, small computational power, limited communication capabilities, noisy sensing, or limited battery power. In this paper, we focus on the last limitation aspect – the limited battery power. In particular, we study how such a limitation influences collective capabilities of the robots to accomplish the delivery task. We concentrate on this single aspect of the robots, and do not limit the other capabilities of the robots. In particular, we assume the robots to have enough memory to store the data, and we are also not interested in the amount of time it takes to finish. We study the delivery task on graphs; for this reason we adapt our terminology to the literature and refer to the robots as *agents*.

P. Flocchini et al. (Eds.): ALGOSENSORS 2013, LNCS 8243, pp. 111–122, 2014.
DOI: 10.1007/978-3-642-45346-5_9, © Springer-Verlag Berlin Heidelberg 2014

Model

We consider undirected, connected, edge-weighted graphs. The weight $w(e)$ of edge e represents the energy required to cross the edge. Therefore, we will sometimes refer to $w(e)$ as the *length* of the edge. By $d(u, v)$ we denote the distance between nodes u and v, i.e., the length of the shortest path from u to v (with respect to the edge weights).

We further consider k mobile agents that are initially placed on vertices of a given graph. Agent i can move along the edges of G. In total, agent i can move along a walk of length at most R_i. The agent can stop anywhere on an edge e. In such a case the travelled distance is proportional to $w(e)$ and to the position of the stop on e.

Furthermore, there are m distinct *sources* $S = \{s_1, \ldots, s_k\} \subset V$, and one *target* $t \in V \setminus S$. Each source contains data that needs to be delivered by the agents to target t. An agent i *collects data from source* s by simply visiting s on its walk. An agent i *collects data from agent* j by meeting agent j (at some location). An agent i visiting target t on its walk *delivers* (or *transfers*) all data that it has collected before.

We study the problem of deciding whether all data (from all sources) can be delivered to the target, i.e., whether there exists a schedule prescribing every agent how to move such that at the end all data is delivered. We call such a schedule *feasible*. In full generality, a schedule describes the movement of an agent in continuous time, assuming that all agents move at unit speed. We will see in a moment, however, that we may concentrate on schedules where at any time at most one agent moves. This then allows us to neglect the travel times and consider the movements of the agents in discrete time steps, where movements happen instantaneously.

We refer to the decision problem of finding a feasible schedule as DATADE-LIVERY. Given just the position of the agents in the network, we also study the related minimization problem of finding the smallest uniform power R for which the agents, when assigned the range R each, can deliver the data to t. We are interested in the computational complexity of the problem, and in approximation and resource-augmented algorithms. We say that an algorithm for the minimization version of DATADELIVERY is ρ-approximate, $\rho > 1$, if it runs in polynomial-time and always finds a feasible schedule for uniform range R such that $R \leq \rho \cdot R^*$, where R^* is the minimum uniform power for which a schedule exists. We say that an algorithm for the decision version of DATADELIVERY with agents' initial ranges R_i is a γ-resource augmented algorithm, $\gamma > 1$, if either the algorithm (correctly) answers that there is no feasible schedule, or it finds a feasible schedule for the modified (augmented) powers $R_i' := \gamma \cdot R_i$.

Related Work

On a very high level, our problem can be seen as a special case of data aggregation in (wireless) sensor networks [10]. There, sensor nodes are deployed in an environment, each possessing some data that they need to route (transmit) over an underlying communication network such that all data eventually arrives in a specific aggregation node. Obviously, the nature in which the data "flows" in the network makes the main difference of data aggregation in sensor networks to our problem.

There has been little previous work on data-aggregation-like problems by mobile agents. Anaya et al. [5] study the *convergecast* problem where a set of mobile agents, deployed in an edge-weighted graph, each possessing certain data and a uniform power R, need to move such that at the end at least one agent knows all data (and every agent travels a distance at most R). The main difference to our problem is that there are no sources and a target where the data need to be delivered. On contrary, in convergecast the "target agent" can be chosen freely to suit the given power constraints. Anaya et al. [5] study the convergecast problem both in the centralized and in the distributed setting. They show that the decision problem is strongly NP-complete, even if G is a tree, provide a linear-time algorithm for the case when G is a line, and a 2-approximation algorithm for the minimization version in general graphs. In the distributed setting, they provide a 2-approximate algorithm for trees and show that this is best possible (even if G is a line).

There is little research on general power-aware computation with mobile agents. A rare example is the study of self-deployment by Heo and Varshney [8]. Arguably, minimizing the total travelled distance (instead of the maximum traveled distance) by any single agent comes close to optimizing individual power-consumption. There is a rich research history accomplishing various tasks (such as pattern formation, exploration, or searching) by mobile agents where the prime optimization goal was the total travelled distance, see e.g. [2,3,6].

Power-aware computation is a relatively new research area. Most of the existing literature focuses on different computational models than mobile robots, e.g., on routing, tracking, and broadcasting in wireless networks [4,9], or on scheduling [1,7]. However, most of these works focus on minimizing the total energy consumption (whereas we focus on leveraging the consumed energy per computational entity).

Important Observations and Further Variants

The nature of the problem allows us to make several crucial observations that limit the space in which we search for feasible strategies. We will argue about the single source case, but the very same observations can be made for the multi-source case as well.

First of all, it is easy to see that no two agents need to move at the same time. Assume that a given instance has a feasible solution and let us consider one. Let us consider the "flow" of the data from s to t in the solution, i.e., consider for every agent that collected the data the path that the agent made after the collection, and the union of all paths of the agents after they collected the information. Thus, this "flow" can be seen as the subgraph of G. It follows that there has to be an s-t path in the subgraph. Obviously, for completing the data delivery task, we can ignore all movements of the agents beyond this path.

Scanning this path from s to t and observing the identity of the agents that are currently active gives a sequence of agents (we do not need to choose more than one agent per position on the path). It is easy to see that the agents then can walk sequentially in this order, and thus we can only consider discrete time steps such that in each time step exactly one agents moves (to an arbitrary position).

It is now also easy to see that without loss of generality, no agent i appears more than once in this sequence: if yes, we can just ignore all agents that appeared in-between the two occurrences of agent i on the s-t path.

These considerations motivate the following natural variant of DATADELIVERY: Find a feasible schedule such that the data is moved from s to t along a *fixed* path (given as part of the input).

Our Results

We first consider the single-source case in Sect. 2, and show that DATADELIVERY is NP-complete in this case, even for the case of uniform ranges R. We then provide a 3-resource augmented algorithm, and a 2-approximation algorithm for the problem. The combination of the ideas of these two algorithms provides a $\min\{3, (1 + \Delta)\}$-resource augmented algorithm, where Δ is the largest ratio of the agent's ranges, i.e., $\Delta := \max_{i,j} \frac{R_i}{R_j}$. We also consider the case when the data needs to be moved along a fixed path P (given as part of the input), and show that also this problem is NP-complete, and that there exists $\gamma^* > 1$ such that there is no γ^*-augmented algorithm, unless $P = NP$. Finding a good approximation or resource-augmented algorithm for this version is left as an open problem. We also consider the special case when G is a line or a tree. If G is a line, we provide a polynomial-time algorithm for the case of uniform ranges. For the general (non-uniform) ranges, we leave the complexity of the problem open (and note that the $\min\{3, (1 + \Delta)\}$-resource augmented algorithm applies). The case when G is a tree translates to the case of a line with general (non-uniform) ranges, and thus remains open as well.

We study the case of multiple sources in Sect. 3. For the constant number sources k and for general graphs, the natural adaptation of the results for single source carry over. For the general number of sources, the problem becomes NP-complete already for trees and for uniform ranges, by a trivial modification of the hardness result for convergecast by Anaya et al. [5].

2   Single Source

In this section we study DATADELIVERY with single source node s. We first show the hardness result.

Theorem 1. *Deciding whether k agents can transfer the information from a given source s to a given target t is (strongly) NP-complete, even for unweighted graphs and for uniform ranges.*

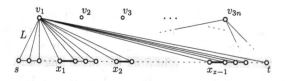

Fig. 1. Illustration of the reduction of 3-Partition to our problem. The horizontally aligned vertices from s to t form the dedicated path P^*. The shaded boxes along P^* are the buckets. Each line connecting v_1 with P^* stands for a path of length L. For simplicity, the lines connecting $v_2, \ldots, v_{3z}$ with P^* are omitted.

Proof. A solution to our problem is a set of walks, one for every agent, whose union forms a subgraph in which s and t are connected. Thus, our problem is obviously in NP – it is easy to check in polynomial time whether the given set of walks satisfy all required conditions.

To show the hardness, we will reduce the 3-Partition problem to our problem: given integers $a_1, a_2, \ldots, a_{3z}$, for some $z \in \mathbb{N}$, and an integer B such that $\sum_i a_i = z \times B$, the 3-Partition problem asks whether there is a partition $S_1, \ldots, S_z$ of the integers such that $\sum_{x \in S_i} x = B$. 3-Partition is NP-complete even if B is polynomially bounded in z, and if for every i, $B/4 < a_i < B/2$.

Given an instance of 3-Partition we construct an unweighted instance of our problem as follows. The graph contains a dedicated s-t path P^* of length $zB + (z - 1)$. The first B edges on the path are called the *first bucket*. At the end of the first bucket, we place an auxiliary agent with range 1. This agent can thus help to transfer the message only along the adjacent edge on the s-t path. After this edge, the *second bucket* starts (containing again B edges), followed by a second auxiliary agent of range 1, and one edge, and the *third bucket* and so on. For every integer a_i we create a vertex v_i and connect it to every vertex of the s-t path by a path of length L. We place an agent to every vertex v_i and set its range to be $L + a_i$, where $L = 2B$. Figure 1 illustrates the reduction.

We now show that there is a solution to 3-Partition if and only if there is a solution in the just described instance of our problem. The "if" part is trivial: Given a solution of Partition, just use the three agents corresponding to the set S_i to move the data within bucket i. Use the auxiliary agents to advance the data on the edge between the buckets. We now argue about the "only if" part. We first show that the data needs to be transported along the dedicated path P^*. The only alternative is to transfer the data from s to vertex v_i, $i = 1, \ldots, z$, and from there to t. This path P_i has length $2L$. Obviously, agent i with range $R_i = L + a_i$ sitting at v_i cannot alone transfer the data along this path. Any other agent j can get to s (by travelling the distance L from v_j) and from there to distance at most $a_j < B/2$ from s on the path P_i. From there, no other agent but v_i can advance the data along the alternative P_i; the agent then can advance the data further to distance $a_j + ((L + a_i) - (L - a_j)) < B/2 + B$, i.e., to a position on P_i that is before vertex v_i. It is easy to see that no agent can further advance the data from there. Thus, the only way to transport the data is to use the dedicated s-t path P^*. Now, because the length of P^* is $zB + (z - 1)$ and because every agent i can advance the data on P^* by at most a_i steps, every solution to our problem needs to use all agents (including the auxiliary ones) in their "full power". Thus, such a solution uses exactly three agents in every bucket, bringing the data collectively from the beginning of the bucket to the end of the bucket. This then gives a solution to the 3-PARTITION.

We can easily modify the reduction for the case of uniform ranges R: just add a path of length $R - R_i$ to every vertex v_i and place the agent i at the end of the path.   □

In the following we show that the hard part of the problem lies in knowing the order in which the agent move (and not in routing the agents through the graph). Namely, we show that given the order in which the agents move, we can solve in polynomial time whether there is a feasible schedule compatible with the given order.

Theorem 2. DATADELIVERY *with single source is solvable in polynomial time if restricted to a fixed order of the agents to move, and if agents can meet only at vertices.*

Proof. For each agent i we can compute a set of feasible "pick-pass" locations, that is, the set of all pairs (x, y) such that i can move to x (to pick up the information from another agent) and move to y (to pass the information to another agent),

$$C_i := \{(x, y) \mid d(i, x) + d(x, y) \le R_i\}.$$

Given an ordered sequence of agents (expressing the order in which they need to advance the data), where each agent appears at most once, we can compute a feasible movement of the agents by looking at the following layered graph. Layer i_k contains the edges of C_{i_k} and a path from s to t in this graph corresponds to a feasible movement of the agents (every agent appears at most once and thus its movement is a single "pick-pass" edge which, by definition, is feasible for its range). Note that, since agents can only meet at the nodes of the graph, this layered graph can be computed in polynomial time.   □

We now present a 3-augmented algorithm for DATADELIVERY with single source on general graphs and with general ranges. Our algorithm first checks whether it is (at all) possible that a feasible schedule exists. For this purpose, consider a *ball* $B(i)$ of radius R_i centered in the initial location of agent i, i.e., the set of all positions (vertices and positions on the edges) at distance at most R_i from i. If there is a feasible schedule, then there is one such that the data travels from s to t along a simple path, carried over by a sequence of ℓ agents $i_1, \ldots, i_\ell$ (and where no agent appears more than once in the sequence). Observe now that (1) s is in the ball of agent i_1 (i_1 is able to reach s to collect the data), (2) the balls of i_j and i_{k+1} intersect (agent i_j collects data from i_j), and (3) t is in the ball of the last agent i_ℓ (agent i_ℓ delivers the data to t). These properties imply the existence of an s-t path in the *connectivity graph*: the vertices are s, t and the agents, and there is an edge between i and j, if the balls $B(i)$ and

$B(j)$ intersect, and where we set $B(s) := \{s\}$ and $B(t) := \{t\}$. We can check the existence of an s-t path in the connectivity graph in polynomial time. If there is no such path, then there is no solution for DATADELIVERY. Otherwise, if there is such a path, we show that there is a feasible schedule for agents with new ranges $R'_i = 3 \cdot R_i$.

The feasible schedule for R'_i can be found in the following way. We first find an s-t path in the connectivity graph; recall that every agent appears at most once in this path. This path induces a natural order on the agents (that appear on the path), and let $i_1, \ldots, i_\ell$ be the order of these agents. For every two agents i_j and i_{j+1}, $j < \ell$, let x_j be an arbitrary vertex in $B(i) \cap B(i+1)$. Define further $x_0 := s$ and $x_\ell := t$. Then, every agent i_j moves as follows: it first goes to x_{j-1}, collects the data there, it goes back to initial position, and from there it goes to x_j. Obviously, with this schedule, the data gets delivered to t. Furthermore, every agent i_j does not travel more than $3 \cdot R_{i_j}$ (as every of its "three" moves are within its range R_i). We have thus proved the following.

Theorem 3. *There is a 3-resource augmented algorithm for* DATADELIVERY *with single source.*

The ideas of the 3-resource augmented algorithm can be adapted to give a 2-approximation algorithm for the optimization variant of DATADELIVERY with single source. Recall that in the optimization version, we are asked to find the minimum uniform range R such that there is a feasible schedule.

We will use the following observations. Consider an optimum solution, i.e., the smallest R^* and a corresponding schedule. Let i_1 be the first agent from the optimum solution to move, i.e., the agent that collects the data from s. Without loss of generality, we may assume that the optimum solution moves agent i_1 to s along a shortest path. This now induces a new instance of the problem: agent i_1 is now located in s, and has range $R' = R^* - d(i_1, s)$, while all other agents remain in their initial positions and with unchanged ranges R^*. By our construction, we know that this instance has a feasible schedule. This then implies that there is a path from i_1 (which sits on node s) to t in the connectivity graph of the modified instance.

The 2-approximation algorithm then works as follows. We first guess the first agent i_1 from the optimum solution that collects the data from s (i.e., technically, we try all possible candidate agents, perform the subsequent steps as explained below, and choose the solution giving the smallest range R among all the candidates). We move agent i_1 to s along a shortest path of length $d = d(i_1, s)$, and compute the smallest R^a such that there is a path from i_1 to t in the connectivity graph of the instance where every agent but i_1 has range R^a, and agent i_1 has range $R^a - d$. By the definition of R^a, we know that $R^* \geq R^a$. Let $i_1, i_2, \ldots, i_\ell, t$ be an i_1-t path in the connectivity graph of the considered instance. Thus, we know that for any $1 \leq j < \ell$, the respective balls intersect, and therefore $d(i_j, i_{j+1}) \leq 2 \cdot R^a$, and furthermore $d(i_\ell, t) \leq R^a$. Observe now that the schedule where agent i_j goes to agent i_{j+1}, $j < \ell$, and agent i_ℓ goes to t, is feasible if we add R^a to the range of every agent. This gives a feasible schedule to the original setting where agent i_1 has not been moved to s, with uniform ranges $2 \cdot R^a$. We can thus return $2 \cdot R^a$ as the solution of the algorithm.

Because $R^* \geq R^a$, i.e., $2 \cdot R^* \geq 2 \cdot R^a$, we obtain that the algorithm is a 2-approximation.

Theorem 4. *There is a 2-approximation algorithm for* DATADELIVERY *with single source.*

Obviously, we can use the ideas of the 2-approximation algorithm for designing an equivalent 2-resource augmented algorithm for the case when the ranges are uniform, i.e., when $R_i = R_j$ for every i, j. The very same algorithm is then $(1+\Delta)$-resource augmented algorithm, where $\Delta = \max_{i,j} \frac{R_i}{R_j}$: It can happen that the algorithm decides for agent i to bring the data (from its initial position) to the initial position of agent j; For this, R_i needs to be increased by additive R_j to be able to do it, which gives the claimed ration $(1 + \Delta)$. Thus, we have the following.

Corollary 1. *There is a* $\min\{3, (1 + \Delta)\}$-*resource augmented algorithm for* DATADELIVERY *with single source.*

We now consider a special case where the delivery of the data needs to happen along a fixed path in G. This is motivated by security reasons when we do not want the data to be delivered in dangerous areas of the environment. We now show that this problem is hard. We present an alternative proof for this case, since this gives us (additionally to the pure hardness result of Theorem 1) hardness for providing arbitrary good γ-resource augmented algorithms.

Theorem 5. *The variant of* DATADELIVERY *in which there is a single source and the data must travel along a fixed path of the graph is NP-hard.*

Proof. We reduce the problem from the restriction of 3-SAT in which every variable appears at most *four* times [11].

The idea of the reduction is as follows (see Fig. 2). Each clause consists of a "gadget" which has some common part with the fixed path. Intuitively speaking, if a clause is satisfied, then the data can travel from left to right along the portion of the path "covered" by that clause. The reduction will ensure that there is a satisfying assignment if and only if the data can travel from left to right through each of the clauses sub-paths.

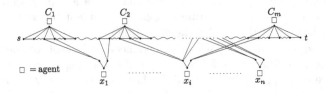

Fig. 2. Overview of the reduction from 3-SAT.

More formally, for each clause $C_j = \{l_{j_1}, l_{j_2}, l_{j_3}\}$, we create a gadget consisting of a graph and an agent with range 5 as shown in Fig. 3 (upper part). For each variable x_i the corresponding gadget is the simple graph plus the agent of range R shown in Fig. 3 (lower part); We shall set $R < 3L$ so that this agent is forced to choose between "true or false". The three *edge literals* in the clause gadget are connected to some vertices of the corresponding variable gadget: The endpoints of an edge for literal x_i (resp., literal $\neg x_i$) are connected to the vertex in the variable gadget of x_i corresponding to *true* (resp., *false*). The overall construction (see Fig. 2) consists in a concatenation of all clause gadgets, where any two consecutive gadgets are connected via a *spline path*, that is, a chain of L edges/agents like the one shown in Fig. 3.

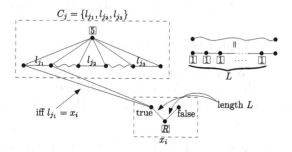

Fig. 3. The gadgets for clauses (upper part) and for variables (lower part) and how they are interconnected. An agent is shown as box with a number inside (its range). The two edges in the variable gadget have length L. Every other edge has unit length and the splines represent a chain of L edges with L additional agents of range 1 each.

We say that an agent *covers* and edge of the path if it traverses that edge from left to right. By setting $R = L + 8$ and $L = 8$ we have that

1. The agent for the clause cannot cover all three literal edges alone, but can cover any two of them. It is impossible for this agent to cover some edge of *another* clause gadget (because of the spline paths between clause gadgets).
2. The agent for a variable x_i can either (1) cover one edge in each of the clauses where x_i appears positive or (2) cover one edge in each of the clauses where x_i appears negated. It is impossible for this agent to cover an edge from a clause where x_i is positive *and* from another clause where x_i is negated.

Note that in the second item we use the fact that every variable appears in at most *four* clauses and, without loss of generality, it appears both positive and negated (so, there are at most three positive occurrences and at most three negated ones).

Claim. For every satisfying truth-assignment there exists a feasible movement of the agents such that the data travels along the path.

Proof (of Claim). If the assignment sets $x_i = true$ (resp., $x_i = false$) then, by Item 2, the variable agent can cover one literal edge in each of the clauses where

x_i appears positive (resp., negated). Since the assignment satisfies all clauses, each clause has one literal edge covered by a variable agent. The remaining two literal edges can be covered by the clause agent of the clause (Item 1). The edges in the spline paths are covered by the corresponding agents.   □

Claim. For every feasible movement of the agents such that the data travels along the path, there exists a satisfying truth-assignment.

Proof (of Claim). In every feasible movement all edges in the path must be covered by some agent. In particular, for each clause, there must be one of the three literal edges that is covered by a variable agent (Item 1). By Item 2 we can obtain a truth-assignment as follows: If the variable agent for x_i covers edge literals of clauses where x_i appears positive, then we set $x_i = true$; Otherwise we set $x_i = false$. By the previous argument, this assignment satisfies all clauses.   □

The two claims above imply the NP-hardness.   □

Note that the proof of hardness can be easily extended to the case of identical ranges. Moreover, with minor modifications, the reduction can be extended to prove that there is no γ-augmented algorithm for this variant of the problem, for some constant $\gamma > 1$.

Corollary 2. *There exists $\gamma > 1$ such that there is no γ-augmented algorithm for* DATADELIVERY *with single source and fixed delivery path, unless P=NP.*

3   Multiple Sources

In this section we consider the version of DATADELIVERY in which the agents have to collect the data from *more than one source* to a common target location.

3.1   A 2-approximation for a Constant Number of Sources with Identical Powers

The 2-approximation algorithm from Theorem 4 can be generalized to the case of a constant number of sources. Intuitively speaking, the algorithm guesses the set of "pick-up" agents that first reach the sources and, if an agent picks up data at more than one source, then it also guesses the *order* in which this is done (this is possible since there is only a constant number of sources).

More formally, in the optimal solution the piece of data at every source s_i travels along some path whose first agent is the *pick-up* agent of that source. Note that an agent can be the pick-up agent of several sources. Given the set P of pick-up agents, each pick-up agent $p \in P$ is then matched to an ordered sequence of sources,

$$s_1^{(p)} \rightarrow s_2^{(p)} \rightarrow \cdots \rightarrow s_{\ell_p}^{(p)},$$

meaning that, in the optimal solution, agent p visits these sources in that particular order (possibly by visiting other locations in between). After being visiting the last source, p will move to some location to pass its data to some other agent i_p. Similarly to the case of a single source, we consider a new instance in which p has moved to its last source $s_{\ell_p}^{(p)}$ and its initial power R has been decreased by the minimum cost of visiting these sources in that particular order:

$$d(p, s_1^{(p)}) + d(s_1^{(p)}, s_2^{(p)}) + \cdots + d(s_{\ell_p-1}^{(p)}, s_{\ell_p}^{(p)}).$$

(Visiting other additional locations between two consecutive sources can only increase this cost.) The new instance is thus feasible and, in particular, the range of p when starting from the last source allows it to move inside the ball of i_p. Thus it can move directly to i_p if provided an extra power of R. Therefore, any path from s to t in the connectivity graph of the modified instance yields a 2-approximation: Each pick-up agent visits all of its sources in the specified order and then moves from the last source to the first agent in the path (by the previous argument this costs at most $2R$). Then the subsequent agents simply bring the collected data directly to the next agent in the path (again the cost is at most $2R$ since these agents have power R also in the modified instance).

The 2-approximation algorithm now suggests itself: Guess the set of pick-up agents and their ordered sequence of sources (there are only constantly many since the number of sources is constant), and then check if the connectivity graph of the modified instance contains a path from s to t (this is indeed the case when the guess in the first step is correct).

Theorem 6. *There is a 2-approximation algorithm for* DATADELIVERY *with a constant number of sources.*

3.2 On Hardness and Approximation of Arbitrary Number of Sources

When there are many sources but each source initially contains an agent, then the problem is NP-hard even for identical powers. This follows from Theorem 4 in [5], which study a conceptually different problem, but the proof of hardness applies also to our problem. We note that one can easily extend the 2-approximation algorithm from Sect. 3.1 for this very special case.

4 Conclusions and Open Problems

In this work we have studied several variants of DATADELIVERY. This problem concerns how a set of energy-constrained agents can collectively move some data from a set of given locations to a common target. The problems turn out to be hard, and the hardness lies (essentially) in finding how and in which order the agents should move to perform this task. On the positive side, we provide algorithms that find solutions which guarantee that no agent uses more than a constant factor the energy required by the optimum. It would be interesting to close the gap between our approximation and resource augmentation algorithms and hardness results. Are there any better algorithms? What is the complexity of DATADELIVERY on special graphs? The problem is open even for trees and for the following simple geometric version: the agents lie on a single line with source and target being the endpoints. We note that if all agents have uniform ranges, then the problem is solvable on the line (but remains open for trees): start with the closest agent (to the source); when advancing the data, never overtake a (previously) unused agent; recursively use the closest agent to advance it further. Finally, it would be interesting to obtain positive results for the variant in which the piece of data must travel along a fixed (given graph), or for the case of an arbitrary number of sources.

References

1. Albers, S., Fujiwara, H.: Energy-efficient algorithms for flow time minimization. ACM Trans. Algorithms **3**(4), 1–17 (2007)
2. Albers, S., Henzinger, M.R.: Exploring unknown environments. SIAM J. Comput. **29**(4), 1164–1188 (2000)
3. Alpern, S., Gal, S.: The Theory of Search Games and Rendezvous, vol. 55. Kluwer Academic, Boston (2002)
4. Ambühl, C.: An optimal bound for the MST algorithm to compute energy efficient broadcast trees in wireless networks. In: Caires, L., Italiano, G.F., Monteiro, L., Palamidessi, C., Yung, M. (eds.) ICALP 2005. LNCS, vol. 3580, pp. 1139–1150. Springer, Heidelberg (2005)
5. Anaya, J., Chalopin, J., Czyzowicz, J., Labourel, A., Pelc, A., Vaxès, Y.: Collecting information by power-aware mobile agents. In: Aguilera, M.K. (ed.) DISC 2012. LNCS, vol. 7611, pp. 46–60. Springer, Heidelberg (2012)
6. Blum, A., Raghavan, P., Schieber, B.: Navigating in unfamiliar geometric terrain. SIAM J. Comput. **26**(1), 110–137 (1997)
7. Chan, H.-L., Edmonds, J., Lam, T.-W., Lee, L.-K., Marchetti-Spaccamela, A., Pruhs, K.: Nonclairvoyant speed scaling for flow and energy. Algorithmica **61**(3), 507–517 (2011)
8. Heo, N., Varshney, P.K.: Energy-efficient deployment of intelligent mobile sensor networks. IEEE Trans. Syst. Man Cybern. (Part A) **35**(1), 78–92 (2005)
9. Li, Q., Aslam, J., Rus, D.: Online power-aware routing in wireless ad-hoc networks. Proceedings of the 7th Annual International Conference on Mobile Computing and Networking, MobiCom 2001, pp. 97–107. ACM, New York (2001)
10. Rajagopalan, R., Varshney, P.K.: Data-aggregation techniques in sensor networks: a survey. IEEE Commun. Surv. Tutorials **8**(4), 48–63 (2006)
11. Tovey, C.A.: A simplified NP-complete satisfiability problem. Discrete Appl. Math. **8**(1), 85–89 (1984)

Approximation Bounds for the Minimum k-Storage Problem

Gianlorenzo D'Angelo[✉], Daniele Diodati, Alfredo Navarra,
and Cristina M. Pinotti

Dipartimento di Matematica e Informatica, Università degli Studi di Perugia,
Perugia, Italy
{gianlorenzo.dangelo,daniele.diodati}@dmi.unipg.it,
{alfredo.navarra,pinotti}@unipg.it

Abstract. Sensor networks are widely used to collect data that are
required for future information retrieval. Data might be aggregated in
a predefined number k of special nodes in the network, called *storage*
nodes, which, for replying to external queries, compress the last received
raw data and send them towards the *sink*. We consider the problem of
locating such storage nodes in order to minimize the energy consumed
for converging the raw data to the storage nodes as well as to converge
the aggregated data to the sink. This is known as the *minimum k-storage
problem*. We first prove that it is *NP*-hard to be approximated within a
factor of $1+\frac{1}{e}$. We then propose a local search algorithm which guarantees
a constant approximation factor. We conducted extended experiments to
show that the algorithm performs very well in many different scenarios.
Further, we prove that the problem is not in APX if we consider directed
links, unless $P = NP$.

1 Introduction

Networks of sensor nodes are usually employed to monitor large areas, collecting data with regular frequency. This large volume of data has to be stored
somewhere for answering to external user queries [8]. Source nodes, which are
responsible for collecting data, can either locally store the data or transmit them
to the *sink*, a powerful node connected to the external world. If data are locally
stored, several problems may arise: (i) data cannot be accumulated for long
periods because nodes are equipped with only limited memory space; (ii) stored
data are lost once the energy of a source node – battery operated – is depleted;
and (iii) searching data for serving query demand results in network-wide communications. Alternatively, source nodes can forward the collected data to the

Research partially supported by the Research Grant 2010N5K7EB 'PRIN 2010' ARS
TechnoMedia (Algoritmica per le Reti Sociali Tecno-mediate), from the Italian Min-
istry of University and Research, and by "Fondazione Cassa di Risparmio della
Provincia dell'Aquila" within project ARISE.

P. Flocchini et al. (Eds.): ALGOSENSORS 2013, LNCS 8243, pp. 123–138, 2014.
DOI: 10.1007/978-3-642-45346-5_10, © Springer-Verlag Berlin Heidelberg 2014

sink. However, communicating data from the source nodes up to the sink makes the network congested, especially if data are transmitted *raw*, that is, uncompressed. In general, source nodes may forward their raw data to some midway nodes, referred to as *storage nodes*. Here, raw data are stored and *aggregated*, i.e., reduced in size, to be transmitted to the sink at the time a query demand from external users is submitted. Such midway nodes must have extra permanent storage, more power battery and more computational capabilities. With this two-tier model, if the number of storage nodes is kept limited, the network becomes less congested at the price of a moderate increase of the network cost. This scheme has been pursued in [16,17] where the problem of selecting a subset of storage nodes so as the overall communication cost is minimized is called *optimal storage placement* problem. When the number of storage nodes is limited by an integer k, we talk about the minimum k-*storage* problem.

Data query is a very important service in networking applications. Thus, the minimum k-storage problem appears also in other contexts, like web caching, peer-to-peer, and database systems.

Related Work. There have been a lot of prior research on data querying models in sensor networks. Initially, no in-network storage was considered: the query was spread out to every sensor by flooding messages and data were returned by following the same path but in the reverse direction [12]. To reduce the communication cost towards the sink, clustering routing protocol have been then considered [19]. Later, the data-centric in-network model has been introduced [18], which stores different data types in different places to facilitate the archiving and retrieving process in the network. Due to the large storage capabilities required by the in-network model, it has been proposed in [7] to store data in a degrading model: fresh-data are stored raw, while long-term data are kept, but compressed. More recently, a two-tier model has been proposed in [17] to ameliorate both the problem of limited storage and communication congestion. The authors formulate the problem as an integer programming problem and propose a 10-approximation rounding algorithm. Differently from this paper, they assume that (i) raw data have size independent from the source node; and (ii) the energy spent for transmitting one unit data between any pair of sensors is proportional to their Euclidean distance. For us, instead, different source nodes may generate data of different size since sensors can monitor different environment aspects. Moreover, we assume that communications follow an underlying network represented by a graph. Each edge of the graph has its own weight that measures the energy required to traverse it.

In [16], the problem is solved assuming that the communication network topology is a directed tree T, rooted at the sink. When a sensor s sends one unit data upwards to the sink the energy cost is fixed, while when a sensor s sends one unit data downwards, the energy cost is proportional to the number of children of s in T. Thus, the energy cost for diffusing the query – which involves communications from the sink downwards to the storage nodes – cannot be ignored, as done in this paper and in [17]. Indeed, we assume that the query cost is negligible. In fact, storage nodes may even simply send query replies in a proactive manner with a predefined query frequency. In [16], the storage placement problem has been solved assuming that the number of storage nodes is either unlimited or limited. Moreover, two tree models are considered: in the fixed tree model, the problem is optimally solved following a dynamic programming approach; in the dynamic tree model, a stochastic analysis of the performance estimation is conducted. Note that, in this paper, the communication network topology is not restricted to any specific class of graphs.

Finally, the minimum k-storage problem is strongly related to the well-known k-median problem [2,10,11]. In fact, if data compressed is assumed to be of negligible size (i.e., there is no cost in sending data from storage nodes to the sink), so as for the energy spent for the queries, and if the sink is assumed to be one of the k selected nodes, then minimum k-storage problem coincides with the minimum k-median problem. Under these assumptions, k-storage minimizes only the energy cost required to send raw data to the storage nodes. This is the classical k-median problem, except that the sink is a special, predefined median. Clearly, the relation with the k-median also suggests affinities with facility location problems [2,10,11]. However, to the best of our knowledge, there is no variant of facility location problems that coincides with our k-storage.

Our Results. First, we prove that it is NP-hard to approximate the metric k-median problem (MMP) within a factor of $1 + \frac{1}{e}$. At the best of our knowledge, in the literature, only a harder variant of MMP, where medians can be chosen only among a subset of nodes, has been shown to be not approximable within a factor of $1 + \frac{2}{e}$ [11]. The obtained result for MMP is then extended to the minimum k-storage problem by means of a polynomial time reduction that preserves approximation. For the case of directed links, we prove that the problem is even not in APX, unless $P = NP$. For the undirected case, a local search algorithm to solve the minimum k-storage problem is then proposed. We show that it guarantees a constant approximation ratio. The algorithm has been also experimentally evaluated by conducting an extended investigation of its performance by varying on different parameters, like graph type and size, number of storage nodes k, raw data size, compression factor of the storage nodes, and accuracy of the solution.

2 Problem Definition and Notation

Let $G = (V, E)$ be a connected graph of n nodes representing a sensor network. Unless differently specified, G is considered undirected. Each node $v \in V$ generates raw data at regular time intervals, with frequency r_d and size $s_d(v)$. Edges of the network have different weights. The energy cost propagation of a message over the edge $\{u, v\} \in E$ is denoted by $w(u, v)$. When communication links are bidirectional (i.e., G is undirected), $w(u, v) = w(v, u)$. Let $d(u, v)$ be the minimum energy cost for propagating a message from u to v.

Each $v \in V$ can be set to serve as *storage* node, i.e., it makes use of higher storage capacity. A solution is a set $S \subseteq V$ of storage nodes such that $|S| \leq k$, for some $k \in \mathbb{N}$.

External users retrieve data, with frequency r_q, from a special storage node $r \in S$, named *sink*. Each node v in V is associated to a storage node, denoted as $\sigma(v) \in S$. Clearly, if $v \in S$, then $\sigma(v) = v$. For replying to a query, a storage node compresses and sends to r the last data generated from its associated nodes. The compressed size of the data produced by a node v becomes $\alpha s_d(v)$, with $\alpha \in [0, 1]$. The compressed data cannot be further compressed if they reach another storage node. A node $v \in V$ is associated to the storage node s that minimizes $r_d s_d(v) d(v, s) + r_q \alpha s_d(v) d(s, r)$, ties are arbitrarily broken. Formally, $\sigma(v) = \arg\min_{s \in S} s_d(v) (r_d d(v, s) + \alpha r_q d(s, r))$.

The total cost per time unit for a set S of storage nodes is given by: $cost(S) = \sum_{v \in V} s_d(v) (r_d d(v, \sigma(v)) + \alpha r_q d(\sigma(v), r))$, where σ is induced by S. The minimum k-*storage* problem (briefly, *MSP*) consists in finding a subset $S \subseteq V$, with $|S| \leq k$ that minimizes $cost(S)$.

We assume that $r_d \geq \alpha r_q$, as otherwise, there is no need for storage nodes and the nodes can send their raw data directly to the sink along the shortest paths. It follows that $\alpha \frac{r_q}{r_d} \leq 1$, and then we can define an instance of the problem with $r'_d = r'_q = 1$ and $\alpha' = \alpha \frac{r_q}{r_d}$. Therefore, we can assume without loss of generality that $r_d = r_q = 1$ and $\alpha \in [0, 1]$. Then, the total cost per time unit is:

$$cost(S) = \sum_{v \in V} s_d(v) (d(v, \sigma(v)) + \alpha d(\sigma(v), r)).$$

It can be then assumed that any solution S to *MSP* has size k, since reducing the number of storage nodes does not decrease its cost.

3   Hardness of Approximation

In this section, we show that it is *NP*-hard to approximate *MSP* within a factor of $1 + \frac{1}{e}$. To this aim, we provide an approximation factor preserving reduction from the k-median problem defined as follows.

Let $G = (V, E)$ be a complete graph, $k \in \mathbb{N}$, and $dist(u, v) \in \mathbb{N}$ be the distance from u to v over the edge that connects them, $(u, v) \in E$. A k-median set for G is a subset $V' \subseteq V$ with $|V'| \leq k$. The *minimum k-median problem* consists in finding a k-median set V' that minimizes the sum of the distances from each vertex to its nearest median, i.e. $\sum_{u \in V} \min_{v \in V'} dist(u, v)$.

In the *metric* minimum k-median problem (briefly, *MMP*) it is assumed that the distance function is symmetric and satisfies the triangle inequality.

In the next theorem, we show that it is *NP*-hard to approximate *MMP* within a factor of $1 + \frac{1}{e}$. This result holds even if the distance function assumes only values in $\{1, 2\}$ and it is of its own interest.

Theorem 1. *It is NP-hard to approximate MMP within a factor $\gamma < 1 + \frac{1}{e}$.*

Proof. Our proof is based on a reduction that preserves approximation from minimum dominating set to *MMP*, and it is similar to the technique used in [10] for the metric uncapacitated facility location problem.

We show that, if there exists an algorithm γ-MMP with approximation factor $\gamma < 1 + \frac{1}{e}$ for MMP, then there exists a $(c \ln n)$-approximation algorithm for the minimum dominating set for some $c < 1$. This implies that it is NP-hard to approximate MMP within a factor γ as it has been shown that it is NP-hard to approximate the minimum dominating set within a factor $c \ln n$ for any $c < 1$ [3,5]. The minimum dominating set problem is defined as follows.

Let $G = (V, E)$ be an undirected graph, a dominating set for G is a subset $V' \subseteq V$ such that for each $u \in V \setminus V'$ there is a $v \in V'$ for which $\{u, v\} \in E$. The *minimum dominating set problem* consists in finding the minimum cardinality dominating set.

Given an instance $G = (V, E)$ of the minimum dominating set problem we build an instance of MMP defined as a graph $G' = (V, E')$, where $E' = V \times V$ and

$$dist(u, v) = \begin{cases} 1 & \text{if } \{u, v\} \in E \\ 2 & \text{otherwise.} \end{cases} \tag{1}$$

We give an approximation algorithm $\mathcal{A}$ for the minimum dominating set problem on G that exploits the algorithm γ-MMP on G'.

Let us suppose for a while that the size k of an optimal dominating set is known. Fixed such a value of k, algorithm $\mathcal{A}$ selects a dominating set V' by repeatedly applying algorithm γ-MMP by using parameter k for the instance of MMP.

Initially, $V' = \emptyset$ and $G_1 = (V_1, E_1)$ with $V_1 = V$ and $E_1 = V_1 \times V_1$. At each iteration $p \geq 1$, the γ-MMP algorithm selects the set S of the k-medians which will be added to the dominating set V', then $\mathcal{A}$ computes the set $\Delta(S) = \bigcup_{v \in S} Adj(v)$, where $Adj(v)$ is the set of neighbors of v in G. Then, it creates a smaller instance G_{p+1} for MMP with $V_{p+1} = V_p \setminus (S \cup \Delta(S))$. The algorithm ends when all nodes are covered.

In order to evaluate the approximation provided for the minimum dominating set problem, observe that if $\mathcal{A}$ terminates after λ iterations, the returned dominating set V' has size at most $\lambda \cdot k$. Then, since k is the size of an optimal dominating set, λ is the approximation factor guaranteed by $\mathcal{A}$ for minimum dominating set.

We now show an upper bound on λ. After the pth iteration of the while loop, let $d_p = |\Delta(S)|$ be the number of neighbors of any medians in G and $i_p = |V_p| - d_p - k$ be the number of remaining nodes. In other words, d_p are the nodes that pay a cost of 1 in the pth instance of MMP and i_p are those that pay a cost of 2. After the first stage of the while loop, by definition $d_1 + i_1 + k = |V_1| = n$. Since k is the size of an optimal dominating set on G, then there exists a solution for MMP on G_1 of cost $n - k$. Therefore, the value of an optimal solution for MMP on G_1 is $OPT \leq n - k$. Algorithm γ-MMP outputs a solution of cost $d_1 + 2i_1 \leq \gamma OPT \leq \gamma(n - k)$. By substitution, $n - k + i_1 \leq \gamma(n - k)$ and finally $i_1 \leq (n - k)(\gamma - 1) \leq n(\gamma - 1)$. After the pth stage of the while loop, $d_p + i_p + k = i_{p-1}$, and $d_p + 2i_p \leq \gamma(i_{p-1} - k)$ and then $i_p \leq n(\gamma - 1)^p$. At the beginning of the last iteration λ we have at most $n(\gamma - 1)^{\lambda-1} = \eta$ uncovered nodes, for some $1 \leq \eta \leq n$, and then, $\lambda - 1 = \log_{(\gamma-1)} \frac{\eta}{n} \leq \log_{(\gamma-1)} \frac{1}{n} = \frac{\ln n}{\ln \frac{1}{\gamma-1}}$.

We know that it is *NP*-hard to approximate the minimum dominating set problem within a factor $(c \ln n)$-approximation, for each $c < 1$. Moreover, for each c' such that $0 < c' < c < 1$, we have $c' \ln n \le c \ln n - 1$, for n sufficiently large. It follows that for each $c' < 1$, $c' \ln n + 1 \le c \ln n < \lambda \le \frac{\ln n}{\ln \frac{1}{\gamma - 1}} + 1$. Therefore, $\frac{1}{\ln \frac{1}{\gamma - 1}} \ge 1$ which implies $\frac{1}{\gamma - 1} \le e$, and hence $\gamma \ge 1 + \frac{1}{e}$.

Since we do not know the size k of the optimal dominating set, we repeat the approximation algorithm based on γ-*MMP* for all the values of k, $1 \le k \le |V|$ and we return the dominating set of minimum size. $\square$

By exploiting Theorem 1, we show that it is *NP*-hard to approximate *MSP* within a factor $\gamma < 1 + 1/e$. The proof is based on a reduction that preserves approximation from *MMP* to *MSP*. Given an instance $G = (V, E)$, k of *MMP*, where $dist(u, v) \in \{1, 2\}$ with $(u, v) \in E$, we build a graph $G' = (V', E')$, where $V' = V \cup I$. Nodes $i_{\{u,v\}} \in I$ are called *intermediate*. For each $\{u, v\} \in E$:

- $\{u, v\} \in E'$ if $dist(u, v) = 1$;
- $\{\{u, i_{\{u,v\}}\}, \{i_{\{u,v\}}, v\}\} \in E'$ if $dist(u, v) = 2$.

The size of raw data is $s_d(i_{\{u,v\}}) = 0$, for each $i_{\{u,v\}} \in I$, and $s_d(v) = 1$, for each $v \in V$. The weight $w(u, v)$ is set to 1 for each $\{u, v\} \in E'$.

Let γ-*MSP* be a γ-approximation algorithm for *MSP*.

Lemma 1. *Let S be the set of storages returned by γ-MSP executed on graph G'. If $S \cap I \neq \emptyset$, starting from S we can generate a solution S' such that $cost(S') \le cost(S)$ by iteratively swapping each storage $i_{\{u,v\}} \in S \cap I$ with u or v.*

Proof. Let $N(u)$ and $N(v)$ be the sets of nodes that send query responses to $i_{\{u,v\}}$, across u and v respectively. If $|N(u)| = |N(v)|$ we can swap the storage in $i_{\{u,v\}}$ with either u or v without changing the cost of the solution. If $|N(u)| \neq |N(v)|$, we can assume without loss of generality that $|N(u)| > |N(v)|$. Hence, we swap the storage in $i_{\{u,v\}}$ with u decreasing the cost of the solution of $|N(u)| - |N(v)| > 0$. $\square$

Theorem 2. *MSP is at least as hard to approximate as MMP.*

Proof. In the following, we provide an approximation algorithm for *MMP*.

For each $r \in V$, we run γ-*MSP* on an instance of *MSP* given by the graph $G' = (V', E')$ defined above and by sink r and $\alpha = 0$. Let $S(r)$ be the set of storages selected by γ-*MSP*, and $OPT_{MSP}(r)$ be the cost of an optimal solution for *MSP* with sink r. We refine solution $S(r)$, obtaining the set $S'(r)$, with the post-processing stage described in the proof of Lemma 1. It follows that $S'(r)$ is an approximate solution for the *MSP* instance with a better performance guarantee than $S(r)$, that is $cost(S'(r)) \le \beta OPT_{MSP}(r)$, with $\beta \le \gamma$.

Let S^* be the subset of medians generated by the algorithm, that is $cost(S^*) = \min_{r \in V} cost(S'(r))$. Let OPT_{MMP} be the cost of an optimal solution for *MMP* on G and let v be a node in an optimal solution for *MMP*. Since a solution for *MSP* with sink v is a feasible solution for the instance of *MMP*, then

$OPT_{MSP}(v) \geq OPT_{MMP}$. However, $OPT_{MSP}(v) \leq OPT_{MMP}$ as there exists an optimal solution for MMP that contains v and then such a solution is feasible for the instance of MSP with sink v. Hence, $OPT_{MSP}(v) = OPT_{MMP}$ and $cost(S^*) \leq cost(S'(v)) \leq \beta OPT_{MSP}(v) = \beta OPT_{MMP} \leq \gamma OPT_{MMP}$, since $cost(S^*) = \min_{r \in V} cost(S'(r)) \leq cost(S'(v))$ and $S'(v)$ is a β-approximate solution for MSP with $\beta \leq \gamma$. Finally, we observe that as in the reduction $\alpha = 0$, then $cost(S^*)$ is the cost of the set of medians S^* in the instance of MMP.   □

Corollary 1. *It is NP-hard to approximate MSP within a factor* $\gamma < 1 + \frac{1}{e}$.

Note that, the above lower bound holds even with unitary edge weights. The reduction can be modified by shortcutting of each intermediate nodes with an edge of weight 2. In this way, we can prove that the above theorem and its corollary hold with unitary s_d but with weights in $\{1, 2\}$.

In the case of directed graphs, the following theorem can be stated.

Theorem 3. *For directed graphs, MSP does not belong to APX, unless* $P = NP$.

4 Local Search Algorithm

A local search algorithm $\mathcal{L}$ for solving MSP on undirected graphs is defined as follows. Each solution is specified by a subset $S \subseteq V$ of exactly k nodes. To move from one feasible solution S to a neighboring one S', we define a *swap* operation between two nodes $s \in S$ and $s' \in V \setminus S$ which consists in adding s' and removing s, that is $S' = S \cup \{s'\} \setminus \{s\}$. In our local search algorithm, we repeatedly check whether any swap move yields a solution of lower cost. In the affirmative case, we apply to the current solution any swap move that improves the solution cost and the resulting solution is set to be the new current solution. This is repeated until, from the current solution, no swap operation decreases the cost, that is, the current solution represents a local optimum.

Let us define $f : (0, 1] \to \mathbb{R}$, $f(\alpha) = \frac{2}{\alpha}$, $g : [0, \frac{1}{2}) \to \mathbb{R}$, $g(\alpha) = \frac{12\alpha}{1-2\alpha}$, and $h : [0, 1] \to \mathbb{R}$,

$$h(\alpha) = \begin{cases} g(\alpha) & \text{if } \alpha = 0 \\ \min\{f(\alpha), g(\alpha)\} & \text{if } \alpha \in (0, \frac{1}{2}) \\ f(\alpha) & \text{if } \alpha \in [\frac{1}{2}, 1]. \end{cases}$$

Theorem 4. *The local search algorithm $\mathcal{L}$ for MSP exhibits a locality gap of at most* $5 + h(\alpha)$.

Proof. The proof follows the scheme of [2], where it is shown that a local search algorithm for MMP provides a 5-approximation. We consider, for any input instance of MSP, the optimal solution, denoted by S^*, and the one provided by $\mathcal{L}$, denoted by S. Clearly, sink r belongs to both S and S^*. For both solutions, each node v is assigned to a storage node according to the definition in Sect. 2. These mappings are denoted as $\sigma(v)$ and $\sigma^*(v)$, respectively. Similarly, C and

C^* denote the total cost of S and S^*, respectively. By definition, S represents a local minimum, that is, it cannot be improved anymore by means of one further step of algorithm $\mathcal{L}$. Clearly, S^* is not known but the methodology still provides a reasonable comparison among the two solutions.

The relation between S and S^* is obtained by considering k ideal swaps, called *crucial swaps*, between the nodes in S and the ones in S^*. Since S is locally optimal, each swap, singularly taken, does not improve the objective function of the resulting solution. Each crucial swap consists in swapping into the solution one node $s^* \in S^*$ and swapping out one node $s \in S$. The main property for each swap is that any element $s^* \in S^*$ will participate in exactly one of these k crucial swaps, and each $s \in S$ will participate in at most two of these k swaps. The case where $s^* \equiv s$ is possible, but still it does not improve the current solution.

The mapping σ can be used to categorize the nodes in S as follows:

- let $\mathcal{O} \subseteq S$ be the set of nodes $s \in S$ that have exactly one node $s^* \in S^*$ with $\sigma(s^*) = s$;
- let $\mathcal{Z} \subseteq S$ be the set of nodes $s \in S$ for which none of the nodes $s^* \in S^*$ have $\sigma(s^*) = s$;
- let $\mathcal{T} \subseteq S$ be the set of nodes $s \in S$ such that s has at least two nodes in S^* assigned to it in the current solution.

The mapping σ provides a matching between a subset $\mathcal{O}^* \subseteq S^*$ and the set $\mathcal{O} \subseteq S$. Hence, if ℓ denotes the number of nodes in $\mathcal{R}^* = S^* \setminus \mathcal{O}^*$, then $|\mathcal{Z} \cup \mathcal{T}| = \ell$ since $|S^*| = |S| = k$. This implies $|\mathcal{T}| \leq \frac{\ell}{2}$, and hence $|\mathcal{Z}| \geq \frac{\ell}{2}$.

We can now construct the crucial swaps as follows: for each node $s^* \in \mathcal{O}^*$, we swap it with $\sigma(s^*)$. Since r belongs to both S and S^*, either it is in $\mathcal{O}$, or it is in $\mathcal{T}$. In both cases, it is swapped with itself, and then other at most ℓ swaps remain to be defined. Each of such swaps moves into the solution a distinct node in $\mathcal{R}^*$, and moves out a node in $\mathcal{Z}$, so that each node in $\mathcal{Z}$ appears in at most two swaps. For those swaps involving nodes in $\mathcal{R}^*$ and $\mathcal{Z}$, we are free to choose any mapping provided that each element of $\mathcal{R}^*$ is swapped in exactly once, and each element of $\mathcal{Z}$ is swapped out once or twice.

For each crucial swap $cs(s, s^*)$ between $s \in S$ and $s^* \in S^*$, let $S' = (S \setminus \{s\}) \bigcup \{s^*\}$. We set the associations of nodes to storage nodes in S' as follows:

- For any v such that $\sigma^*(v) = s^*$, v is associated to s^*, since $s^* \in S'$;
- For any v such that $\sigma^*(v) \neq s^*$ and $\sigma(v) = s$, v is associated to $\sigma(\sigma^*(v))$;
- Any other v remains associated to $\sigma(v)$.

In order to proceed with the proof, we need the following lemma which proves that $\sigma(\sigma^*(v)) \in S'$.

Lemma 2. *If* $\sigma^*(v) \neq s^*$, *then* $\sigma(\sigma^*(v)) \neq s$.

Proof. By contradiction, let $\sigma(\sigma^*(v)) = s$. We know that $s \in \mathcal{O}$. In fact, each node swapped out by a crucial swap is either in $\mathcal{Z}$ or $\mathcal{O}$. However the former case is not possible since by the definition of $\mathcal{Z}$ the nodes is $\mathcal{Z}$ have no associated nodes of S^*. Since $s \in \mathcal{O}$, it is mapped by function σ to exactly one element in $\mathcal{O}^*$, and we build a crucial swap by swapping s with that one element. Hence, $\sigma^*(v) = s^*$ which contradicts the hypothesis.   □

We can now evaluate the cost of the obtained solution S' after a crucial swap. By the above discussion, S' differs from S only for elements v such that $\sigma^*(v) = s^*$, and those for which $\sigma^*(v) \neq s^*$ but $\sigma(v) = s$. Moreover, $cost(S) \leq cost(S')$ since S is locally optimal. Hence, $0 \leq cost(S') - cost(S) \leq$

$$\leq \sum_{v:\sigma^*(v)=s^*} s_d(v) \cdot [(d(v,\sigma^*(v)) + \alpha d(\sigma^*(v),r)) - (d(v,\sigma(v)) + \alpha d(\sigma(v),r))] + \quad (2)$$

$$+ \sum_{\substack{v\,:\,\sigma^*(v)\,\neq\,s^* \\ \sigma(v)\,=\,s}} s_d(v) \cdot [(d(v,\sigma(\sigma^*(v))) + \alpha d(\sigma(\sigma^*(v)),r)) - (d(v,\sigma(v)) + \alpha d(\sigma(v),r))]$$

$$(3)$$

By summing up over all crucial swaps $cs(s,s^*)$, the term (2) of the inequality becomes

$$\sum_{cs(s,s^*)} \sum_{v:\sigma^*(v)=s^*} s_d(v) \cdot [(d(v,\sigma^*(v)) + \alpha d(\sigma^*(v),r)) - (d(v,\sigma(v)) + \alpha d(\sigma(v),r))] =$$

$$\sum_{v\in V} s_d(v) \cdot [(d(v,\sigma^*(v)) + \alpha d(\sigma^*(v),r)) - (d(v,\sigma(v)) + \alpha d(\sigma(v),r))] = C^* - C.$$

In fact, $\bigcup_{s^*\in S^*}\{v : \sigma^*(v) = s^*\} = V$, and, for any $s_1^*, s_2^* \in S^*$, $\{v : \sigma^*(v) = s_1^*\} \cap \{v : \sigma^*(v) = s_2^*\} = \emptyset$, implies $\sum_{cs(s,s^*)} \sum_{v:\sigma^*(v)=s^*} m(v) = \sum_{s^*\in S^*} \sum_{v:\sigma^*(v)=s^*} m(v) = \sum_{v\in V} m(v)$, for any function m.

From the term (3) of the inequality, we first consider $d(v,\sigma(\sigma^*(v))) + \alpha d(\sigma(\sigma^*(v)),r)$ which can be upper bounded by

$$d(v,\sigma^*(v)) + \alpha d(\sigma^*(v),r) + (1+\alpha)d(\sigma^*(v),\sigma(\sigma^*(v))),$$

since, by the triangle inequality, $d(v,\sigma(\sigma^*(v))) \leq d(v,\sigma^*(v)) + d(\sigma^*(v),\sigma(\sigma^*(v)))$ and $d(\sigma(\sigma^*(v)),r) \leq d(\sigma^*(v),\sigma(\sigma^*(v))) + d(\sigma^*(v),r)$, see Fig. 1 for a visualization of these bounds.

We now give two different upper bounds on $d(\sigma^*(v),\sigma(\sigma^*(v)))$, for different values of α which imply two different upper bounds on the term (3) of the inequality.

Bound 1. Clearly, $d(\sigma^*(v),\sigma(\sigma^*(v))) \leq d(\sigma^*(v),r)$ since $r \in S$ and $\sigma^*(v)$ is associated to $\sigma(\sigma^*(v))$, hence obtaining for $\alpha > 0$:

$$d(v,\sigma^*(v)) + \alpha d(\sigma^*(v),r) + (1+\alpha)d(\sigma^*(v),r) \leq$$

$$(x+1)(d(v,\sigma^*(v)) + \alpha d(\sigma^*(v),r)),$$

with $x = \frac{1+\alpha}{\alpha}$, that is $(1+\alpha)d(\sigma^*(v),r) \leq x\alpha d(\sigma^*(v),r) \leq x d(\sigma^*(v),r)$.

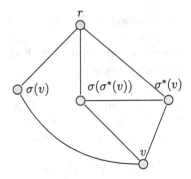

Fig. 1. Upper bounds on distance $d(\sigma^*(v), \sigma(\sigma^*(v)))$

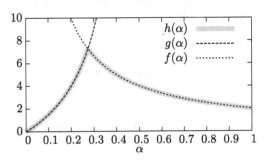

Fig. 2. The two upper bound functions to the locality gap.

From the above bounding, by summing up over all the crucial swaps:

$$\sum_{cs(s,s^*)} \sum_{\substack{v\,:\,\sigma^*(v)\,\neq\,s^* \\ \sigma(v)\,=\,s}} s_d(v) \cdot [(d(v, \sigma(\sigma^*(v))) + \alpha d(\sigma(\sigma^*(v)), r))$$

$$-(d(v, \sigma(v)) + \alpha d(\sigma(v), r))] \le$$

$$\le \sum_{cs(s,s^*)} \sum_{\substack{v\,:\,\sigma^*(v)\,\neq\,s^* \\ \sigma(v)\,=\,s}} s_d(v) \cdot (x+1)(d(v, \sigma^*(v)) + \alpha d(\sigma^*(v), r)) \le$$

$$2(x+1) \sum_{v \in V} s_d(v) \cdot (d(v, \sigma^*(v)) + \alpha d(\sigma^*(v), r)) = 2(x+1)C^*$$

Where the last inequality is due to the fact that each $s \in S$ appears in at most two crucial swaps.

Hence for any $\alpha > 0$, $0 \le cost(S') - cost(S) \le C^* - C + 2(x+1)C^* = C^*(2x+3) - C$, that is, $\frac{C}{C^*} \le 2x + 3 = 5 + \frac{2}{\alpha}$.

Bound 2. Since solution S contains both $\sigma(\sigma^*(v))$ and $\sigma(v)$, then (see Fig. 1):
$d(\sigma^*(v), \sigma(\sigma^*(v))) + \alpha d(\sigma(\sigma^*(v)), r) \le d(\sigma^*(v), \sigma(v)) + \alpha d(\sigma(v), r)$, hence

$d(\sigma^*(v), \sigma(\sigma^*(v))) \leq d(\sigma^*(v), \sigma(v)) + \alpha d(\sigma(v), r) \leq d(\sigma^*(v), v) + d(\sigma(v), v) + \alpha d(\sigma(v), r) \leq d(\sigma^*(v), v) + \alpha d(\sigma^*(v), r) + d(\sigma(v), v) + \alpha d(\sigma(v), r)$. The penultimate inequality holds by the triangle inequality (see Fig. 1), and in the last one we added $\alpha d(\sigma^*(v), r)$.

By summing up over all crucial swaps,

$$\sum_{cs(s, s^*)} \sum_{\substack{v \, : \, \sigma^*(v) \neq s^* \\ \sigma(v) = s}} s_d(v) \cdot [d(\sigma^*(v), v) + \alpha d(\sigma^*(v), r) +$$

$$+ (1 + \alpha)(d(\sigma^*(v), v) + \alpha d(\sigma^*(v), r) + d(\sigma(v), v) + \alpha d(\sigma(v), r))$$

$$- (d(v, \sigma(v)) + \alpha d(\sigma(v), r))] =$$

$$\sum_{cs(s, s^*)} \sum_{\substack{v \, : \, \sigma^*(v) \neq s^* \\ \sigma(v) = s}} s_d(v) \cdot [(2 + \alpha)(d(\sigma^*(v), v) + \alpha d(\sigma^*(v), r)) +$$

$$\alpha(d(\sigma(v), v) + \alpha d(\sigma(v), r))] \leq$$

$$2 \sum_{v \in V} s_d(v) \cdot [(2 + \alpha)(d(\sigma^*(v), v) + \alpha d(\sigma^*(v), r)) + \alpha(d(\sigma(v), v) + \alpha d(\sigma(v), r))]$$

$$= 2(2 + \alpha)C^* + 2\alpha C$$

Hence obtaining, $0 \leq cost(S') - cost(S) \leq C^* - C + 2(2 + \alpha)C^* + 2\alpha C = (5 + 2\alpha)C^* + (2\alpha - 1)C$. If $\alpha < \frac{1}{2}$, then $2\alpha - 1 < 0$, and $(1 - 2\alpha)C \leq (5 + 2\alpha)C^*$, that is $\frac{C}{C^*} \leq 5 + \frac{12\alpha}{1 - 2\alpha}$. $\qquad\square$

Theorem 4 provides two upper bounds to the locality gap given by $5 + f(\alpha)$ and $5 + g(\alpha)$. Functions f, g, and h are plotted in Fig. 2. Function f is monotonic decreasing, while g is monotonic increasing, in their intervals of definition. We have that $f(\alpha) = g(\alpha)$ for $\alpha = \frac{1}{6}(\sqrt{7} - 1) \approx 0.274$ where $f(\alpha) = g(\alpha) < 7.3$. For all the other values of α, one of the two functions is always below such a threshold, that is the approximation ratio is always below 12.3.

Actually, algorithm $\mathcal{L}$ is not yet an approximation algorithm, as the number of iterations needed to find a local optimum solution might be superpolynomial. To fix this problem, as in [2], we can change the stopping condition of $\mathcal{L}$ so it finishes as soon as it finds an approximate local optimum solution, i.e., when the solution S is such that every neighboring solution S' of S has $cost(S') > (1 - \epsilon)cost(S)$, for some $\epsilon > 0$. This leads to at most $\frac{\log(\frac{cost(S_0)}{cost(S^*)})}{\log(\frac{1}{1 - \epsilon})}$ iterations, where S_0 is the initial solution selected in the first iteration of $\mathcal{L}$. This is polynomial in the size of the input.

Corollary 2. *There exists an $\frac{1}{1 - \epsilon}(5 + h(\alpha))$-approximation algorithm MSP for any $\epsilon \in (0, 1)$.*

Finally, by following the arguments in [2], the algorithm can be improved by allowing t simultaneous swaps. For such algorithm, the analysis given in Theorem 4 can be extended by defining the crucial swaps in a way that each storage in the local optimal solution appears in at most $\frac{t+1}{t}$ swaps. This leads to a locality gap of $h'(\alpha)$, where $h' : [0,1] \to \mathbb{R}$,

$$h'(\alpha) = \begin{cases} g'(\alpha) & \text{if } \alpha = 0 \\ \min\{f'(\alpha), g'(\alpha)\} & \text{if } \alpha \in (0, \frac{t}{t+1}) \\ f'(\alpha) & \text{if } \alpha \in [\frac{t}{t+1}, 1], \end{cases}$$

$f' : (0,1] \to \mathbb{R}$, $f'(\alpha) = 1 + \frac{t+1}{t}\frac{1+2\alpha}{\alpha}$ and $g' : [0, \frac{t}{t+1}] \to \mathbb{R}$, $g' = \frac{(3+\alpha)t+2+\alpha}{(1-\alpha)t-\alpha}$. To give an idea on the improvement provided by this method, we computed the maximum value of the upper bounds on the approximation ratio for $t = 2, 3, 4$, which is less than 8.67, 7.78 and 7.05, respectively. It follows that for $t \geq 2$ and any value of α, our algorithm improves over the 10-approximation algorithm provided in [17].

5   Experimental Analysis

In this section, we experimentally analyze the performance of algorithm $\mathcal{L}$. We implemented the algorithm in C++ and compared the values of the found solutions with those of an optimal one computed by solving the following integer program with the GLPK solver [9].

$$\begin{aligned} \min \ & \textstyle\sum_{v,s \in V} x_{vs} \cdot s_d(v)(d(v,s) + \alpha d(s,r)) \\ \text{s.t.} \ & \textstyle\sum_{s \in V} x_{vs} = 1 && \text{for each } v \in V \\ & x_{vs} \leq y_s && \text{for each } v, s \in V \\ & \textstyle\sum_{s \in V} y_s \leq k, \ y_r = 1 \\ & y_s, x_{vs} \in \{0,1\} && \text{for each } v, s \in V. \end{aligned}$$

Our experiments are performed on a workstation equipped with a 3.60 GHz Intel processor and 24 GB of main memory. The program has been compiled with GNU g++ compiler 4.4.3 under Linux. We used the following datasets.

Random Geometric Graphs (RGG). These graphs simulate sensors thrown at random in a two-dimensional space [14]. By throwing n points in a unit square uniformly at random, the probability that no nodes are inside a circle of diameter $d = \sqrt{\frac{\rho \log n}{n}}$ is smaller than $n^{-\frac{\rho}{4}}$, hence, for $\rho \geq 4$ and large n, this probability is very low [15]. Therefore, to generate the graphs we proceeded as follows. First, nodes are generated and a uniformly random position in a unit size square is associated to each of them. An edge between nodes u and v is added to the graph if the Euclidean distance between u and v is at most $\frac{1}{2}\sqrt{\rho\frac{\log n}{n}}$. Edge weights are set to the Euclidean distances. We set $\rho = 4$ and $n \in \{100, 300, 1000\}$. Similar instances, for $n = 100$ have been used in [17].

Barabasi-Albert Graphs (BA). These graphs have been proven to model many real-world networks such as the Internet, the World Wide Web, citation graphs, and some social networks [1]. A Barabási–Albert topology is generated by iteratively adding one node at a time, starting from a given connected graph with at least two nodes. A newly added node is connected to any other existing nodes with a probability that is proportional to the degree that the existing nodes already have. Hence, the more connected a node is, the more likely it is to receive new connections. We set $n \in \{100, 300, 1000\}$ and the weight of each edge is a value sampled from a uniform distribution defined in the range $[40, 60]$.

OR Library (PMED). We used the dataset for *MMP* of the OR Library [4], a collection of test datasets for several problems. The dataset consists in 40 instances named PMED1, PMED2,..., PMED40, whose sizes range from $n = 100$ to 900. This dataset has been used for several experimental studies on *MMP* [13].

Erdős-Rényi Random Graphs (ER). In these graphs, given a fixed number of nodes n and a parameter p, there exists an edge between two nodes u and v with probability p. The probability p represents the density of the graph, that is the ratio between the number of edges in the graph and that in a complete graph of n nodes. We generated these graphs with the same parameters of [6] and [13], that is $n = 100$ with $p = 0.1$, and $n = 150$ with $p = 0.05$. The weight of each edge is randomly taken in the range $[40, 60]$.

The sink node is chosen uniformly at random among the nodes of the graph. As motivated in Sect. 2, we set r_d and r_q to 1. For each of the above graphs, we made a test configuration by setting $\alpha \in \{0.0, 0.1, \ldots, 1\}$ and 30 values of k in the interval $\{1, \ldots, n\}$ with step $\lfloor n/30 \rfloor$. In the cases of $n = 1000$, the sampling for k has been considered for 10 values. The value of $s_d(v)$ is randomly taken from a uniform distribution in the range $[1, 10]$, independently for each $v \in V$. For each of the above test configuration, we generated 5 different graphs in order to compute average values and standard deviations of the measured performance. Finally, we set the parameter ϵ in Corollary 2 to 0.005, 0.01, and 0.1 and the number t of simultaneous swaps to 1. We made this last choice with the aim to analyzing the worst-case behavior of $\mathcal{L}$.

Analysis. In Fig. 3.left, we plot the approximation ratio obtained by our algorithm on a RGG graph of 300 nodes as a function of k, assuming that $\alpha = 0.1$ (for any other value of α, $\mathcal{L}$ exhibits better behavior). In the diagram, we plot three functions, one for each tested value of ϵ. In Fig. 3.right, we report the number of iterations needed for achieving such a ratio in the same setting. This represents the computational time required by the algorithm in a machine-independent way. In our machines, each iteration requires about 0.25 s. in average for graphs of 300 nodes (See Table 1 for the computational time). As expected, the ratio decreases with ϵ, and in particular, for $\epsilon = 0.005$ it is always less than 1.108, that is, the obtained solution is at most 10.8 % worse than the optimal one and it is below the theoretical lower bound (i.e. $1 + 1/e > 1.367$). On the other hand, decreasing ϵ increases the number of iterations. When k is small, the approximation ratio is reduced, in fact for $k < 100$, it is less than 1.07. However, this required up to 18 iterations. We obtain better results both in terms of approximation ratio and number of iterations with large values of k. In fact, for $k > 250$, the ratio is below 1.05 and it is obtained with at most 2 iterations.

In Fig. 4, we plot the approximation ratio and the number of iterations obtained on a RGG graph of 300 nodes as a function of α, assuming that $k = 21$

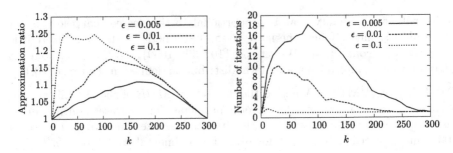

Fig. 3. Random geometric $n = 300$, $\alpha = 0.1$

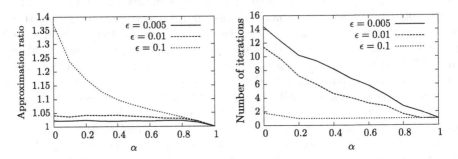

Fig. 4. Random geometric $n = 300$, $k = 21$

(for any other value of k, $\mathcal{L}$ exhibits better behavior). For small values of ϵ the ratio is very small. In detail, it is at most 1.023 for $\epsilon = 0.005$ and 1.042 for $\epsilon = 0.01$. For $\epsilon = 0.1$ we observe a decreasing tendency. This is in contrast with the theoretical upper bound plotted in Fig. 2 (which is however confirmed if functions s_d and w are constant). The maximum value measured is 1.38, obtained when $\alpha = 0$. For $\epsilon = 0.005$, the good values of the approximation ratio required more iterations if α is small. This is due to the fact that if α approaches 1, then the usage of storage nodes does not significantly decrease the objective function and hence the first feasible solution already has a good approximation ratio. In particular, for $\alpha = 1$ the usage of storage nodes does not decrease the energy costs.

For the other settings, varying on the number of nodes, the values of α and k and the type of graph, the overall tendencies are similar. This is also confirmed by the plots of Fig. 5, where we compare the performance of the algorithm in different graph types. In detail, in Fig. 5 we report the average values of the graphs with 100 nodes, $\alpha = 0.1$ and $\epsilon = 0.005$, as a function of k and classified by graph type. We do not observe any significant difference with respect to the type of graph. We note that in these cases the approximation ratio is smaller than the previously reported ones.

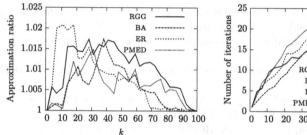
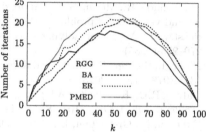

Fig. 5. Graph type comparison $n = 100$, $\alpha = 0.1$, $\epsilon = 0.005$

Table 1. Average computational time required for each iteration of $\mathcal{L}$ when $k = n/2$. Note that the computational time of the iterations in the extreme cases, i.e. $k = 1$ or $n = k$ is always < 0.0001.

Graph type	n	k	Time per iteration (s)
RGG	100	50	0.0121
	300	150	0.3274
	1000	500	15.0086
BA	100	50	0.0122
	300	150	0.3291
	1000	500	14.2448
PMED	100	50	0.0116
	300	150	0.3191
	900	450	10.9442
ER	100	50	0.0137
	150	75	0.0392

6   Conclusions

The minimum k-storage problem has been considered from the theoretical and experimental points of view. The obtained results show that, in undirected graphs, the problem is not approximable within a factor of $1 + \frac{1}{e}$. On the other hand, it admits a constant-factor polynomial-time approximation algorithm based on local search technique which performs very well in practical scenarios. For the case of directed graphs, the problem has been proved to not belong to APX.

References

1. Albert, R., Barabási, A.-L.: Statistical mechanics of complex network. Rev. Mod. Phys. **74**, 47–97 (2002)
2. Arya, V., Garg, N., Khandekar, R., Meyerson, A., Munagala, K., Pandit, V.: Local search heuristics for k-median and facility location problems. SIAM J. Comput. **33**(3), 544–562 (2004)

3. Bar-Yehuda, R., Moran, S.: On approximation problems related to the independent set and vertex cover problems. Discrete Appl. Math. **9**(1), 1–10 (1984)
4. Beasley, J.: A note on solving large p-median problems. Eur. J. Oper. Res. **21**, 270–273 (1985)
5. Dinur, I., Steurer, D.: Analytical approach to parallel repetition. Technical Report 1305.1979, arXiv (2013)
6. Galvao, R.D., ReVelle, C.: A lagrangean heuristic for the maximal covering location problem. Eur. J. Oper. Res. **88**(1), 114–123 (1996)
7. Ganesan, D., Greenstein, B., Estrin, D., Heidemann, J.S., Govindan, R.: Multiresolution storage and search in sensor networks. Trans. Storage **1**(3), 277–315 (2005)
8. Gehrke, J., Madden, S.: Query processing in sensor networks. IEEE Perv. Comp. **3**(1), 46–55 (2004)
9. GLPK. GNU Linear Programming Kit. http://www.gnu.org/software/glpk
10. Guha, S., Khuller, S.: Greedy strikes back: improved facility location algorithms. J. Algorithms **31**(1), 228–248 (1999)
11. Jain, K., Mahdian, M., Saberi, A.: A new greedy approach for facility location problems. In: Proceedings of the 34th ACM Symp. on Theory of Computing (STOC), pp. 731–740 (2002)
12. Madden, S., Franklin, M.J., Hellerstein, J. M., Hong, W.: The design of an acquisitional query processor for sensor networks. In: SIGMOD Conference, pp. 491–502 (2003)
13. Nagarajan, C., Williamson, D.P.: An experimental evaluation of incremental and hierarchical k-median algorithms. In: Pardalos, P.M., Rebennack, S. (eds.) SEA 2011. LNCS, vol. 6630, pp. 169–180. Springer, Heidelberg (2011)
14. Penrose, M.: Random Geometric Graphs. Oxford University Press, Oxford (2003)
15. Raab, M., Steger, A.: balls into bins - a simple and tight analysis. In: Rolim, J., Serna, M., Luby, M. (eds.) RANDOM 1998. LNCS, vol. 1518, pp. 159–170. Springer, Heidelberg (1998)
16. Sheng, B., Li, Q., Mao, W.: Optimize storage placement in sensor networks. IEEE Trans. Mob. Comput. **9**(10), 1437–1450 (2010)
17. Sheng, B., Tan, C.C., Li, Q., Mao, W.: An approximation algorithm for data storage placement in sensor networks. In: Proceedings of the International Conference on Wireless Algorithms, Systems and Applications (WASA), pp. 71–78. IEEE, New York (2007)
18. Shenker, S., Ratnasamy, S., Karp, B., Govindan, R., Estrin, D.: Data-centric storage in sensornets. Comp. Commun. Rev. **33**(1), 137–142 (2003)
19. Tilak, S., Abu-Ghazaleh, N.B., Heinzelman, W.R.: A taxonomy of wireless microsensor network models. Mob. Comp. Commun. Rev. **6**(2), 28–36 (2002)

Counting in Anonymous Dynamic Networks: An Experimental Perspective

Giuseppe Antonio Di Luna[1]([⊠]), Silvia Bonomi[1],
Ioannis Chatzigiannakis[2], and Roberto Baldoni[1]

[1] Cyber Intelligence and Information Security Research Center and
Dipartimento di Ingegneria Informatica, Automatica e Gestionale Antonio Ruberti,
Universitá degli Studi di Roma La Sapienza, Via Ariosto, 25, 00185 Roma, Italy
{baldoni,bonomi,diluna}@dis.uniroma1.it
[2] Computer Technology Institute and Press "Diophantus" (CTI), Patras, Greece
ichatz@cti.gr

Abstract. Counting is a fundamental problem of every distributed system as it represents a basic building block to implement high level abstractions. In anonymous dynamic networks, counting is far from being trivial as nodes have no identity and the knowledge about the network is limited to the local perception of the process itself. Moreover, nodes have to cope with continuous changes of the topology imposed by an external adversary. A relevant example of such kind of networks is represented by wireless sensor networks characterized by the dynamicity of the communication links due to possible collisions or to the presence of duty-cycles aimed at battery preservation. In a companion paper [14], a leader-based algorithms to count the number of processes in an anonymous dynamic network, namely $\mathcal{A}_{NoK}$, has been proposed. Such algorithm employs a technique that mimics an energy transfer from the anonymous nodes to the leader to converge to an exact count of the number of nodes having no knowledge on the dynamic network. Unfortunately $\mathcal{A}_{NoK}$ is an *unconscious* counting algorithm, i.e., the algorithm eventually converges to the exact count but there is no node in the network that is able to detect when this happens. In this paper, we define a new algorithm, called $\mathcal{A}_{NoK}^*$, by augmenting $\mathcal{A}_{NoK}$ with a termination heuristic that allows the leader to decide when it should output the current count and we provide an experimental evaluation, for both $\mathcal{A}_{NoK}$ and $\mathcal{A}_{NoK}^*$, considering different types of dynamic graphs.

1 Introduction

Networks of tiny artifacts will play a fundamental role in the computational environments and applications of tomorrow. Networked embedded sensors and mobile devices will produce a constant flow of data between the real world and modern and traditional networks such as information, communication and social networks. Such hyperconnected dynamic environments create very challenging system models where what was trivially solvable in a static system, is now far from being trivial. What is becoming apparent is that in such environments,

P. Flocchini et al. (Eds.): ALGOSENSORS 2013, LNCS 8243, pp. 139–154, 2014.
DOI: 10.1007/978-3-642-45346-5_11, © Springer-Verlag Berlin Heidelberg 2014

theory and models for static distributed systems do not capture anymore the new kind of applications that are emerging. As a result, over the last years, dynamic distributed systems have attracted a lot of interest from the relevant research community (see e.g., [7,8,24]).

A critical issue in designing such hyperconnected dynamic infrastructures is security and trust, especially when artifacts exchange crucial information that needs to be protected [10]. It is evident that contemporary networks have significant difficulties dealing with third-party tracking and monitoring online, much of it spurred by data aggregation, profiling, and selective targeting. Terms like information security, data confidentiality and integrity, entity authentication and identification need to be considered [26]. A promising approach for addressing these problems is to incorporate privacy in the design and models of such future systems by guaranteeing the *anonymity* of the artifacts.

In this paper, we consider the problem of counting the number of nodes in a network, without revealing any information about their identities or providing information about the network state. Counting is among the most fundamental problems of distributed computation and it is a key function for network management and control, and the vast number of papers appearing in the relevant literature is a clear indication of its importance. A large part of these studies deals with causes of dynamicity such as failures and changes in the topology that eventually stabilize [15]. However, the low rate of topological changes that is usually assumed is unsuitable for reasoning about truly dynamic networks. We envision future networks with highly dynamic changes: connected artifacts may become immediately unreachable after they have been received a message from them. We consider recent theoretical models for dynamic networks in which the topology may change arbitrarily from round to round. In some models (e.g.,[18]), edges - representing communication among hyperconnected artifacts - are changed at each round by an adversary, that is forced to modify edges in such a way that the network is always connected. In other models (e.g., [5,12]), edges appear by following a random distribution where certain properties of the dynamic network hold with high probability. Under these assumptions taken for granted, theoretical results indicate that we can design protocols for distributed tasks that are robust, scalable and that terminate.

In this work we remove fundamental assumptions made by previous theoretical models: (a) we avoid any assumption on the network knowledge: nodes do not know the size (or an upper bound) n of the network, or any other metrics [14,19]; (b) we also avoid any assumption on providing unique identities (ids) to the artifacts: nodes execute identical programs and in symmetric networks it is impossible to count the nodes unless a leader is not introduced [19]; (c) we do not require the network to be connected at each time instance (e.g., 1-Interval connectivity property [18,21]). We believe that the resulting mode of operation is more suitable for future hyperconnected environments, where privacy is incorporated in the model. Under this mode of operation, we propose a new distributed algorithm, namely $\mathcal{A}^*_{NoK}$, that defines a termination heuristic in order to provide estimates on the size of the anonymous network.

$\mathcal{A}^*_{NoK}$ is built starting from the *unconscious* counting algorithm $\mathcal{A}_{NoK}$ introduced in the companion paper [14]. $\mathcal{A}_{NoK}$ guarantees that the leader eventually converges to the exact count but it has no way to detect when this happens (unconsciousness). In this paper, we equip the leader of $\mathcal{A}_{NoK}$ with an heuristics as terminating condition. In this new algorithm, namely $\mathcal{A}^*_{NoK}$, the leader uses the heuristics to guess when it can conclude the count and output a value.

Both the algorithms exploits the energy-transfer technique to count the exact number of nodes. Informally the technique works as follows: each node v_i is assigned with a fixed energy charge, stored in the variable e_{v_i}, and during each round it discharges itself by disseminating energy around to its neighbors i.e., e_{v_i} decreases of a value $k \leq e_{v_i}/2$, then this quantity k is equally split among the neighbors of v_i and this value is added to v_i's neighbors variable. The leader acts as a sink collecting energy (i.e., energy is not transferred by the leader to neighbors). Our technique enforces, at each round, a global invariant on the sum of energy among networks' nodes (i.e., $\sum_{v_i} e_{v_i} = \#nodes$), that resorts to the fact that energy is not created or destroyed in the anonymous dynamic distributed system (energy conservation property). Considering the behavior of the nodes, the energy is eventually transferred to the leader and stored there. The leader measures the energy received to count the size of the network.

We follow a detailed experimental approach and investigate the performance of both $\mathcal{A}_{NoK}$ and $\mathcal{A}^*_{NoK}$ (Sects. 4.1 and 4.2) in the presence of an *oblivious adversary* that rearranges the edges of the communication graph without a strategy and acts according to some probability distribution that may also disconnect the network[1]. In particular, we consider different random evolving graph models in order to identify the error rate of the algorithm (the number of times the termination heuristic fails) as well as the efficiency for terminating the computation (the difference from the actual size of the network and the value output by the termination heuristic). We also look into networks that are periodically disconnected as the artifacts duty-cycle (Section 4.4).

For the case of densely connected anonymous networks, $\mathcal{A}^*_{NoK}$ terminates always correctly. In cases where the network experiences regular partitions, $\mathcal{A}^*_{NoK}$ provides estimates on the size whose accuracy varies according to the degree of disconnection of the network (see Sect. 4.2). Longer periods of network disconnections bring to lower accuracy in counting. Let us finally remark that $\mathcal{A}^*_{NoK}$ is able to answer to predicates such as *"does the network contain more than T nodes?"* (i.e., $|V| \geq T$)in a number of rounds lesser than the one needed by the base $\mathcal{A}_{NoK}$ algorithm presented in [14] as shown in Section 4.3.

2 Related Work

The question concerning which problems can be solved in a distributed system where all processors use the same algorithm and start from the same state has a long history, with its roots dating back to the seminal work of Angluin [3], who

---

[1] In [14] we addressed the problem of counting in the presence of a worst-case adversary that is constrained to preserve only the connectivity of the network.

investigated the problem of establishing a "center". She was the first to realize the connection with the theory of graph coverings, which was going to provide, in particular with the work of Yamashita and Kameda [25], several characterizations for problems that are solvable under certain topological constraints. Other well-known studies on unknown networks have dealt with the problems of robot-exploration and map-drawing of an unknown graph [2,13,22] and on information dissemination [6]. Sakamoto [23] studied the "usefulness" of initial conditions for distributed algorithms (e.g. leader or knowing n) on anonymous networks by presenting a transformation algorithm from one initial condition to another. Fraigniaud *et al.* [16] assumed a unique leader in order to break symmetry and assign short labels as fast as possible. Recently, Chalopin *et al.* [9] have studied the problem of naming anonymous networks in the context of snapshot computation. Finally, Aspnes *et al.* [4] studied the relative powers of reliable anonymous distributed systems with different communication mechanisms: anonymous broadcast, read-write registers, or read-write registers plus additional shared-memory objects.

Distributed systems with worst-case dynamicity were first studied in [21] by introducing the 1-interval connectivity model. They studied flooding and routing problems in asynchronous communication and allowed nodes to detect local neighborhood changes. Under the same model, [18] studied the problem of counting for networks where nodes have unique IDs and provided an algorithm that requires $O(n^2)$ rounds using $O(\log n)$ bits per message. In [19], the authors studied the problem of anonymous counting in this worst-case dynamicity model and provided an algorithm where given that the nodes know an upper bound on the maximum degree that will ever appear, the nodes obtain an upper bound on the size of the network. In [20] the 1-interval connectivity assumption is replaced by other less restrictive *temporal connectivity* conditions that only require that *another causal influence occurs within every time-window of some given length*. They introduce several novel metrics for capturing the speed of information spreading in a dynamic network and provide terminating algorithms for fast propagation of information under continuous disconnectivity.

Let us finally remark that the counting problem has been also investigated also in the context of gossip-based protocols where the dynamicity of the system is governed by a sampling oracle [17] that, at each round, provides each node with a set of neighbors, obtained by randomly sampling network members. This is similar to have a oblivious adversary. However, in the model of [17] the number of neighbors is known before the send step, in our model this is actually not possible.

3   Background

System Model. A *dynamic network* is a network whose topology changes along time due to possible failures of nodes or communication links. We consider computations executed in discrete *synchronous* rounds, controlled by a fictional global clock accessible to all the nodes. Thus, all nodes have access to the current round number via a local variable that we usually denote by r. A dynamic network is modeled by a *dynamic graph* $G(r) = (V, E(r))$, where V is a set of

n nodes (or processors) and $E : \mathbb{N} \rightarrow \mathcal{P}(E')$, where $E' = \{\{u, v\} : u, v \in V\}$, is a function mapping a round number $r \in \mathbb{N}$ to a set $E(r)$ of bidirectional links drawn from E' [18]. Intuitively, a dynamic graph G is an infinite sequence $G(1), G(2), \ldots$ of *instantaneous graphs*, whose edge sets are subsets of E' chosen by an *adversary*. The set V is assumed throughout this work to be *static*, that is it remains the same throughout the execution.

Nodes in V are *anonymous*, i.e. they have no identifier. At each round r, the local view of a node v, denoted as $l_v(r)$, is defined by the multi set containing all the states of processes that are neighbors of v at round r (i.e. all the local variables maintained by the neighbors of v at round r).

Nodes in the network communicate by sending and receiving messages via *anonymous broadcast*; in every round r, each node u generates a single message $m_u(r)$ to be delivered to all its current neighbors in $N_u(r) = \{v \mid \{u, v\} \in E(r)\}$.

Oblivious Dynamic Graph Adversaries. In order to model the dynamicity of the network graph, we consider the following four adversary models:

1. **$G(n, p)$ Graph[11]:** at the beginning of each round r the set of edges is emptied and then for any pair of processes $u, v \in V$, the edge uv is created according to a given probability p. Let us recall that in the $G(n, p)$ graph model, there exist a connectivity threshold t, depending on the number of nodes n, such that if probability p is above the threshold, then $G(n, p)$ is connected with very high probability.
2. **Edge-Markovian (EM) Graph[12]:** at each round r, edges are modified according to the following rules:
   (a) For each edge $uv \in E(r-1)$, uv is removed from $E(r)$ with a probability p_d (i.e., *death probability*).
   (b) For each edge $uv \notin E(r-1)$, uv is created and inserted in $E(r)$ with a probability p_b (i.e., *birth probability*).
   Clearly, connectivity of the graph at each round depends on p_d and p_b.
3. **Duty-cycle based Graph:** at round r_0 the dynamic graph has a fixed, connected, topology. Each node follows a duty cycling phase during which, if at a given round r_i the node is awake it can receive and send messages according the topology of r_0 to any neighboring node that is also awake. While when at round r_j it is in sleep mode, all adjacent edges are removed from the graph. The presence of the duty cycle essentially brings some dynamicity in the graph since not all edges will be set at each round. This model constructs evolving graphs that reflect realistic deployments of resource constraint devices. Remark that this model does not guarantee that the graph will be connected at each round.

Energy-Transfer Technique. In [14], we presented a technique, called *energy-transfer*, for counting the size of the network overcoming the lack of identities and the constantly dynamic environment. Such technique is inspired to the physical conservation of energy invariant and we abstract the notion of energy by assigning to each process a variable (representing an energy charge). In particular, each node is assigned a fixed energy charge, and during each round it discharges itself by disseminating it around to its neighbors. The leader acts as a sink collecting energy (i.e., energy is not transferred by the leader to neighbors). The technique enforces, at each round, an invariant on the sum of energy among networks' nodes: energy is not created or destroyed. Considering the behavior of the nodes, the energy is eventually transferred to the leader and stored there. The leader measures the energy received to count the size of the network. Interestingly, this technique is very simple to implement and depends on very limited information about the attributes of a given network (i.e., node ids, channel ids etc). [14] introduced a series of algorithms applying the *energy-transfer technique* that either assume knowledge on certain aspects of the network (e.g., an upper bound on node degree) in order to terminate the computation (i.e. *conscious* algorithm), or do not make any additional assumption but do not terminate as the $\mathcal{A}_{NoK}$ algorithm (i.e. *unconscious* algorithm where the computation converges to the correct count, but nodes are not able to detect when to terminate).

We remark that the results in [14], especially those concerning the absence of any knowledge assumption, represent an interesting feasibility point, even if they cannot be used in practice since the leader is not able to verify any terminating condition and thus it is not able to provide an answer to the counting problem. In this paper, we present an algorithm, namely $\mathcal{A}_{NoK}^*$, obtained by the basic $\mathcal{A}_{NoK}$ one, in which we define a terminating condition based on the definition of an heuristic and we show that it enables an accurate count.

4   Unconscious Counting Algorithms with Termination Heuristic

4.1   The Unconscious No-Knowledge Algorithm $\mathcal{A}_{NoK}$

The No-Knowledge Algorithm ($\mathcal{A}_{NoK}$) presented in [14] works in the following way: each non-leader node v starts, at round r_0, with energy quantity $e_v = 1$ and it transfers half of its current energy to the neighbors. However, v has no knowledge about the network and thus it cannot know the exact number of neighbors in r before receiving messages, but it can only guess such number. Thus, v supposes to have d neighbors and it broadcasts a quantity of energy $\frac{1}{2d}$ (as if there are really d neighbors). Then v starts to collect messages transmitted by its neighbors at the beginning of the round and it stores such messages in a local variable S_{msg}. At the end of the round, v updates its energy e_v to $\frac{1}{2} + (d - |S_{msg}|)\frac{1}{2d} + \sum_{\forall m \in S_{msg}} m$ to preserver the quantity of energy over all the network.

Notice that, if the real number of neighbors at round r is lower than the estimation (i.e., $|N_v(r)| \leq d$) then the global energy conserved among all the processes is still constant (this is due to the compensation done by v at the end of the round based on the effective number of received messages). On the contrary, if the number of neighbors is greater than the estimation (i.e., $|N_v(r)| > d$) then, there is the release of a *local surplus* of energy. As an example, consider the case where v has energy e_v the estimation of neighbors is $d = 2$ and the real number of neighbors is $N_v(r) = 8$. When v sends $\frac{e_v}{4}$ to each neighbors, the total amount of energy transferred is twice the energy stored by v (i.e., the energy transferred is $8 \times \frac{e_v}{4} = 2e_v$ while node v had only e_v residual energy). However, since v adjusts its local residual energy considering the number of received messages, it follows that its residual energy will become negative and globally the energy is still preserved.

The local surplus of positive/negative energy could create, in the leader, a temporary value of energy e that is greater than $|V|$ or negative. Moreover, the adversary could change, at each round, the degree of nodes in order to avoid the convergence of the leader. To overcome these issues each processes stores locally the highest number of neighbors it has ever seen and it uses such number as estimation of its degree d. node v can increase d at most $|V| - 1$ times, from 1 to $|V|$. This implies that worst case adversary cannot create an infinite surplus of local energy. Since the conservation of energy is not violated and the local surplus of energy is finite, it is straightforward to prove that the leader has to converge to the value $|V|$ and the adversary could delay this convergence only a finite number of times. Intuitively, the adversary cannot delay too much its moves, because when the energy stored in $V \setminus \{v_l\}$ is less than a certain value, the local surplus of energy that it could create, even in the worst case, it is not enough to change the leader count. So, if at each round r the leader counts $\lceil e_{v_l} \rceil$, it is possible to prove that there exists a round r^* after which the leader will always count the correct value despite the move of the adversary [14].

Unfortunately, looking to the number of consecutive rounds in which the leader outputs always the same count is not sufficient to provide a terminating condition as such number can always be influenced by the adversary. As a consequence, the leader cannot detect convergence. In fact, let us suppose that the leader stops when the increment of energy at round r is below a threshold t. It is always possible to have a network of size $t + 1$ where each node have a residual charge of t. So each increment on the leader energy is below the termination threshold but the residual energy on the network is greater than 1, so if the leader terminates it will miss one node.

4.2   Termination Heuristic: $\mathcal{A}^*_{NoK}$

In this section, we will present the heuristic added to the basic $\mathcal{A}_{NoK}$ to obtain the new algorithm $\mathcal{A}^*_{NoK}$ working in an anonymous network with No Knowledge assumption and having a termination condition. The heuristic is used by the leader to decide at which time the current count can be considered as the final one. The heuristic is based on the assumption that the dynamicity of the graph is governed by a random process (i.e., a graph where links change according to a uniform probability distribution) and it considers the notion of *flow observed by the leader*.

At each round r, the leader v_l will receive a fraction of energy from all its neighbors. So the flow of energy to the leader at round r can be expressed as:

$$\Phi^r(v_l) = \sum_{\forall v \in N_{v_l}(r)} \frac{e_v(r)}{2d_v^{max}(r)}$$

where $e_v(r)$ is the energy of v at round r and $d_v^{max}(r)$ is the maximum number of neighbors that node v has so far. After a sufficient number of rounds, the estimation of the flow observed by the leader is

$$\Phi^r(v_l) = \sum_{\forall v \in N_{v_l}(r)} \frac{e_v(r)}{2d_v^{max}(r)} \simeq \frac{|N_{v_l}(r)|}{2d_{avg}^{max}(r)} \overline{e_v(r)}$$

where $2d_{avg}^{max}(r)$ is the average of the maximum degrees seen by nodes in G at round r and $\overline{e_v(r)}$ is the average of the energy kept by all non-leader nodes at round r.

Let us remark that, in the absence of the leader, the energy is always balanced among nodes in the network and let us recall that the leader is the only node absorbing energy. As a consequence, nodes being neighbors of the leader could have less energy than others as they transferred part of their energy to the leader without receiving nothing from it. Due to the assumption about the probabilistic nature of the edges creation process and considering the functioning of $\mathcal{A}_{NoK}$, those non-leader nodes will tend to have a similar quantity of energy as they will balance energy surplus. Thus, the leader can estimate $\overline{e_v(r)} \simeq \frac{|V| - e_{v_l}(r-1)}{|V|}$.

Due to the assumption about the probabilistic nature of the edges creation process, the leader will see almost the same maximum number of neighbors as the other nodes. Thus, $2d_{v_l}^{max}(r) \simeq 2d_{avg}^{max}(r)$. Thus, substituting we have

$$\Phi^r(v_l) \simeq \frac{|N_{v_l}(r)|}{2d_{v_l}^{max}(r)} \frac{|V| - e_{v_l}(r-1)}{|V|}$$

from which we obtain

$$|\bar{V}(r)| \simeq \frac{\rho(r)e_{v_l}(r-1)}{\rho(r) - \Phi^r(v_l)}$$

where $|\bar{V}(r)|$ is estimation of the number of processes in the network done by the leader at round r and $\rho(r) = \frac{|N_{v_l}(r)|}{2d_{v_l}^{max}(r)}$.

Let $k = \lceil e_{v_l}(r) \rceil$ be the number representing the count done by the leader at round r, and let $\Delta(r) = |\bar{V}(r)| - e_{v_l}(r)$ be difference between the network size estimated with the energy flow and the energy currently stored at the leader. We can finally define a termination condition as follows: $\rceil$ as the count outputted by the leader at round r, as long as $\lceil e_{v_l}(r) \rceil$ remains stable, the leader computes the average $\overline{\Delta}$ of $\Delta(r)$ over the last k rounds and if after k consecutive rounds the quantity $\lceil e_{v_l}(r+k) + \overline{h} \rceil$ is equal to k and $\lceil e_{v_l}(r+k) \rceil = k$ the counting procedure terminates and the leader outputs k.

5   Performance Evaluation

Simulator. In order to run our experiments, we developed a JAVA simulator using the Jung library [1] to keep track of the graph data structure. Each process v is seen as a node in the graph and it exposes an interface composed of two methods: the first one allowing to send a message for round r and the second one allowing to deliver messages for the round r. Moreover, each node has associated a queue q_v storing the messages that it has to receive. The simulation is done trough a set of threads; a thread T_j takes a node from a list l_m containing all of nodes to be examined in this round, removes it from the list and invokes the method send message. T_j also takes the message m generated by v, and adds it to the queues of $N_v(r)$. When l_m is empty, a different set of threads is activated to deliver messages. T_j takes a node v from a list l_d and manage the delivery of all messages in q_v that v received during the current round. When all the messages in the queues are delivered to all the processes, the round terminates and the topology can be modified according to the dynamicity model considered and a new round can start.

Metrics and Parameters. We investigate three key performance metrics:

- **Convergence Time Distribution:** the convergence time is defined by the first round at which the algorithm outputs the correct value. In the following, we studied the probability distribution of the convergence time to show the average latency of the algorithms before reaching a correct count.
- **Flow Based Gain Δ:** such metrics represents the difference measured by the leader between the size estimated through the flow and the the size estimated trough the energy stored inside the leader (i.e., $\Delta(r) = |\bar{V}(r)| - e_{v_l}(r)$).
- **Error Frequency ρ:** we measured the percentage of uncorrect termination obtained while adopting the heuristics-based termination condition defined in Sect. 4.2.

The above metrics have been evaluated by varying the following parameters:

- **Dynamicity Model:** we considered different types of oblivious dynamic graph adversaries to evaluate the factors impacting every metrics (see Sect. 3 for a formal description).
- **Edges Creation Probability p:** such probability governs the graph dynamicity according to the specific model considered ($G_{(n,p)}$ or Edge-Markovian).

We have evaluated the performance of the algorithms under different metrics in networks comprised of $\{10, 100, 1000\}$ nodes. When not explicitly stated, tests are the results of 1000 independent runs.

5.1 Evaluation of $\mathcal{A}_{NoK}$

We implemented and tested $\mathcal{A}_{NoK}$ on both $G_{(n,p)}$, Edge-Markovian and Duty-cycle-based graphs. Let us first consider the case of $G_{(n,p)}$ graphs and let us recall that the connectivity threshold t is defined according to the number of nodes in the graph (i.e., $t = \frac{ln(|V|)}{|V|}$). We evaluate our algorithm for several probability p. In particular, for any probability greater than $2t$, we consider only

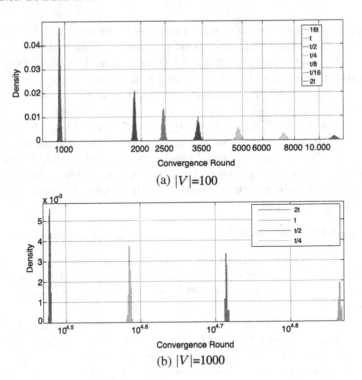

Fig. 1. $\mathcal{A}_{NoK}$ Convergence time distribution for $G_{(n,p)}$ graphs with different edge creation probabilities p.

connected graph instances, i.e., at each step, we check the connectivity and in case of disconnected graph we sample a new random graph. For probabilities smaller than $2t$ we allow disconnected graph instances.

Figure 1 shows $\mathcal{A}_{NoK}$ convergence time distribution when the algorithm runs on $G_{(n,p)}$ graphs. As expected the convergence time becomes worse when we consider disconnected instances. However, it is worth notice that the algorithm is able to converge to the correct count even in presence of disconnected instances. Moreover, the increment of convergence time is inversely proportional to p and there is an increment of the distribution variance due to the presence of disconnected instances.

When considering Edge-Markovian graphs, we set the probability of creating an edge as in the $G_{(n,p)}$ graphs and we fixed the probability of deleting an edge to 0.25 (i.e., $p_d = 0.25$ and $p_b = f(t)$).

Figure 2 shows $\mathcal{A}_{NoK}$ convergence time distribution; as we can see, it is comparable to $G_{(n,p)}$ graph one. In addition, the persistence of edges across rounds (due to $p_d \leq 1$) mitigates the low values of edge creation probability. As a consequence, the convergence is faster than the pure $G_{(n,p)}$.

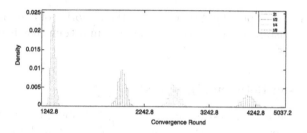

Fig. 2. $\mathcal{A}_{NoK}$ Convergence time distribution for edge-Markovian graphs with different edge creation probabilities p_b and $|V| = 100$.

5.2 Evaluation of $\mathcal{A}^*_{NoK}$

In the following, we evaluate the $\mathcal{A}^*_{NoK}$ algorithm on both $G_{(n,p)}$ and Edge-Markovian graphs. Figure 3 shows several measures related to the heuristic correctness. In particular, in addition to the error frequency ρ, we measured also the average error and maximum error done, by the heuristic, in terms of number of nodes missed with respect to the real number of nodes in the graph. We omit from the Figure some probabilities since they always terminate correctly ($p \geq \frac{t}{2}$ in case of $G_{(n,p)}$ graphs and $p_b \geq \frac{t}{4}$ for the Edge-Markovian). In case of disconnected topologies, i.e., $p \leq \frac{t}{4}$ for the $G_{(n,p)}$ or $p_b \leq \frac{t}{8}$ for the Edge-Markovian, we have that the percentage of counting instances terminating correctly is smaller that 100 % and it becomes proportionally worse with the decrease of p. Moreover, it is possible to see a bimodal behavior of the heuristic when it fails: two cases are frequent in the experiments (i) the heuristic forces the termination in the first rounds of the counting process with the consequence of having the leader outputting a count much smaller than the real number of processes and (ii) the heuristic fails when the energy accumulated by the leader is close to the current network size. In all our experiments we have not found a case in which the heuristics forces the termination in a case different from this two. Moreover in the table we indicate the *Convergence Detection Time*, that is the number of rounds after the first convergence that the heuristics employs to correctly ter-

Model		$G_{(n,p)}$							Edge-Markovian $p_d = 0.25$
p		$\frac{t}{4}$		$\frac{t}{8}$	$\frac{t}{16}$	$\frac{t}{32}$	$\frac{t}{8}$	$\frac{t}{16}$	$\frac{t}{32}$
$|V|$	10	100	1000	100	100	100	100	100	100
ρ	22%	3%	2%	19%	25%	84%	30%	68%	76%
Average Error	2,02	8,96	1	9	44,5	41,4	1	3,12	11,8
Max Error in Nodes	8	96	1	99	99	99	1	99	99
σ of Error	2,1166	27,4	0	27,4	48,3	48,8	1	14,23	29,73
Convergence Detection Time Average	10,2	100	1000	100	100	100	100	100	100
Convergence Detection Time Max	40	100	1000	100	100	100	100	100	100
Convergence Detection Time Min	10	100	1000	100	100	100	100	100	100

Fig. 3. Evaluation of the Termination Correctness ρ. The results are the outcome of 500 experiments

minate the count. It is possible to see that in the majority of experiments, even on disconnected instances the heuristic converges in a time that is equal to the size of the network.

5.3 Comparison Between $\mathcal{A}_{NoK}$ and $\mathcal{A}^*_{NoK}$

The flow could be used to estimate the size of $|V|$ obtaining a faster count. Figure 4 shows the evolution of Δ, i.e., difference measured by the leader between the size estimated through the flow and the the size estimated trough the energy stored inside the leader, both from a temporal perspective 4(a) and from the energy perspective 4(b).

The value Δ reaches the maximum when the energy at the leader is approximately half of the network size; in this case, when the network is connected (i.e., $p \geq t$), the use of the heuristic allows the leader to predict, correctly, the presence of at least others 17 nodes.

So, on connected instances our approach could be useful to answer faster to predicates likes $|V| \geq t$. In addiction, the flow-based estimation continues to perform well on non-connected instances only until a certain threshold, then the gain obtained with the flow drops to one or two nodes more than the ones estimated by the energy.

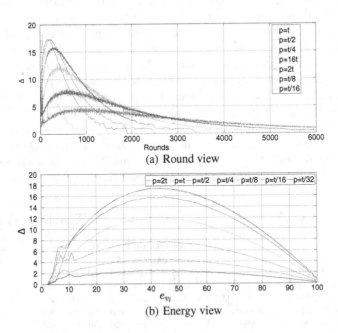

(a) Round view

(b) Energy view

Fig. 4. Difference between the size estimated with the flow ($\mathcal{A}^*_{NoK}$) and the size estimated by looking to the energy stored at the leader ($\mathcal{A}_{NoK}$) in a $G_{n,p}$ network of $|V|=100$.

Moreover the figures show why the termination heuristics works bad on instances where $p \leq \frac{t}{4}$, we can see that Δ falls behind the threshold of 1, both when the energy in the leader is low, and when the energy in the leader is approaching the value $|V|$ this could lead to two possible misbehavior, terminating after few rounds from the start, so with a value that could be sensibly distant from the value of $|V|$ or it could terminate near $|V|$, when Δ falls again behind 1.

Figure 4(a) shows the behavior of Δ along time. In particular,

- when the network is connected (i.e., $p \geq t$), the counting done by the leader fast approaches half of the network size (i.e., the maximum value for Δ). The energy-based count approaches the actual size with an exponential time; this is visible from the exponential decay of Δ. This behavior is present also when $p < t$, even tough there is a slower decay of Δ that obviously reflects a slower approach to the actual size.
- for values of $p \leq t$ the curves show a high variance. This is due to the presence of disconnected topologies that introduce a variance in the convergence time for which the magnitude is proportional to the inverse of p. This high variance in convergence is due to the high variance of the flow that the leader will see during the execution.

The same behavior can be observed in Edge-Markovian graphs (cfr. Figure 5). The presence of more edges in the edge-markovian graph affects positively the Δ measures since it is less prone to the value of p. It is possible to notice a slightly low maximum value for the edge-markovian process, 17 against 17.3 of the $G_{(n,p)}$ graph.

We run also tests with larger graphs ($|V| = 1000$) but we omit them here since curves exhibit the same behavior of those shown in Figs. 4 and 5, notably in this case the maximum delta is about 170 nodes.

5.4   Duty Cycle

In order to test the adaptiveness of our heuristic, we run $\mathcal{A}^*_{NoK}$ on regular topologies: rings and chains. Over those topologies, we simulate a duty-cycle of 80 %. Each node independently sleeps for 20 % of the time and during this period links of sleeping nodes are deleted. Considering a ring topology with $|V| = 100$, the average convergence time is around 26986 rounds for 100 experiments, for the chain the convergence time is on average 70000 rounds . We also tested random $G_{(n,p)}$ topologies where $p = 2t$, in this case the average over 200 experiments shows a convergence time of 1059 rounds. The most noticeable phenomenon is that on graphs with duty-cyle both the termination heuristic and the size estimation perform really bad: on rings and chains the termination heuristics always fails and on random graphs it fails on the 23 % of the instances.

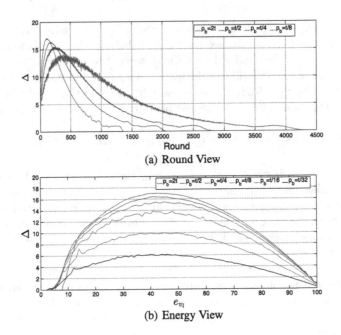

(a) Round View

(b) Energy View

Fig. 5. Difference between the size estimated with the flow $(\mathcal{A}^*_{NoK})$ and the size estimated by looking to the energy stored at the leader $(\mathcal{A}_{NoK})$ for Edge Markovian network with $p_d = 0.25$ of $|V|=100$.

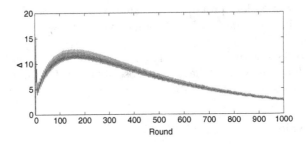

Fig. 6. Difference between the size estimated with the flow $(\mathcal{A}^*_{NoK})$ and the size estimated by looking to the energy stored at the leader $(\mathcal{A}_{NoK})$ of 200 runs for duty cycle and random graph with $|V|=100$.

6   Conclusion

In this work we presented a practical algorithm based on a paradigm that mimics an energy transfer from nodes to the leader, namely $\mathcal{A}^*_{NoK}$, obtained by the algorithm $\mathcal{A}_{NoK}$ presented in [14]. This is done by equipping the leader of $\mathcal{A}_{NoK}$ with a termination heuristic. Both algorithms has been implemented and tested. Experiments show that $\mathcal{A}^*_{NoK}$ terminates correctly on dense graphs and it has acceptable error rate (i.e., the termination heuristic fails just a few times returning a wrong count) on disconnected graph instances; however, the error rate becomes high when we consider sparse and extremely disconnected graph instances or regular topologies. An interesting point revealed by the analysis is that when the heuristic fails, it exhibits a bimodal behavior: either the count output by the terminating heuristic has a small difference with the actual network size or this difference is on the order of the size of the network.

This interesting feature has to be further investigated to understand if and how it is possible to design better heuristics. Thanks to the concept of energy-flow, $\mathcal{A}^*_{NoK}$ could answer faster to predicates like $|V| \geq T$.

Acknowledgments. This work has been partially supported by the TENACE project (MIUR-PRIN 20103P34XC) and Supported in part by the project "Foundations of Dynamic Distributed Computing Systems" (FOCUS) which is implemented under the "ARISTEIA" Action of the Operational Programme "Education and Lifelong Learning" and is co-funded by the European Union (European Social Fund) and Greek National Resources.

References

1. Jung. http://jung.sourceforge.net/ (2013)
2. Albers, S., Henzinger, M.R.: Exploring unknown environments. SIAM J. Comput. **29**(4), 1164–1188 (2000)
3. Angluin., D.: Local and global properties in networks of processors (extended abstract). In: Proceedings of the Twelfth Annual ACM Symposium on Theory of Computing, STOC '80, pp. 82–93. ACM (1980)
4. Aspnes, J., Fich, F.E., Ruppert, E.: Relationships between broadcast and shared memory in reliable anonymous distributed systems. Distrib. Comput. **18**(3), 209–219 (2006)
5. Avin, C., Koucký, M., Lotker, Z.: How to explore a fast-changing world (cover time of a simple random walk on evolving graphs). In: Aceto, L., Damgård, I., Goldberg, L.A., Halldórsson, M., Ingólfsdóttir, A., Walukiewicz, I. (eds.) ICALP 2008, Part I. LNCS, vol. 5125, pp. 121–132. Springer, Heidelberg (2008)
6. Awerbuch, B., Goldreich, O., Vainish, R., Peleg, D.: A trade-off between information and communication in broadcast protocols. J. ACM (JACM) **37**(2), 238–256 (1990)
7. Baldoni, R., Bertier, M., Raynal, M., Tucci-Piergiovanni, S.: Looking for a definition of dynamic distributed systems. In: Malyshkin, V.E. (ed.) PaCT 2007. LNCS, vol. 4671, pp. 1–14. Springer, Heidelberg (2007)
8. Casteigts, A., Flocchini, P., Quattrociocchi, W., Santoro, N.: Time-varying graphs and dynamic networks. CoRR, abs/1012.0009 (2010)
9. Chalopin, J., Métivier, Y., Morsellino, T.: On snapshots and stable properties detection in anonymous fully distributed systems (extended abstract). In: Even, G., Halldórsson, M. (eds.) SIROCCO 2012. LNCS, vol. 7355, pp. 207–218. Springer, Heidelberg (2012)
10. Chen, J., Wu, J.: A survey on cryptography applied to secure mobile ad hoc networks and wireless sensor networks. In: Chen, J., Wu, J. (eds.) Handbook of Research on Developments and Trends in Wireless Sensor Networks: From Principle to Practice. IGI Global, Hershey (2010)

11. Clementi, A.E.F., Pasquale, F., Monti, A., Silvestri, R.: Communication in dynamic radio networks. In: Proceedings of the Twenty-Sixth Annual ACM Symposium on Principles of Distributed Computing, PODC '07, pp. 205–214. ACM, New York (2007)
12. Clementi, A.E.F., Macci, C., Monti, A., Pasquale, F., Silvestri, R.: Flooding time in edge-markovian dynamic graphs. In: Proceedings of the Twenty-Seventh ACM Symposium on Principles of Distributed Computing, PODC '08, pp. 213–222. ACM, New York (2008)
13. Deng, X., Papadimitriou, C.H.: Exploring an unknown graph. In: Proceedings of the 31st Annual Symposium on Foundations of Computer Science, pp. 355–361. IEEE (1990)
14. Di Luna, G., Baldoni, R., Bonomi, S., Chatzigiannakis, I.: Counting on anonymous dynamic networks through energy Transfer. Technical report, Dipartimento di Ingegneria Informatica, Automatica e Gestionale Antonio Ruberti., 2013. The paper will appear at ICDCN 2014 with the title "Conscious and Unconscious Counting on Anonymous Dynamic Networks"
15. Dolev, S.: Self-stabilization. MIT Press, Cambridge (2000)
16. Fraigniaud, P., Pelc, A., Peleg, D., Pérennes, S.: Assigning labels in unknown anonymous networks. In: Proceedings of the Nineteenth Annual ACM Symposium on Principles of Distributed Computing, pp. 101–111. ACM (2000)
17. Kempe, D., Dobra, A., Gehrke, J.: Gossip-based computation of aggregate information. In: FOCS, pp. 482–491 (2003)
18. Kuhn, F., Lynch, N., Oshman, R.: Distributed computation in dynamic networks. In: Proceedings of the 42nd ACM Symposium on Theory of Computing, STOC '10, pp. 513–522. ACM, New York (2010)
19. Michail, O., Chatzigiannakis, I., Spirakis, P.G.: Brief announcement: naming and counting in anonymous unknown dynamic networks. In: Aguilera, M.K. (ed.) DISC 2012. LNCS, vol. 7611, pp. 437–438. Springer, Heidelberg (2012)
20. Michail, O., Chatzigiannakis, I., Spirakis, P.G.: Causality, influence, and computation in possibly disconnected synchronous dynamic networks. In: Baldoni, R., Flocchini, P., Binoy, R. (eds.) OPODIS 2012. LNCS, vol. 7702, pp. 269–283. Springer, Heidelberg (2012)
21. O'Dell, R., Wattenhofer, R.: Information dissemination in highly dynamic graphs. In: Proceedings of the 2005 Joint Workshop on Foundations of Mobile Computing, DIALM-POMC '05, pp. 104–110. ACM, New York (2005)
22. Panaite, P., Pelc, A.: Exploring unknown undirected graphs. In: Proceedings of the Ninth Annual ACM-SIAM Symposium on Discrete Algorithms, pp. 316–322. Society for Industrial and Applied Mathematics (1998)
23. Sakamoto, N.: Comparison of initial conditions for distributed algorithms on anonymous networks. In: Proceedings of the Eighteenth Annual ACM Symposium on Principles of Distributed Computing, PODC '99, pp. 173–179. ACM (1999)
24. Scheideler, C.: Models and techniques for communication in dynamic networks. In: Alt, H., Ferreira, A. (eds.) STACS 2002. LNCS, vol. 2285, pp. 27–49. Springer, Heidelberg (2002)
25. Yamashita, M., Kameda, T.: Computing on an anonymous network. In: Proceedings of the Seventh Annual ACM Symposium on Principles of Distributed Computing, PODC '88, pp. 117–130. ACM, New York (1988)
26. Zhou, Y., Fang, Y., Zhang, Y.: Securing wireless sensor networks: a survey. IEEE Communi. Surveys Tutorials 10(3), 6–28 (2008)

Station Assignment with Applications to Sensing

Antonio Fernández Anta[1], Dariusz R. Kowalski[2],
Miguel A. Mosteiro[3,4(✉)], and Prudence W.H. Wong[2]

[1] Institute IMDEA Networks, Madrid, Spain
antonio.fernandez@imdea.org
[2] Department of Computer Science, University of Liverpool, Liverpool, UK
{D.Kowalski,pwong}@liverpool.ac.uk
[3] Department of Computer Science, Kean University, Union, NJ, USA
[4] Universidad Rey Juan Carlos, GSyC, Madrid, Spain
mmosteir@kean.edu

Abstract. We study an allocation problem that arises in various scenarios. For instance, a health monitoring system where ambulatory patients carry sensors that must periodically upload physiological data. Another example is participatory sensing, where communities of mobile device users upload periodically information about their environment. We assume that devices or sensors (generically called *clients*) join and leave the system continuously, and they must upload/download data to static devices (or *base stations*), via radio transmissions. The mobility of clients, the limited range of transmission, and the possibly ephemeral nature of the clients are modeled by characterizing each client with a *life interval* and a *stations group*, so that different clients may or may not coincide in time and/or stations to connect. The intrinsically shared nature of the access to base stations is modeled by introducing a maximum *station bandwidth* that is shared among its connected clients, a client *laxity*, which bounds the maximum time that an active client is not transmitting to some base station, and a *client bandwidth*, which bounds the minimum bandwidth that a client requires in each transmission. Under the model described, we study the problem of assigning clients to base stations so that every client transmits to some station in its group, limited by laxities and bandwidths. We call this problem the *Station Assignment* problem. We study the impact of the rate and burstiness of the arrival of clients on the solvability of Station Assignment. To carry out a worst-case analysis we use a typical adversarial methodology: we assume the presence of an adversary that controls the arrival and departure of clients. The adversary is limited by two parameters that model the rate and the burstiness of the stations load (hence, limitting the rate and burstiness of the client arrivals). Specifically, we show upper and lower bounds on the rate and burstiness of the arrival for various client arrival schedules and protocol classes. The problem

This work was supported in part by the Comunidad de Madrid (S2009TIC-1692), the Spanish MICINN/MINECO (TEC2011-29688-C02-01), the National Natural Science Foundation of China (61020106002), the National Science Foundation (CCF-0937829, CCF-1114930), and Kean University UFRI grant.

P. Flocchini et al. (Eds.): ALGOSENSORS 2013, LNCS 8243, pp. 155–169, 2014.
DOI: 10.1007/978-3-642-45346-5_12, © Springer-Verlag Berlin Heidelberg 2014

has connections with Load Balancing and Scheduling, usually studied using competitive analysis. To the best of our knowledge, this is the first time that the Station Assignment problem is studied under adversarial arrivals.

Keywords: Station Assignment · Periodic sensing · Health monitoring systems · Participatory sensing · Continuous adversarial dynamics

1   Introduction

We study a dynamic allocation problem that arises in various scenarios where data sensed using mobile devices has to be gathered using one of many static access points available. Examples include *wearable health-monitoring systems*, where ambulatory patients carry physiological sensors, and the data gathered must be periodically uploaded, and *participatory sensing*, where communities of mobile device users upload periodically information about their environment. We call this problem *Station Assignment.*

We study the Station Assignment problem assuming a continuous arrival of mobile devices, called *clients*, who have to upload data to static devices, called *base stations*, via radio transmissions. The mobility of clients, the limited range of transmission, and the possibly limited time that they will be uploading to a given station, are modeled by assigning to clients a *life interval* and a *stations group*. That is, as clients move, the set of stations that they could possibly upload data may change. We model this event as a client that departs and a new client that arrives. The intrinsically shared nature of the base stations is modeled by introducing a maximum *station bandwidth* that a station can share among its connected clients, a client *laxity*, which bounds the maximum time that an active client is not transmitting to some base station, and a *client bandwidth*, which bounds the minimum bandwidth that a client requires in each transmission.

Under the described model, we study the problem of assigning clients to base stations so that every client transmits to some station in its group, limited by laxities and bandwidths. We study the impact of the frequency of client arrivals on the solvability of Station Assignment. Specifically, we show upper and lower bounds on the rate and burstiness of the client arrivals for solvability of Station Assignment under various client arrival schedules and protocol classes. To carry out a worst-case analysis, we use a typical adversarial methodology: we assume the presence of an adversary that controls the arrival and departure of clients. The adversary is limited by two parameters that model the rate and burstiness of the arrival. We also study the connections of this problem with online load balancing and scheduling, usually studied using competitive analysis. To the best of our knowledge, this is the first time when the Station Assignment problem is studied under adversarial arrivals.

2   Adversarial Model and Problem Definition

Model. We consider a Mobile Radio Network composed of a set S of base stations, or simply **stations** for short and a set C of **clients** that want to transmit packets to some station. Throughout we denote $n \triangleq |C|$ and $m \triangleq |S|$. The time is assumed to be slotted and the time domain is $\mathbb{N}$. Each time slot is long enough to transmit one packet.

Each client $c \in C$ has the following characterization.

- A **life interval**, which is the set $\tau_c = [a, b] \subseteq \mathbb{N}$ of consecutive slots in which c is active.
- A **stations group**, which is the set $S_c \subseteq S$ of stations to which c may transmit packets.
- A **laxity** $w_c \in \mathbb{N}$, $0 < w_c \leq |\tau_c|$, such that $c \in C$ must transmit to some station in S_c at least once within every w_c consecutive time slots in τ_c. In this work we assume the laxity to be some value w, which is the same for all clients.
- A **bandwidth** $b_c \in \mathbb{R}^+$ that models a resource requirement (such as frequency bandwidth).

On the other hand, each station $s \in S$ has the following characterization.

- A **bandwidth** $B_s \in \mathbb{R}^+$, which limits the sum of the bandwidth of the clients transmitting to s. In this work we assume the station bandwidth to be some value B, which is the same for all stations.

We refer to the set of stations (with their parameters) as the **system** and to the set of clients (with their parameters) as the **client arrival schedule**.

To carry out a worst-case analysis, we consider adversarial client arrival schedules where the adversary is limited as follows. For any $C' \subseteq C$, let $S(C') = \bigcup_{c \in C'} S_c$. For a given pair of values $\rho > 0$ and $\beta \geq 0$ (that limit the rate and burstiness of the stations load, which in turn limits the arrival/departure of clients), we say that a client arrival schedule is (ρ, β)-**admissible** if the following conditions hold:

$$\forall C' \subseteq C : \forall T = [t, t'] \subseteq \mathbb{N} : \sum_{c \in C'} b_c \frac{|\tau_c \cap T|}{w} \leq |T||S(C')|\rho B + \beta \qquad (1)$$

$$\forall c \in C : b_c \leq B . \qquad (2)$$

The first condition (1) restricts the load of the stations for any set of clients C' and any time interval T. In particular, given any C' and any T, the total bandwidth requested by the clients in C' (specifically, $\sum_{c \in C'} b_c |\tau_c \cap T|/w$) has to be no larger that a fraction ρ of the bandwidth that can be provided by the stations that can serve the clients in C' ($S(C')$) plus a constant term β (that allows for some burstiness). The second condition (2) imposes that the requested bandwidth b_c of each client must be no larger that the bandwidth B of each station. Naturally, if some client had a request of bandwidth larger than

B it would be impossible to satisfy it. Adversarial methodology characterized as above is typically used for performing worst-case analysis of the considered problem [3, 10, 11].

Problem. The **Station Assignment problem** is defined as follows. For a given system and admissible client arrival schedule, for each time slot $t \in \mathbb{N}$ schedule a set of clients to transmit to each station in t, so that

1. Each client $c \in C$ transmits to some station in S_c at least once within each w consecutive time slots in τ_c using a bandwidth b_c;
2. For each station $s \in S$ the sum of the bandwidths of the clients transmitting to s in any time slot is at most B.

Protocols. We consider the following classes of protocols, commonly used in scheduling literature.

- A Station Assignment algorithm is called *irrevocable* if for each client c all the transmissions of c are to the same station s. We say that the algorithm *irrevocably assigns* the client c to station s.
- A Station Assignment algorithm is called *online* if the information about any client c is revealed to the algorithm only at the arrival time of c.
- A Station Assignment algorithm is called *improvident* if the algorithm does not know when a client will leave the system.

3   Our Results

The results presented in this work are summarized in Tables 1 and 2. The tables are organized by the system characteristics (columns) and the rows are further subdivided by double lines into comparable settings for which upper and lower bounds are presented. Lower bounds are for impossibility whereas upper bounds are for solvability.

Table 1. Summary of bounds on problem solvability for offline protocols.

b_c	S_c	Arrival time	Offline protocol class	β	ρ	Theorem
Identical	Identical	Identical	Any	$> mwB\left(\frac{n/(mw)}{\lceil n/(mw)\rceil} - \rho\right)$ $n = \lceil \frac{mwB\rho+\beta}{B}\rceil$	$> \frac{n/(mw)}{\lceil n/(mw)\rceil}$	1
Identical	Identical	Identical	Even assignment	$\le mwB\left(\frac{n/(mw)}{\lceil n/(mw)\rceil} - \rho\right)$	$\le \frac{n/(mw)}{\lceil n/(mw)\rceil}$	2
Distinct	Identical	Identical	Any	$-$	$> 1/2$	3
Distinct	Identical	Identical	Any	$> mB(1/m + 1/2 - \rho)$	$-$	4
Any	Identical	Identical	Balance station-bandwidth usage	$< mwB(1/2 - \rho)$	$< 1/2$	5
Distinct	Distinct	Identical	Any	$> mwB\left(1/(mw) - \rho\right)$	$> 1/(mw)$	6
Any	Any	Any	Any	$\le mwB(1/(mw) - \rho)$	$\le 1/(mw)$	7

Table 2. Summary of bounds on problem solvability for online protocols.

b_c	S_c	Arrival time	τ_c	Offline protocol class	β	ρ	Theorem
1	Distinct	Distinct	Open	Irrevocable	$> mB\left(\frac{1}{\ln m} - \rho\right)$	$> \frac{1}{\ln m}$	9
1	Distinct	Distinct	Distinct	Irrevocable improvident randomized	$> mB\left(\frac{3}{\sqrt{2m}} - \rho\right)$	$> \frac{3}{\sqrt{2m}}$	10
1	Distinct	Distinct	Distinct	Irrevocable improvident deterministic	$> mB\left(\frac{1}{\sqrt{2m}} - \rho\right)$	$> \frac{1}{\sqrt{2m}}$	10
$b \geq \rho B$	Any	Any	Open	Irrevocable improvident	$< \rho B$	$\leq \frac{1}{1+\sqrt{2m}}$	11

We study offline Station Assignment under various model assumptions, starting from a more optimistic one where all clients have the same bandwidth and the same stations group, and removing gradually assumptions making the model more pessimistic and, hence, realistic. Studying different models gives insight on what the inherent challenges of Station Assignment are, c.f., Table 1.

We start considering adversarial client arrival schedules where all clients have the same stations group and bandwidth. Then, Theorem 1 shows that for each $\beta > mwB\left((n/(mw))/\lceil n/(mw)\rceil - \rho\right)$, where $n = \lceil (mwB\rho + \beta)/B\rceil$, there exists a (ρ, β)-admissible client arrival schedule such that no Station Assignment algorithm can solve the problem, even if all clients arrive simultaneously and have the same life interval. Given that it must be $\beta \geq 0$, this lower bound for non-solvability implies also a lower bound of $\rho > (n/(mw))/\lceil n/(mw)\rceil$. Corollary 1 shows a stronger bound on β that holds for any positive ρ. Under the same conditions, Theorem 2 shows that the offline algorithm that distributes the clients evenly solves Station Assignment, for any (ρ, β)-admissible client arrival schedule that matches those bounds on β and ρ.

Then, we move to a class of client arrival schedules where clients may have different bandwidths, although the stations group is still the same for all. In this scenario, Theorem 3 shows that, for each $\rho > 1/2$, there exists a (ρ, β)-admissible client arrival schedule such that no Station Assignment algorithm can solve the problem, even if all clients must arrive simultaneously. Changing the adversarial client arrival schedule slightly, Theorem 4 shows a bound of $\beta > mB(1/m + 1/2 - \rho)$ for the same conditions. This bound implies a bound on ρ as well, but it is subsumed by Theorem 3. Under the same conditions, Theorem 5 shows that an algorithm that (somehow) balances the station-bandwidth usage solves Station Assignment, for any (ρ, β)-admissible client arrival schedule such that $\beta < mwB(1/2 - \rho)$ and $\rho < 1/2$.

The last class of client arrival schedules we consider in our offline analysis does not restrict station groups or bandwidths. Theorem 6 shows that, for each $\beta > mB(1/m - \rho)$ (which implies $\rho > 1/m$ because $\beta \geq 0$) there exists a (ρ, β)-admissible client arrival schedule such that Station Assignment cannot be solved by any algorithm, even if all clients arrive at the same time. Theorem 7 matches those bounds, showing that if $\rho \leq 1/(mw)$ and $\beta \leq mwB\left(1/(mw) - \rho\right)$ the Station Assignment problem is solvable using any algorithm for *any* client arrival schedule.

Moving to online protocols, c.f., Table 2, by showing a reduction from Load Balancing [7], we prove in Theorem 9 that for any irrevocable algorithm, that is, algorithms where the station-client assignments are final, there is a client arrival schedule such that if $\beta > mB(1/\ln m - \rho)$ the Station Assignment problem is not solvable. Again, the lower bound implies a lower bound of $\rho > 1/\ln m$ because $\beta \geq 0$. If the algorithm is additionally improvident, that is, the departure time of clients already in the system is not known in advance, then Theorem 10 shows lower bounds of $\beta > mB\left(3/\sqrt{2m} - \rho\right)$ and $\beta > mB\left(1/\sqrt{2m} - \rho\right)$ for randomized and deterministic algorithms respectively. Those bounds imply that if $\rho > 3/\sqrt{2m}$ and if $\rho > 1/\sqrt{2m}$ respectively the Station Assignment problem cannot be solved. Finally, Theorem 11 shows that, when all clients have the same bandwidth $b \geq \rho B$ and do not depart, even if the station groups and arrival times are different, if $\rho \leq 1/(1 + \sqrt{2m})$ and $\beta < \rho B$ the algorithm that distributes clients evenly (restricted to station group) solves Station Assignment.

We also show in Theorem 8 (not included in Table 2) a lower bound on ρ for non-solvability with irrevocable algorithms that applies to systems with distinct station-bandwidths. Corollary 2 shows that instantiating Theorem 8 on a system where all stations have the same bandwidth B, the lower bound on ρ for non-solvability is $\rho > 1/(1 + \ln m)$.

4   Related Work

Adversarial queuing was introduced in [3,11], applied to store-and-forward networks, to measure stability of buffers and packet latency of dynamically injected packets. Later, there were approaches to apply it in the context of wireless networks: modelled as time-varying channels [4], radio channels with collisions [2,12], or SINR networks [15]. In a single-hop radio channels with collisions, more detail competitive analysis of dynamic and stochastic traffic was performed [9]. The difference between this line of research and our work is that it considered simple packet forwarding requests without additional scheduling constraints.

In [7], Azar, Broder, and Karlin studied a load balancing problem where a set of tasks that arrive and depart in time (temporary, as opposed to permanent when tasks do not depart) have to be assigned to a set of machines. Each task has an associated weight that represents the load that the processing of such task adds to a machine. Additionally, each task has an associated subset of machines that may process the task (restricted assignment). Upon arrival, a task must be assigned to a machine immediately and cannot be transferred to another machine later. The machine starts processing the task immediately and continues until the task departs. An assignment algorithm selects a machine to assign each task upon arrival. In the online version the algorithm does not know future arrivals or departures, whereas an offline algorithm has complete knowledge. The cost of an assignment of a given input is the maximum load of

Table 3. Competitive ratios of load balancing problem.

	Unknown duration	Known duration	Permanent
Identical	$2 - o(1)$ [7,13]	$2 - o(1)$ [7,13]	$2 - \epsilon$ [14]
Related	$\Theta(1)$ [8]	$\Theta(1)$ [8]	$\Theta(1)$ [8]
Restricted	$O(\sqrt{m})$ [8]	$O(\log mT)$ [8]	$\Theta(\log m)$ [5]
	$\Omega(\sqrt{m})$ [7]		

the machines for such assignment. The authors study the competitive ratio of an online algorithm with respect to an offline one as the supremum over all inputs of the cost ratio. Specifically, for the greedy online algorithm that assigns each task to the least loaded machine, they show matching upper and lower bounds of $((3m)^{2/3}/2)(1+o(1))$ on the competitive ratio, and a lower bound of $\Omega(\sqrt{m})$ for any deterministic or randomized algorithm. The lower bound is matched in [8]. Variants of the problem include relaxing the constraint such that the duration of a job is known on arrival (temporary) or the job never departs (permanent). Another direction of relaxation includes making all machines to be available for all jobs (identical or related). Table 3 gives a summary of the results.

In [1], Alon et al. studied a similar model for permanent tasks. They consider two cases: (i) the tasks have associated weights and can be assigned to any machine (unrestricted), (ii) the tasks have unit weights and can be assigned only to a subset of the machines (restricted). They provide an ϵ-approximation scheme for the L_p norm of the loads. Interestingly, for the restricted unit-weights model, they show that there exists an assignment that is optimal for all norms. For further references on dynamic online scheduling and load balancing, see the chapters [6,16].

5   Analysis of Offline Protocols

In this section, we study the impact of ρ and β on the offline solvability of Station Assignment.

5.1   Unique Stations Group and Client Bandwidth

We start with a very optimistic scenario (for Station Assignment algorithms) where all clients have the same stations group and the same bandwidth. We show a lower bound for non-solvability that holds even under those optimistic conditions. Given that $\beta \geq 0$ by definition, the bound obtained implies a lower bound on ρ.

Theorem 1. *Given a system of m stations each with bandwidth B, even if all clients must have the same stations group and the same bandwidth, for any $\beta > mwB \left(\frac{n/(mw)}{\lceil n/(mw) \rceil} - \rho \right)$, where $n \geq \lceil (mwB\rho + \beta)/B \rceil, n \in \mathbb{Z}^+$, there exists a (ρ, β)-admissible client arrival schedule such that no algorithm can solve the Station Assignment problem, even if all clients must have the same life interval.*

Proof. Consider a client arrival schedule of n clients, for any $n \geq \lceil (mwB\rho + \beta)/B \rceil, n \in \mathbb{Z}^+$, with the same bandwidth $b = (mwB\rho + \beta)/n$ and the same life interval of length w. Such schedule is (ρ, β)-admissible because, for any $n' \leq n$ and any subinterval T of the life interval of the clients (i.e., $|T| \leq w$), it holds $n'b\frac{|T|}{w} \leq nb\frac{|T|}{w} = (mwB\rho + \beta)\frac{|T|}{w} \leq m|T|B\rho + \beta$, and $b = \frac{mwB\rho+\beta}{n} \leq \frac{mwB\rho+\beta}{\lceil (mwB\rho+\beta)/B \rceil} \leq B$. However, by the pigeonhole principle, there is at least one station and one slot for which the sum of bandwidths of the clients assigned to the station in the slot is at least $\lceil n/(mw) \rceil b = \lceil n/(mw) \rceil (mwB\rho + \beta)/n$. Replacing $\beta > mwB ((n/(mw))/\lceil n/(mw) \rceil - \rho)$, the latter is bigger than B. □

Given that the client arrival schedule is adversarial, by choosing the station group to be a singleton in the above proof, that is $m = 1$, and the laxity $w = 1$, the lower bound obtained becomes $\beta > B(1 - \rho)$, which implies that if $\rho > 1$ the Station Assignment is not solvable. We assume that $\rho \leq 1$ throughout the rest of the paper. This result can also be used to show that, for some higher values of β, Station Assignment is not solvable for any $\rho > 0$.

Corollary 1. *Given a system of m stations each with bandwidth B, even if all clients must have the same stations group and the same bandwidth, if $\rho > 0$ and $\beta \geq nB/\lceil n/(mw) \rceil$, where $n = \lceil (mwB\rho+\beta)/B \rceil$, there exists a (ρ, β)-admissible client arrival schedule such that no algorithm can solve the Station Assignment problem, even if all clients must have the same arrival time.*

Proof. Let $\rho > 0$, from Theorem 1 it is enough to prove the claim that $\beta > mwB \left(\frac{n/(mw)}{\lceil n/(mw) \rceil} - \rho \right)$, where $n = \lceil (mwB\rho + \beta)/B \rceil$. This holds if $\beta \geq mwB\frac{n/(mw)}{\lceil n/(mw) \rceil} = \frac{nB}{\lceil n/(mw) \rceil}$. □

Now we show a matching upper bound for solvability in the same optimistic scenario. That is, all clients have the same stations group and bandwidth.

Theorem 2. *Given any (ρ, β)-admissible client arrival schedule of n clients, such that all clients have the same bandwidth, the same station group of size $m > 0$, and the same arrival time, if $\beta \leq mwB \left(\frac{n/(mw)}{\lceil n/(mw) \rceil} - \rho \right)$, the algorithm that assigns clients evenly among stations and intervals of w times slots solves the Station Assignment problem on any system of at least m stations each with bandwidth B.*

Proof. Let b be the client bandwidth. In order to show the claim, it is enough to show it for the initial w time slots after the arrival of the clients, given that, if some client departs, the bandwidth usage of the assigned station is reduced. Note that the life interval of all clients is at least w, by the definition of laxity. Given that the assignment of clients is even, the station most used has at most $\lceil n/(mw) \rceil$ clients assigned per slot. Hence, in order to prove the claim, it is enough to prove $\lceil \frac{n}{mw} \rceil b \leq B$. Due to admissibility (Eq. (1)) for w slots (i.e., $|T| = w$), we know that $nb \leq mwB\rho+\beta$. Replacing this bound on b, it is enough to show that $\lceil \frac{n}{mw} \rceil \frac{mwB\rho+\beta}{n} \leq B$. Replacing the bound on β, it can be seen that the inequality holds. □

5.2   Unique Stations Group and Distinct Client Bandwidth

We now consider a less optimistic scenario where the client bandwidths may be different. Theorems 3 and 4 show lower bounds for non-solvability on ρ and β respectively.

Theorem 3. *Given a system of m stations each with bandwidth B, even if all clients must have the same station group, for any $\rho > 1/2$, there exists a (ρ, β)-admissible client arrival schedule such that no algorithm can solve the Station Assignment problem, even if all clients must have the same life interval.*

Proof. Consider a client arrival schedule of $mw + 1$ clients with the same station group S and the same life interval of length w. One of the clients, call it x, has bandwidth $b = (\rho - \delta)mwB$ for some value δ such that $1/2 < \delta < \rho$ and $\rho - 1/(mw) \leq \delta < (\rho mw - 1)/(mw - 1)$. Each of the remaining mw clients has bandwidth δB. Such schedule is (ρ, β)-admissible because, for any subset of $n \leq mw + 1$ clients that includes x, Eq. (1) becomes $\forall T : |T| \leq w :$ $((n-1)\delta B + (\rho - \delta)mwB)\frac{|T|}{w} \leq |T|m\rho B + \beta$, which is true because $n-1 \leq mw$ and $\beta \geq 0$. On the other hand, if we consider the $n \leq mw$ clients that do not include x, Eq. (1) becomes $\forall T : |T| \leq w : n\delta B\frac{|T|}{w} \leq |T|m\rho B + \beta$, which is true because $n \leq mw$, $\beta \geq 0$, and $\delta < \rho$. Finally, Eq. (2) also holds because $\delta B < \rho B \leq B$ because $\rho \leq 1$, and $(\rho - \delta)mwB \leq B$ for $\delta \geq \rho - 1/(mw)$. However, given that there are $mw + 1$ clients, due to the pigeonhole principle two clients have to be assigned to the same slot of the same station. Then, there is a slot in some station such that the sum of the assigned clients is either $2\delta B > B$ or $\delta B + (\rho - \delta)mwB > B$ because $\delta < (\rho mw - 1)/(mw - 1)$.   □

The following theorem shows a lower bound on β for this scenario. The proof uses an adversarial client arrival schedule similar to the schedule used in the proof of Theorem 3. The details are left to the full version of this paper.

Theorem 4. *Given a system of m stations each with bandwidth B, even if all clients must have the same station group, for any $\beta > mB(1/m + 1/2 - \rho)$, there exists a (ρ, β)-admissible client arrival schedule such that no algorithm can solve the Station Assignment problem, even if all clients must have the same arrival time.*

Now we show an upper bound for solvability for the same scenario. That is, the stations group is unique among clients but the bandwidth may be different.

Theorem 5. *Given any (ρ, β)-admissible client arrival schedule, such that all clients have the same station group of size $m > 0$ and the same arrival time, if $\beta < mwB(1/2 - \rho)$, there exists a polynomial time algorithm that computes an assignment of clients to stations that solves the Station Assignment problem on any system of at least m stations each with bandwidth B. The transmission schedule of such assignment is periodic with period w.*

Proof. Consider a (ρ, β)-admissible client arrival schedule where all clients have the same station group, arrive simultaneously, and all have laxity w. Let the time slot of clients arrival be labeled as 1. We will focus on the first interval of slots $[1, w]$. Notice that all clients that arrive at time 1 stay active during such interval, given that by definition $\forall c \in C : w_c \leq |\tau_c|$. We will show how to assign each client to one station and one slot within this window, so that no station is overloaded in any slot. The assignment in all the subsequent intervals $[iw + 1, (i+1)w]$, for each integer $i > 0$, is identical. Let $\mathcal{A}$ be any initial assignment of each client to one of the m stations and one of the w slots. Let $B_{\max}$ be the maximum bandwidth used in $\mathcal{A}$ in any slot, and let (s, i) be some station-slot pair with such bandwidth usage in the assignment $\mathcal{A}$. If $B_{\max} \leq B$, we are done. Otherwise, given that $B_{\max} > B$, station s has more than one client assigned in slot i, since otherwise the client arrival schedule would violate Eq. (2). If the sum of the bandwidth used on some pairs (s', j) and (s'', k) is at most B, consider another assignment $\mathcal{A}'$ where the clients assigned to (s', j) and (s'', k) in $\mathcal{A}$ are now all assigned to (s', j), and the clients assigned to (s, i) in $\mathcal{A}$ are now split between (s, i) and (s'', k). Repeat the procedure above until the sum of bandwidth usage in each two station-slot pairs is at least B, or $B_{\max} \leq B$. In the latter case we are done. Otherwise, adding in pairs, the total bandwidth used throughout all stations and slots is at least $mwB/2$. But, according to Eq. (1), the total bandwidth used must be at most $mw\rho B + \beta < mwB/2$. Which is a contradiction.   □

A similar bound can be obtained if clients never depart, even if they arrive at different times.

5.3   Distinct Stations Group and Client Bandwidth

Now we consider the harshest scenario where clients may have different station groups and different bandwidths. Given that $\beta \geq 0$ by definition, the bound obtained implies that if $\rho > 1/m$ the problem is not solvable.

Theorem 6. *Given a system of m stations each with bandwidth B, for each $\beta > mwB(1/(mw) - \rho)$, there exists a (ρ, β)-admissible client arrival schedule such that no algorithm can solve the Station Assignment problem, even if all clients must have the same life interval.*

Proof. Consider a client arrival schedule of $n + 1$ clients, where $n = amw$, for some integer $a \geq 1$, such that $n \geq (mwB\rho + \beta - B)/B$. The first n clients have a singleton station group so that, for each station s_i, $i = 1, 2, \ldots, m$, the number of clients with station group $\{s_i\}$ is aw. The bandwidth of each of these n clients is $b = (mwB\rho + \beta - B)/n$. There is one additional client x with station group M and bandwidth B. All the $n + 1$ clients in the client arrival schedule have the same life interval of length $w \geq 1$). Such client arrival schedule is (ρ, β)-admissible because, for any subinterval T such that $|T| \leq w$, the total bandwidth of any subset of $n' \leq n + 1$ clients is, if x is included then $((n' - 1)b + B)\frac{|T|}{w} = ((n' - 1)\frac{mwB\rho + \beta - B}{n} + B)\frac{|T|}{w} \leq (mwB\rho + \beta)\frac{|T|}{w} \leq |T|mB\rho + \beta$. Otherwise, if x is not included, and hence $n' \leq n$, the total bandwidth is $n'b\frac{|T|}{w} = n'\frac{mwB\rho+\beta-B}{n}\frac{|T|}{w} = \frac{n'}{aw}B\rho|T| + n'\frac{\beta-B}{n}\frac{|T|}{w} \leq \left\lceil\frac{n'}{aw}\right\rceil|T|B\rho + \beta$. Therefore, Eq. (1) holds. Additionally, replacing the expression of n in b, it can be seen that $b \leq B$. Thus, Eq. (2) holds for all clients. However, for any assignment, there must be at least one slot of one station with bandwidth usage $B + ab = B + \frac{n}{m}b = B + \frac{n}{mw}\frac{mwB\rho+\beta-B}{n} = B(1+\rho) + \frac{\beta-B}{mw}$, which is bigger than B for $\beta > mwB(1/(mw) - \rho)$. $\qquad\square$

Now we show a matching upper bound for solvability for the same strict scenario. That is, both, the stations group and bandwidth, may be different among clients.

Theorem 7. *Given any (ρ, β)-admissible client arrival schedule, if $\beta \leq mwB(1/(mw) - \rho)$, the Station Assignment problem can be solved on any system of m stations each with bandwidth B.*

Proof. Consider an assignment of a given (ρ, β)-admissible client arrival schedule. Consider the set $C' \subseteq C$ of clients that are active at any given time step t in such assignment. Because the client arrival schedule is (ρ, β)-admissible, making $|T| = w$ in Eq. (1) and using that $w \leq |\tau_c|$, it must be $\sum_{c \in C'} b_c \leq w|S(C')|\rho B + \beta \leq wm\rho B + \beta$. Replacing in the latter the upper bound on β, we have that $\sum_{c \in C'} b_c \leq B$. Thus, no station can have a bandwidth usage bigger than B. $\qquad\square$

6   Analysis of Online Protocols

In this section, we present bounds for irrevocable improvident online protocols.

6.1   Lower Bounds for Non-solvability

We show now that irrevocable algorithms do not always solve the problem. Theorem 8 applies to a more general model where the station bandwidths may be different. The corollary that follows instantiates the result on a model where the station bandwidth is unique. The proof is left to the full version of this paper.

Theorem 8. *For any system of m stations, where station s has bandwidth B_s, any $\beta \geq 0$, and for each irrevocable online algorithm $\mathcal{A}$, there is a station labeling $\{s_1, \ldots, s_m\}$ and a (ρ, β)-client arrival schedule such that, if $\rho > B_{s_m}/\left(B_{s_m} + \sum_{j=1}^{m-1}\left(\sum_{i=j}^{m} B_{s_i} - \max_{j \in [j,m]} B_{s_j}\right)\frac{1}{m-j+1}\prod_{k=2}^{m-j}\left(1 - \frac{1}{k}\right)\right)$, $\mathcal{A}$ cannot solve the Station Assignment problem.*

Corollary 2. *For any system of m stations each with bandwidth B, and for each irrevocable algorithm $\mathcal{A}$, and for any $\rho > 1/(1+\ln m)$ and $\beta \geq 0$, there is a (ρ, β)-client arrival schedule such that $\mathcal{A}$ cannot solve the Station Assignment problem.*

Proof. Replacing all bandwidths in the lower bound of ρ in Theorem 8 by B, we get $\rho > \left(\sum_{j=1}^{m}\frac{1}{j}\right)^{-1} = H_m^{-1} > \frac{1}{1+\ln m}$. $\qquad\square$

Observe that for the above proof to work it is not needed that an irrevocable algorithm assigns a client to a station forever. It is enough that it assigns it for $m + w$ steps to reach the same result.

The following theorem for irrevocable algorithms relates β and ρ for the case where the bandwidth of all stations is the same. Given that $\beta \geq 0$ by definition, the bound implies that if $\rho > 1/\ln m$, the Station Assignment problem is not solvable.

Theorem 9. *For any system of m stations, such that all stations have the same bandwidth B, and for each irrevocable algorithm $\mathcal{A}$, there is a (ρ, β)-client arrival schedule such that, if $\beta > mB(1/\ln m - \rho)$, then $\mathcal{A}$ cannot solve the Station Assignment problem.*

Proof. Consider an adaptive adversary that decides the clients that arrive according to the actions of $\mathcal{A}$. The adversarial client arrival schedule is the following. Let $w = 1$. For each client c, it is $b_c = 1$. The life interval of all clients is open ended. That is, upon arrival, clients stay active forever. Clients arrive in batches. A new batch of clients arrive after the previous batch has been irrevocably assigned by algorithm $\mathcal{A}$. Time is conceptually divided in m **rounds**, which are enumerated sequentially as $1, 2, \ldots, m$. A new round starts when a new batch of clients arrive. The number of clients arriving in each round is $\rho B + \beta/m$. (We omit ceilings and floors throughout the proof for clarity.) All clients arriving in the same round i have the same stations group S_i. Starting from the whole set of stations S in the first round, the stations group for each new round is reduced by one station. We say that such station is **removed**. Thus, for round 1 the stations group has size m, for round 2 the size is $m - 1$, and so on until round m when the stations group has size 1. For any round $r > 1$, the station removed is the station with the smallest number of clients assigned.

First we notice that the client arrival schedule defined is (ρ, β)-admissible. For this purpose, it is enough to show that the property is preserved after each batch of arrivals. Consider any round $i = 1, \ldots, m$. Let C_j be any subset of clients with stations group S_j, for $j = 1, \ldots, i$. We know that $|C_j| \leq \rho B + \beta/m$. So, in Eq. (1), the ρB term can be applied to the station removed in round j, and putting together all the β/m terms they add up to $i\beta/m \leq \beta$.

We show now that, with the above client arrival schedule, the sum of the bandwidths of the clients assigned to the station in S_m is more than B. Let the number of clients arriving in each round be called $X = \rho B + \beta/m$. In round 1 the overall number of clients is X. Given that the station removed is the one with the smallest number of clients, in round 2 the overall number of clients assigned to stations in S_2 is at least $X(1 - 1/m) + X$. Likewise, in round 3, the overall number of clients assigned to stations in S_3 is at least $((X(1 - 1/m) + X)(1 - 1/(m - 1)) + X$. Inductively, the number of clients assigned to the station in S_m is at least

$$\left(\ldots \left(\left(X \frac{m-1}{m} + X \right) \frac{m-2}{m-1} + X \right) \frac{m-3}{m-2} \ldots \right) \frac{1}{2} + X = X \left(\frac{1}{m} + \frac{1}{m-1} + \cdots + \frac{1}{2} + 1 \right) >$$

$X \ln m$. That is, the total bandwidth of the clients assigned to the station in S_m is at least $\ln m(\rho B + \beta/m)$. Thus, if $\beta > mB(1/\ln m - \rho)$ the claim follows. □

The following theorem shows that the restriction on ρ for solvability with irrevocable assignments is stronger for improvident algorithms. Theorem 10 shows that, for randomized online algorithms, if $\beta > mB\left(3/\sqrt{2m} - \rho\right)$ the Station Assignment problem is not solvable, and if $\beta > mB\left(1/\sqrt{2m} - \rho\right)$ the Station Assignment problem is not solvable online deterministically. Given that $\beta \geq 0$ by definition, the bound implies that if $\rho > 3/\sqrt{2m}$, or if $\rho > 1/\sqrt{2m}$ respectively, the problem is not solvable.

Theorem 10. *For any set of m stations each with bandwidth B, the following holds, even if all clients must have the same bandwidth:*

1. *For any $m \geq 5$ and $\beta > mB\left(3/\sqrt{2m} - \rho\right)$, there exists a (ρ, β)-admissible client arrival schedule such that no online irrevocable improvident randomized algorithm can solve Station Assignment.*
2. *For any $m \geq 1$ and $\beta > mB\left(1/\sqrt{2m} - \rho\right)$, there exists a (ρ, β)-admissible client arrival schedule such that no online irrevocable improvident deterministic algorithm can solve Station Assignment.*

Proof. If $\beta > mB(1-\rho)$, the claim follows from Theorem 1. So, for the rest of the proof we assume that $\beta \leq mB(1 - \rho)$.

For the Load Balancing problem, where computing tasks have to be assigned to servers, the proof of Theorem 3.3 in [7] shows a sequence of unit-weight tasks such that, the maximum (over the servers) off-line load at all times is 1, and the competitive ratio of any randomized irrevocable improvident algorithm is at least $(\sqrt{2m}/3)(1+o(1))$. (The theorem is stated in asymptotic notation, but the bound obtained in the proof is the expression given here.) We reuse such adversary mapping tasks to clients, servers to stations and weights/loads to bandwidths. Let the bandwidth of such clients be instead $\rho B + \beta/m$ and the laxity $w = 1$. This client arrival schedule is (ρ, β)-admissible because *(i)* $\beta \leq mB(1 - \rho)$ and then $\rho B + \beta/m \leq B$, and *(ii)* the maximum off-line bandwidth at all times on each station is at most $\rho B + \beta/m$. However, the bandwidth used at some station is at least $(\sqrt{2m}/3)(\rho B + \beta/m)$, which is larger than B if $\beta > mB\left(3/\sqrt{2m} - \rho\right)$, which is feasible for $m > 9/2$. The same argument can be used for deterministic algorithms and competitive ratio of $\sqrt{2m}$. $\quad\square$

6.2   Upper Bounds for Solvability

The following theorem applies to a setting where the station bandwidth is unique, but the station group may be different for each client.

Theorem 11. *For any system of m stations each with bandwidth B, there exists an online algorithm, such that if $\rho \leq 1/(1 + \sqrt{2m})$, $\beta < \rho B$, all clients have the same bandwidth $b \geq \rho B$ and laxity $w = 1$, and never depart, the Station Assignment problem is solved.*

Proof. Let S be the set of stations in the system and, for any subset of stations $S' \subseteq S$, let C' be the set of clients $C' = \{c | S_c = S'\}$. Using that $b \geq \rho B$ and $\beta < \rho B$, the following properties arise from admissibility.

Property 1. $\forall S' \subseteq S : |C'| \leq |S'|$.

That is, for each station group of x stations, there are at most x clients with that station group.

Property 2. $\forall S'' = \{S'|S' \subseteq S\} : |\cup_{S' \in S''} C'| \leq |\cup_{S' \in S''} S'|$.

That is, for any set of station groups, the maximum number of clients with those station groups is at most the size of the union of those groups.

Consider the online algorithm that, for each client c, assigns c to the station $s \in S_c$ with the largest available bandwidth, breaking ties arbitrarily. We show that, under the assumptions of the theorem, this algorithm solves the problem. For the sake of contradiction, assume that some station s_i is overloaded. That is, s_i has some integer number k of clients assigned such that $k > 1/\rho$. We show that then the number of clients in the system must be more than m, which is not possible according to Property 1.

We compute a lower bound on the number of clients that should be in the system in order to have more than $1/\rho$ clients in s_i. For clarity, we label the clients assigned to s_i in the order in which they were assigned. Client 1 is the first one and, hence, does not require any other clients to be in the system before. For each client $c = 2, 3, \ldots, k$, we identify clients that must have been assigned before c to some station. We *allocate* some of those clients to each c. In order to avoid over-counting, sometimes we may *reallocate* some clients, so that each client in the system is allocated to at most one client.

Assume that, for each client $c \in [2, k]$, we can allocate $c - 1$ "new" clients. Then, overall, we will have $1 + 2 + \cdots + k - 1$ allocated clients which yields a lower bound of $k(k-1)/2 > m$ clients in the system proving the claim. The details of the allocation procedure follow.

For each client $c = k, k - 1, \ldots, 2$ in s_i, we know that there must be at least $c - 1$ clients in each station in S_c, because the algorithm distributes clients evenly in S_c. If the clients in one or more of the stations in S_c have not been allocated yet, we choose one of those stations arbitrarily and allocate the $c - 1$ clients assigned to that station to c. If the clients in all stations in S_c have been already allocated to some client, assume that there is at least one client $c' \in [c + 1, k]$ such that the clients assigned to some station $s_j \in S_{c'}$ have not been allocated. Then, we reallocate $c - 1$ clients from c' to c, and we allocate the $c' - 1$ clients in s_j to c'.

We show now that if the latter assumption is false, Property 2 has been violated. For the sake of contradiction, assume that, at the point of allocating clients for some client c, the clients in all stations in $\cup_{c'=c+1}^{k} S_{c'}$ have been already allocated to some client in $[c + 1, k]$. This implies that $k - c = |\cup_{c'=c+1}^{k} S_{c'}|$. Thus, if $S_c \subseteq \cup_{c'=c+1}^{k} S_{c'}$, Property 2 is violated.

7   Conclusions

This paper presented worst-case (adversarial) analysis of scheduling periodic communication between base stations and mobile clients. We considered various classes of scheduling settings and protocols, and provided limitations on feasible mobility patterns given in the form of upper and lower bounds on client injection rates and burstiness. The obtained variety of results is a promising starting point for further study of more complex scheduling settings in the proposed mobility model, including the settings motivated by sensor and local wireless network applications.

References

1. Alon, N., Azar, Y., Woeginger, G.J., Yadid, T.: Approximation schemes for scheduling. In: SODA, pp. 493–500 (1997)
2. Anantharamu, L., Chlebus, B.S., Rokicki, M.A.: Adversarial multiple access channel with individual injection rates. In: Abdelzaher, T., Raynal, M., Santoro, N. (eds.) OPODIS 2009. LNCS, vol. 5923, pp. 174–188. Springer, Heidelberg (2009)
3. Andrews, M., Awerbuch, B., Fernández, A., Leighton, F.T., Liu, Z., Kleinberg, J.M.: Universal-stability results and performance bounds for greedy contention-resolution protocols. J. ACM **48**(1), 39–69 (2001)
4. Andrews, M., Zhang, L.: Scheduling over a time-varying user-dependent channel with applications to high-speed wireless data. J. ACM **52**(5), 809–834 (2005)
5. Aspnes, J., Azar, Y., Fiat, A., Plotkin, S.A., Waarts, O.: On-line load balancing with applications to machine scheduling and virtual circuit routing. In: STOC, pp. 623–631 (1993)
6. Azar, Y.: On-line load balancing. In: Online Algorithms, pp. 178–195 (1996)
7. Azar, Y., Broder, A.Z., Karlin, A.R.: On-line load balancing. Theor. Comput. Sci. **130**(1), 73–84 (1994)
8. Azar, Y., Kalyanasundaram, B., Plotkin, S.A., Pruhs, K., Waarts, O.: On-line load balancing of temporary tasks. J. Algorithms **22**(1), 93–110 (1997)
9. Bienkowski, M., Jurdzinski, T., Korzeniowski, M., Kowalski, D.R.: Distributed online and stochastic queuing on a multiple access channel. In: Aguilera, M.K. (ed.) DISC 2012. LNCS, vol. 7611, pp. 121–135. Springer, Heidelberg (2012)
10. Blesa, M.J., Calzada, D., Fernández, A., López, L., Martínez, A.L., Santos, A., Serna, M.J., Thraves, C.: Adversarial queueing model for continuous network dynamics. Theor. Comput. Syst. **44**(3), 304–331 (2009)
11. Borodin, A., Kleinberg, J.M., Raghavan, P., Sudan, M., Williamson, D.P.: Adversarial queuing theory. J. ACM **48**(1), 13–38 (2001)
12. Chlebus, B.S., Kowalski, D.R., Rokicki, M.A.: Adversarial queuing on the multiple-access channel. In: Proceedings of the 25th ACM Symposium on Principles of Distributed Computing (PODC), pp. 92–101 (2006)
13. Graham, R.L.: Bounds on multiprocessing timing anomalies. Bell Syst. Tech. J. **45**, 1563–1581 (1966)
14. Karger, D.R., Phillips, S.J., Torng, E.: A better algorithm for an ancient scheduling problem. J. Algorithms **20**, 400–430 (1996)
15. Kesselheim, T.: Dynamic packet scheduling in wireless networks. In: PODC, pp. 281–290 (2012)
16. Pruhs, K., Sgall, J., Torng, E.: Online scheduling. In: Leung, J. (ed.) Handbook of Scheduling: Algorithms, Models and Performance Analysis (Chapter 15), pp. 15-1–15-41. CRC Press, Boca Raton (2004)

On Local Broadcasting Schedules
and CONGEST Algorithms in the SINR Model

Fabian Fuchs$^{(\boxtimes)}$ and Dorothea Wagner

Institute of Theoretical Informatics, Karlsruhe Institute of Technology (KIT),
Karlsruhe, Germany
{fabian.fuchs,dorothea.wagner}@kit.edu

Abstract. We consider the distributed construction of a deterministic local broadcasting schedule in the SINR model of interference. During the execution of such a schedule each node should be able to transmit one message to its neighbors. Our construction requires only $\mathcal{O}(\Delta \log n)$ time slots, where Δ is the maximum node degree in the network and n the number of nodes. We prove that the length of the constructed schedule is asymptotically optimal, i.e. of length $\mathcal{O}(\Delta)$. Considering the simulation of $\mathcal{CONGEST}$ algorithms in the SINR model, our deterministic schedule achieves a runtime of $\mathcal{O}(\tau \Delta^2 + \Delta \log n)$ time slots, where τ is the original runtime in the $\mathcal{CONGEST}$ model. We show that there is a lower bound of $\Omega(\Delta^2)$ for the simulation of each one of the τ rounds, hence our simulation is optimal apart from the logarithmic factor. If we restrict the knowledge of the nodes and let the maximum node degree Δ be unknown, we can prove that at least $\Omega(\mathrm{D} + \tau \Delta^2)$ time slots are required to simulate synchronized $\mathcal{CONGEST}$ algorithms in the SINR model of interference, where D is the diameter of the network. For our algorithms we assume location information to be given. Regarding the case without location information we argue that a deterministic algorithm to compute local broadcasting schedules by Derbel and Talbi [ICDCS'10], which requires transmission power adaption, needs messages of size $\mathcal{O}(\log n)$ to simulate $\mathcal{CONGEST}$ algorithms. This is a logarithmic factor less than stated by the authors.

1 Introduction

Local broadcasting is one of the most fundamental task in wireless networks. In contrast to global broadcasting, where one message must be spread over the whole network, in the problem of local broadcasting each node must send one message only to all direct neighbors. In wireless networks usually only a fraction of all nodes can broadcast simultaneously due to the signal interference of multiple transmissions. Hence local broadcasts must be coordinated in order to avoid too high interference. Since interference is modeled relatively realistic in the SINR model (Signal-to-Interference-and-Noise-Ratio model, cf. Sect. 2), the problem of finding a local broadcasting schedule must be tackled by algorithms designed for this model, whereas for many other models such as the message-passing based $\mathcal{CONGEST}$ or $\mathcal{LOCAL}$ models [1] the broadcasting problem does

P. Flocchini et al. (Eds.): ALGOSENSORS 2013, LNCS 8243, pp. 170–184, 2014.
DOI: 10.1007/978-3-642-45346-5_13, © Springer-Verlag Berlin Heidelberg 2014

not occur as interference-free communication is assumed (cf. Sect. 2). Thus, in these models message reception is guaranteed regardless of other transmissions.

However, wireless technology is becoming more and more ubiquitous and hence distributed computing in a wireless context—along with the SINR model—received increasing attention in recent research. Local broadcasting is a fundamental problem in the SINR model that can be used as a building block to solve higher-level problems. Hence it is quite well studied and can be solved in $\mathcal{O}(\Delta \log n)$ time slots [2] (where Δ is the maximum number of nodes in any transmission region of the network) if Δ is known. Further results will be discussed in Sect. 1.1. Due to the vast amount of algorithms designed for message-passing models, one particularly interesting application of local broadcasting is to simulate algorithms designed for message-passing models in the SINR model.

For complex algorithms it may be more effective to invest some time in a preprocessing step in order to achieve faster local broadcasting. In fact, this can be beneficial and both Derbel and Talbi [3] and Jurdzinski and Kowalski [4] achieve—using different methods and assumptions—local broadcasting in $\mathcal{O}(\Delta)$, which is optimal due to a trivial lower bound[1] For Derbel's and Talbi's approach such a preprocessing requires $\mathcal{O}(\Delta \log n)$ time slots while Jurdzinski's and Kowalski's approach requires $\mathcal{O}(\Delta \log^3 n)$ slots. Inspired by both approaches we describe how to construct a deterministic local broadcasting schedule with optimal length $\mathcal{O}(\Delta)$ and preprocessing time of $\mathcal{O}(\Delta \log n)$ time slots. We use distributed node coloring proposed by Derbel and Talbi [3] to construct an infeasible local broadcasting schedule and combine it with the concept of dilution by Jurdzinski and Kowalski [4], which enables us to achieve feasibility of the schedule while increasing the length of the schedule only by a constant factor. We require the nodes to know an upper bound on the number of nodes n, the maximum node degree Δ in the network, their own ID, and location information. We do not require carrier sensing and restrict ourselves to uniform and non-adjustable transmission powers.

Our deterministic local broadcasting algorithm differs from the previously mentioned algorithms in various ways. In contrast to the distributed node coloring by Derbel and Talbi [3] we do not require the nodes to tune their transmission power, while they require the nodes to tune the transmission power by a constant factor. With regard to the backbone structure constructed by Jurdzinski and Kowalski [4] the method described in this work is faster by a polylogarithmic factor.

Using the local broadcasting schedule to simulate algorithms (with original runtime τ) designed for the $\mathcal{CONGEST}$ model, we achieve a runtime of $\mathcal{O}(\tau \Delta^2 + \Delta \log n)$ time slots in the SINR model. Regarding the case that nodes do not know the global maximum degree, we show a lower bound of $\Omega(\tau \Delta^2 + D)$ (with diameter D) on the runtime in the SINR model for the simulation of synchronized $\mathcal{CONGEST}$ algorithms.

---

[1] As only one transmission can be received in a time slot, Δ nodes in a transmission region require $\Omega(\Delta)$ time slots to transmit to one (shared) neighbor.

Finally, we argue that the local broadcasting based on a coloring described in [3] is capable of simulating message-passing algorithms with messages that are by a factor of $\log n$ smaller than stated. This results in an approach that is capable of simulating $\mathcal{CONGEST}$ algorithms in $\mathcal{O}(\tau \Delta^2 + \Delta \log n)$ using messages of size $\mathcal{O}(\log n)$. This is as fast as the deterministic local broadcasting schedule described in this work, however, note that they assume the nodes to tune their transmission power by a constant factor, while we require location information to be given.

1.1 Related Work

A few years ago, the SINR model has only been considered for basic communication problems in wireless networks such as connectivity [5,6], link scheduling [7], or local broadcasting [2]. However, it recently attracted considerable attention even in the distributed computing community. There are now initial works considering distributed computing problems in the SINR model, for example distributed node coloring [3,8], independent sets [8] or dominating sets [9].

However, due to the complexity of analyses in the SINR model, it is reasonable to use local broadcasting as a building block in order to run more evolved distributed computing algorithms on wireless networks. By simulating a round-based message-passing environment through local broadcasting even complex distributed algorithms such as for example all-pairs shortest paths [10] or graph partition [11] designed for the message-passing-based $\mathcal{CONGEST}$ model can be made available in the SINR model.

The simulation of message-passing algorithms in radio networks (in which a message is successfully received if the receiver is silent and only one of its neighbors is transmitting) has first been studied by Alon et al. in [12]. They propose a separate simulation of each round of the message-passing algorithm. Among other results they proved a bound of $\Theta(\Delta^2)$ for the case that each node transmits a different message to each of its neighbors. The lower bound translates to the SINR model with a slightly modified proof (see Sect. 4.1), while the upper bound has not yet been reached. Kuhn et al. [13] proposed an abstract interface— an abstract MAC layer—that enables easier models (i.e., message-passing based models) to be executed in more realistic models for wireless communication. However, they did only describe an implementation of the abstract MAC layer by local broadcasting in the radio network model, which does not account for global interference.

Local broadcasting in the SINR model has first been studied by Goussevskaia et al. in [2]. They considered local broadcasting with known and unknown competition (which is the number of nodes within a certain region around the node) in asynchronous networks and propose two randomized algorithms for the asynchronous SINR model with runtimes of $\mathcal{O}(\Delta \log n)$ and $\mathcal{O}(\Delta \log^3 n)$ for known and unknown competition. Yu et al. [14] improve the approximation ratio for the unknown competition by a logarithmic factor to $\mathcal{O}(\Delta \log^2 n)$ and propose two algorithms for the synchronized model (with synchronous and asynchronous wake-up) that make use of carrier sensing and thereby achieve local broadcasting in $\mathcal{O}(\Delta \log n)$ time slots. In [15] Yu *et al.* improve the algorithm for asynchronous time slots and unknown competition further to $\mathcal{O}(\Delta \log n + \log^2 n)$ and provide a lower bound of $\Omega(\Delta + \log n)$ for randomized algorithms in this model. Halldórsson and Mitra [16] provide an algorithm with the same running time of $\mathcal{O}(\Delta \log n + \log^2 n)$ in the same model, that is slightly simpler and more robust. They also provide an algorithm that achieves a running time of $\mathcal{O}(\Delta + \log^2 n)$ per round of local broadcasting with the assumption that acknowledgments are received freely.

The first result that achieves local broadcasting in the synchronized SINR model in $\mathcal{O}(\Delta)$ after a preprocessing stage of $\mathcal{O}(\Delta \log n)$ time slots is from Derbel and Talbi [3]. They transfer a distributed node coloring algorithm proposed by Moscibroda and Wattenhofer [17] to the SINR model and, by tuning the transmission power during the coloring step, achieve a deterministic local broadcasting schedule of length $\mathcal{O}(\Delta)$ that is feasible in the SINR model. A second result by Jurdzinski and Kowalski [4], which assumes the location to be known to the nodes, achieves the optimal runtime of $\mathcal{O}(\Delta)$ for local broadcasting without requiring the capability of nodes to tune their transmission power. However, the preprocessing stage requires $\mathcal{O}(\Delta \log^3 n)$ time slots. The authors introduce the concept of dilution (cf. Sect. 2.2) and build a deterministic backbone structure that enables communication to the backbone in $\mathcal{O}(\Delta)$ and local broadcasts from within the backbone in constant time. This backbone structure also enables local broadcasting in $\mathcal{O}(\Delta)$. For an overview of related results, see Table 1.

Table 1. Local broadcasting results for the SINR model. Ordered chronologically by appearance with separation in algorithms with and without preprocessing.

Publication	Assumptions	Runtime
Goussevskaia *et al.* [2]	Asynchron model (async), Δ	$\mathcal{O}(\Delta \log n)$
Goussevskaia *et al.* [2]	Async	$\mathcal{O}(\Delta \log^3 n)$
Yu *et al.* [14]	Async	$\mathcal{O}(\Delta \log^2 n)$
Yu *et al.* [14]	Sync. model, carrier sense (c.s.)	$\mathcal{O}(\Delta \log n)$
Yu *et al.* [15]	Async	$\mathcal{O}(\Delta \log n + \log^2 n)$
Halldórsson & Mitra [16]	Async, c.s. or free ACKs	$\mathcal{O}(\Delta + \log^2 n)$

Publication	Assumptions	Runtime +prepr.
Derbel & Talbi [3]	Sync, Δ, tune transmission power	$\mathcal{O}(\Delta) + \mathcal{O}(\Delta \log n)$
Jurdzinski & Kowalski [4]	Sync, Δ, location	$\mathcal{O}(\Delta) + \mathcal{O}(\Delta \log^3 n)$
This work	Sync, Δ, location	$\mathcal{O}(\Delta) + \mathcal{O}(\Delta \log n)$

1.2 Structure

The rest of this paper is structured as follows. In the next section, we describe required models and state some basic definitions. In Sect. 3, the construction of the deterministic local broadcasting schedule is described and we show its feasibility in the SINR model. Afterwards we consider the simulation of $\mathcal{CONGEST}$ algorithms in the SINR model in Sect. 4. We conclude this work with some final remarks in Sect. 5.

2   Model and Definitions

We consider a wireless network consisting of n nodes, that are placed arbitrarily on the Euclidean plane. The global maximum number of nodes within a transmission region is called the maximum degree of any node in the network and denoted by Δ. We usually assume that all nodes in the network know their ID and an upper bound $\tilde{n}$ on n, with $\tilde{n} \leq n^c$ for some constant $c \geq 1$. As the upper bound influences our results only by a constant factor we usually write n even though only $\tilde{n}$ may be known by the nodes.

In the geometric SINR model a transmission from node v to node w is successful iff the SINR condition holds:

$$\frac{\frac{P_v}{\text{dist}(v,w)^\alpha}}{\sum_{u \in \mathcal{I}} \frac{P_u}{\text{dist}(u,w)^\alpha} + N} \geq \beta \tag{1}$$

where P_v (P_u) denotes the transmission power of node v (u), α is the attenuation coefficient[2] depending on the network environment, the SINR-threshold $\beta \geq 1$ is a hardware-defined constant, N is the environmental noise and $\mathcal{I}$ is the set of nodes sending simultaneously with v. We assume uniform transmission powers, hence $P_v = P$ for each node v.

Based on the SINR condition the maximum transmission range of each node is $(\frac{P}{N\beta})^{1/\alpha}$. However, as soon as only one other node in the network transmits simultaneously, this transmission range cannot be achieved anymore. Having only one transmission in the whole network is clearly not desired, hence we define the maximum transmission range R_T such that twice the amount of noise can be tolerated: $R_T = (\frac{P}{2N\beta})^{1/\alpha}$. Note that this is a usual assumption and consistent with [3]. We do not exactly require twice the amount of noise, any constant factor $b > 1$ would also be sufficient. The area that is within the transmission range of a node v is denoted by D_T^v.

2.1   Simulating $\mathcal{CONGEST}$ Algorithms in the SINR Model

Let us first introduce the $\mathcal{CONGEST}$ model of distributed computation [1] briefly. This model focuses on the effects of congestion in distributed networks. Algorithms in the $\mathcal{CONGEST}$ model enforce a $\mathcal{O}(\log n)$ limitation on the maximum message size, while messages can only be sent to neighboring nodes. Note that with one message only a constant number of node IDs in the range $[0, \ldots, n]$ can be transmitted in this model. Hence, unlike in the $\mathcal{LOCAL}$ model which allows messages of unlimited sizes but restricts the runtime to a constant number of rounds [1] only a small fraction of the possibly obtained information can be made known to neighbors in reasonable time.

For a simulation of algorithms designed for the $\mathcal{CONGEST}$ model of distributed computation in the geometric SINR model we require the following properties to hold:

---

[2] The higher α is, the faster the signal fades. Usual values are $\alpha \in [2, 6]$.

– **Locality:** The neighbors of each node v must be reachable in our model, i.e., in the nodes transmission area D_T^v.
– **Disambiguity:** Each message is intended to one receiver, which is specified in the message by the receivers ID.
– **Synchronization:** Two neighbors are not allowed to be in different rounds of the $\mathcal{CONGEST}$ algorithm.

For the simulation to be successful we require that one or more transmission per sender-receiver-pair must be feasible in the SINR model of interference with high probability (w.h.p.—at least probability $1 - \frac{1}{n^c}$ for a constant $c > 0$) in each round of the $\mathcal{CONGEST}$ algorithm. Note that by disambiguity messages that are overheard by a node but not intended for it are discarded upon reception. This is not required in any part of our algorithms but increases clarity of the required properties. We usually assume the network to be connected, hence synchronization in combination with connectivity implies that all nodes must be in the same round of the $\mathcal{CONGEST}$ algorithm.

2.2 Dilution and Backbone Structure

In accordance with [4] we call a partition of the 2-dimensional plane in boxes of size $\gamma \times \gamma$, where $\gamma = R_T/\sqrt{2}$, the *pivotal grid* G_γ. Note that the dimensions of the box are such that all nodes within the same box are within each others transmission radius. Formally each box includes its bottom and left side but does not include its top and right side. We assume box $C(i, j)$ to be the box with lower left coordinates $(i, j) \in \mathbb{R}^2$. A node with position (x, y) is in box $C(i, j)$ iff $\lfloor \frac{x}{\gamma} \rfloor = i$ and $\lfloor \frac{y}{\gamma} \rfloor = j$.

A *local broadcast schedule* can be seen as an assignment of 0/1-bitstrings to nodes indicating in which time slots the node is allowed to broadcast. In the deterministic schedule constructed in this work, however, each node sends only once throughout an execution of the schedule. Hence we can simply store the number of the time slot instead of a 0/1 bitstring.

In order to combine geometric information with local broadcast schedules, we use the concept of *dilution* as introduced in [4]. For a constant δ, which determines the distance between two active transmissions and will be defined later, we assign each node v local coordinates $(l_x^v, l_y^v) = (\lfloor \frac{x}{\gamma} \rfloor \mod \delta, \lfloor \frac{y}{\gamma} \rfloor \mod \delta) = (i \mod \delta, j \mod \delta)$. This ensures that nodes in the same box of G_γ share the same local coordinates. Now, we can *dilute* a local broadcast schedule by a factor of δ^2 by allowing each node v with local coordinates (l_x^v, l_y^v) to send in time slot $t\delta^2 + l_x^v \delta + l_y^v$ iff v was allowed to send in time slot t in the original schedule.

3 Deterministic Local Broadcasting Schedule

One main approach for wireless transmission scheduling problems is to find a graph coloring and then use this coloring to decide when and for how long each node is allowed to transmit a message. This can be done by simply associating each color with a time slot. Let us first consider the simpler protocol model, in which a transmission is successful iff in the interference range (which often equals the transmission range) of the receiver only one node is transmitting at a given time. Even in this simpler model a node coloring which ensures that two nodes are assigned different colors if they are within each others transmission range is not sufficient to directly build a feasible transmission schedule as depicted in Fig. 1. However, for the protocol models this can be overcome by using a distance-2-coloring (i.e., a coloring which ensures unique colors within each transmission region D_T).

Due to the global nature of interference in the SINR model, finding some sort of agreement about transmission schedules (i.e., medium access) is required for deterministic local broadcasting schedules. In the case of coloring in the SINR model, even the more refined coloring that achieves unique colors within each transmission region is not sufficient as shown in Fig. 1(b). However, schedules can be made feasible if the node coloring ensures unique colors in an area larger than the transmission region. Unfortunately finding such a coloring is not possible if we cannot reach nodes outside the transmission region. Finding a coloring can be made possible by tuning the nodes transmission power to reach a larger transmission region, cf. [3], investing time in $\Omega(\mathrm{D})$ (given the network is connected), or having additional knowledge such as location information or knowledge about the topology. As computation of the diameter requires $\Omega(n)$ time slots [18], we restrict ourselves to some additional knowledge. In this work we consider location information to be known by each node. In the following theorem we show that we can distributedly construct a feasible local broadcasting schedule based on the location information and a given node coloring, even if the coloring does not ensure unique colors within each transmission region D_T. Note that such a coloring is easy to compute within $\mathcal{O}(\Delta \log n)$ time slots even in the SINR model [3]. If not noted otherwise we assume such a coloring.

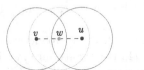

(a) Unique colors within distance R_T

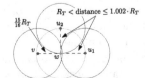

(b) Unique colors within each transmission region

Fig. 1. Using a coloring as depicted on the left does not yield a feasible local broadcasting schedule in the protocol model as the transmission from v to w is not feasible as according to the coloring u and v transmit simultaneously. However, the coloring on the right corresponds to a local broadcasting schedule that is feasible in the protocol model. Still it is not feasible in the SINR model as the SINR constraint is violated (at least for $\alpha \leq 6$).

Theorem 1. *Given a network of nodes in which each node knows its location, the color assigned by a coloring using at most $c_{max} = \mathcal{O}(\Delta)$ colors, and c_{max}*

itself. Then we can distributedly compute a local broadcasting schedule that is feasible under the SINR model of interference with length in $\mathcal{O}(\Delta)$.

In order to prove the theorem we first show that such a coloring is a local broadcasting schedule in which at most one node sends in each box of the pivotal grid G_γ (Lemma 1), and then prove that we can achieve a feasible schedule by applying dilution to this schedule (Lemma 2).

Lemma 1. *Given a network in which each node has a unique color within distance R_T. This implies a local broadcast schedule in which in each slot at most one node is transmitting in each box of the pivotal grid G_γ.*

Proof. As each node knows the number c of its color and a shared upper bound $c_{\max}$ on the number of colors assigned to the nodes in the network we can assign each color to one of $c_{\max}$ time slot. Consider a node v within box $C(i,j)$ and color c. Since the diameter of each box is exactly R_T, the coloring ensures that there is no other node within box $C(i,j)$ that has color c.

We extend Proposition 1 in [4] by explicitly giving a formula to compute the constant δ (depending only on α) that enables us to prove feasibility of a δ-diluted schedule in the SINR model of interference for $\alpha > 2$. For $\alpha = 2$ we can also achieve feasibility, however for $\delta \in \mathcal{O}(\log n)$, which is now additionally dependent on n. This leads to an increase in the schedule length of a multiplicative factor of $\delta^2 \in \mathcal{O}(\log^2 n)$.

Lemma 2. *Let $\alpha > 2$ and $\delta = \left(\dfrac{8\,P \sum_{k=1}^{\infty} \frac{1}{k^{\alpha-1}}}{N\,\gamma^\alpha} \right)^{1/\alpha} + 3$. Then a local broadcasting schedule $\mathcal{S}$ in which at most one node in each box of the pivotal grid G_γ transmits in each time slot can be made feasible in the SINR model of interference with a constant increase in the schedule length.*

The case $\alpha = 2$ is considered after the proof.

Proof. Let length$(\mathcal{S})$ be the length of the local broadcasting schedule $\mathcal{S}$. In order to achieve a feasible schedule, we dilute the schedule $\mathcal{S}$ by a constant δ^2 and obtain a feasible schedule $\mathcal{S}'$ with length$(\mathcal{S}') = \mathcal{O}(\text{length}(\mathcal{S}) \cdot \delta^2) = \mathcal{O}(\text{length}(\mathcal{S}))$. In this schedule $\mathcal{S}'$ a node v with local coordinates (l_x^v, l_y^v) sends in time slot $t\delta^2 + l_x^v \delta + l_y^v$ if and only if the node would have sent in time slot t of schedule $\mathcal{S}$.

Let us now consider an arbitrary time slot of schedule $\mathcal{S}'$, a node v that transmits a message in this time slot, and another node w that is within the transmission region of v. Let $C(i,j)$ be the box in which v is located and accordingly $(l_x^v, l_y^v) = (i \mod \delta, j \mod \delta)$ the local coordinates of v. We claim that w can successfully receive the message sent by v and hence—as we considered an arbitrary sender, receiver and time slot—this schedule is feasible in the SINR model. To show this claim we bound the interference received by w from simultaneously transmitting nodes by first upper bounding the number of simultaneously transmitting nodes within certain distances and then computing an upper bound on the interference of all those nodes on w.

The application of δ-dilution ensures that only nodes u with local coordinates $(l_x^u, l_y^u) = (i \bmod \delta, j \bmod \delta) = (l_x^v, l_y^v)$ transmit simultaneously with v. Note that local coordinates are shared by all nodes in the same box. Hence we call boxes that have nodes with the same local coordinates as v, i.e. boxes that are also allowed to send in the considered time slot, *active*. Due to the cyclicity of the modulo operator, δ-dilution results in a grid of active boxes with distance $\xi := (\delta - 1)\gamma$ between each two active boxes, as depicted in Fig. 2. Note that according to Lemma 1 at most one node in each active box transmits in each time slot.

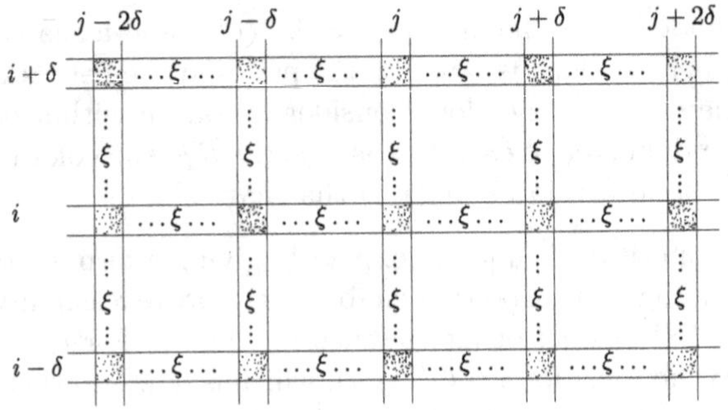

Fig. 2. Grid cells of G_γ that are active simultaneously to a transmission originating from box $C(i,j)$. Note that in order to increase readability $\xi := (\delta - 1)\gamma$.

Let us now examine how many active boxes there are at specified distances. We consider the boxes in so-called rings, which actually are the border layer of active boxes of a square centered at the box $C(i,j)$. In the situation of Fig. 2 all depicted nodes in columns $j - \delta$, j and $j + \delta$ except for $C(i,j)$ itself are in boxes of ring level 1 from $C(i,j)$. It can be observed that in each ring of level $k \geq 1$, exactly $8k$ active boxes can be accommodated. Also, each node in level k has distance at least $k((\delta - 3)\gamma)$ from w ($\delta - 3$ since w can be at most 2 boxes away from v).

Using this relation we can now upper bound the interference received by w from all nodes sending simultaneously with v, which are at most $8k$ nodes from each ring level k. Hence the interference at w is at most

$$\sum_{\substack{u \in V \setminus \{v\}, \\ u \text{ sending simultaneously with } v}} \frac{P}{\text{dist}(u, w)^\alpha} \leq \sum_{k=1}^{\infty} (8k) \frac{P}{(k(\delta - 3)\gamma)^\alpha} \tag{2}$$

$$\leq \sum_{k=1}^{\infty} \frac{8\,P\,k}{k^\alpha (\delta - 3)^\alpha \gamma^\alpha} \leq \frac{8\,P}{(\delta - 3)^\alpha \gamma^\alpha} \left(\sum_{k=1}^{\infty} \frac{1}{k^{\alpha-1}} \right) \leq N \tag{3}$$

where the first equation follows from applying the considerations about the ring levels and the last equation follows by insertion of δ. Note that the sum, which is the generalized harmonic number of order $(\alpha - 1)$, evaluates to a value lower than 6 for $\alpha > 2.2$ and is in $\mathcal{O}(1)$ for any $\alpha > 2$ [19].

Evaluating the SINR at node w yields

$$\text{SINR}_w = \frac{\frac{P}{\text{dist}(v,w)^\alpha}}{\sum\limits_{\substack{u \in V\setminus\{v\}, \\ u \text{ sending simultaneously with } v}} \frac{P}{\text{dist}(u,w)^\alpha} + N} \geq \frac{\frac{P}{R_T^\alpha}}{2N} \geq \beta$$

where the first inequality follows from $\text{dist}(v,w) \leq R_T$ and Eq. 3 and the last inequality follows from the definition of the transmission range $R_T = (\frac{P}{2N\beta})^{1/\alpha}$. This concludes the proof for $\alpha > 2$.

We will now briefly consider the case of $\alpha = 2$.

Corollary 1. *Let $\alpha = 2$ and $\delta = \left(\frac{8P \sum_{k=1}^n \frac{1}{k^{\alpha-1}}}{N\gamma^\alpha} \right)^{1/\alpha} + 3$. Then a local broadcasting schedule $\mathcal{S}$ in which at most one node in each box of the pivotal grid G_γ transmits in each time slot can be made feasible in the SINR model of interference with a factor $\delta^2 \in \mathcal{O}(\log^2 n)$ increase in the schedule length.*

Proof. Note that we changed the sum introduced in Eq. 2 from $\sum_{k=1}^\infty$ to $\sum_{k=1}^n$. This is possible as at most n non-empty ring levels exist. Since the distance of the levels increases it holds that

$$\sum_{\substack{\text{non-empty ring levels } k}} \frac{1}{k^{\alpha-1}} \leq \sum_{k=1}^n \frac{1}{k^{\alpha-1}} \tag{4}$$

and hence the resulting sum $\sum_{k=1}^n \frac{1}{k}$ can be evaluated to $\mathcal{O}(\log n)$[19]. This implies $\delta \in \mathcal{O}(\log n)$ and finally $\text{length}(\mathcal{S}') = \mathcal{O}(\text{length}(\mathcal{S}) \cdot \delta^2) = \mathcal{O}(\text{length}(\mathcal{S}) \cdot \log^2 n)$ as claimed in the corollary.

A pseudo code of the procedure described above is given in Algorithm 1. First an initial schedule is computed by distributed node coloring, then this schedule is diluted in order to obtain a schedule that is feasible in the SINR model. We can see that the algorithm itself is very simple. For a definition of the parameters cf. Sect. 2. Note that regarding δ neither the ceiling nor limiting the sum at n affects our theoretic results.

4 Simulating CONGEST Algorithms in SINR

Using the deterministic local broadcasting schedule constructed in Sect. 3, CONGEST algorithms with a runtime in $\mathcal{O}(\tau)$ can be simulated in $\mathcal{O}(\tau\Delta^2 + \Delta \log n)$ for $\alpha > 2$. This can be done by first computing the local broadcasting schedule in $\mathcal{O}(\Delta \log n)$ and then simulating the algorithm using so-called single-round-simulation as introduced by Alon et al. [12]. This requires Δ executions of the local broadcasting schedule for each round of the message-passing algorithm.

Algorithm 1. Distributed computation of a feasible local broadcasting schedule at node v

---

Require: location information (x^v, y^v), α, N, β, P, Δ, n

$\quad c \leftarrow$ color assigned by distributedNodeColoring(Δ, n, α, N, β, P) (e.g., [3])

$\quad \delta \leftarrow \left\lceil \left(\frac{8\,\mathrm{P} \sum_{k=1}^{n} \frac{1}{k^{\alpha-1}}}{\mathrm{N}\,\gamma^{\alpha}} \right)^{1/\alpha} \right\rceil + 3$ // dilution constant

$\quad (l_x^v, l_y^v) \leftarrow (\lfloor \frac{x^v}{\gamma} \rfloor \bmod \delta, \lfloor \frac{y^v}{\gamma} \rfloor \bmod \delta)$ // local coordinates

$\quad \text{active\_slot} \leftarrow \delta^2 c + \delta l_x^v + l_y^v$

We restrict ourselves to the simulation of general $\mathcal{CONGEST}$ algorithms in most parts of our work. In this model a node can send a different message of size $\mathcal{O}(\log n)$ to each neighbor in each round (cf. Sect. 2.1). However the methods transfer to the simulation of algorithms designed for similar models, for example if the same message is sent to all neighbors or if differently-sized messages are used. In particular for messages of arbitrary size s in a message-passing algorithm, the message size during simulation in the SINR model is $O(s + \log n)$. If unlike in the $\mathcal{CONGEST}$ model the same message is sent to each neighbor the runtime of the simulation decreases to $\mathcal{O}(\tau \Delta + \Delta \log n)$.

4.1 The Maximum Node Degree and the Simulation of (Synchronized) $\mathcal{CONGEST}$ Algorithms

Regardless of which local broadcasting strategy we use to simulate the rounds of the message-passing algorithm, all nodes must know the maximum number of time slots required to simulate one round of the message-passing algorithm. This number is needed so that each node can determine the time slot in which all nodes should finish with a certain round of the $\mathcal{CONGEST}$ algorithm. In the case of our local broadcasting schedule the number of slots required per round is $r = \Delta(\delta^2 \cdot c_{\max}) \in \mathcal{O}(\Delta^2)$, where $c_{\max}$ is the number of colors used by the node coloring.

So far we assumed the global maximum node degree Δ to be known to all nodes. In this section we will show that without an upper bound on the maximum node degree we cannot simulate a synchronized message-passing algorithm in less than $\Omega(D + \tau\Delta^2)$ time slots, where D is the diameter of the network. In order to show this results, let us briefly consider a lower bound on the number of time slots required to simulate one round of a general message-passing algorithm. Such a lower bound has already been stated by Alon *et al.* in [12] for the radio network model. However, it does not directly transfer to the SINR model. Note that we show the lower bound for message-passing models that allow to send a different message to each neighbor in each round (which is consistent with the assumptions of Alon *et al.*). This includes the general $\mathcal{CONGEST}$ model.

Lemma 3. *One round of a message-passing algorithm cannot be simulated in less than $\Omega(\Delta^2)$ time slots, where Δ is the maximum node degree of all nodes in the network.*

Proof. Assume a graph with all nodes within one transmission radius R_T and let this graph consist of two clusters S_l, S_r of the same (geometric) diameter d. Let those clusters be at least η times the diameter apart from each other and $\eta > 1$ be chosen such that $\frac{P}{(\eta d)^\alpha} - \frac{P}{((\eta+2)d)^\alpha} < N$ (note that the left part tends towards 0 for increasing values of η). Such clusters are shown in Fig. 3.

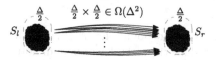

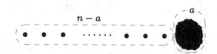

Fig. 3. Two clusters of same diameter within one transmission region. The distance between the clusters is more than η times the diameter of the cluster.

Fig. 4. The network is constructed such that a nodes are in the cluster on the right. For $a > 2$ the maximum node degree Δ occurs in the cluster on the right and must be communicated through the network. The transmission range is such that on the left part at most two nodes are within each others transmission range.

Let us only consider the transmission from the left cluster to the right cluster. Each node in the left cluster must transmit one different message to each node in the right cluster. This yields $\frac{\Delta}{2} \times \frac{\Delta}{2} \in \Omega(\Delta^2)$ inter-cluster-transmissions.

We will now show that at most one inter-cluster transmission can occur in one time slot. Let $v \in S_l$ be in the left cluster and $w \in S_r$ be in the right cluster. Assume v transmits to w in time slot t and assume another node u transmits to any other node in the same time slot. There are 2 cases: u can either be in S_l or S_r. In both cases u transmits simultaneously to v and we show that w cannot successfully receive v's message due to a SINR of less than 1. Let $u \in S_l$, then the SINR constraint (cf. Sect. 2) evaluates to

$$\frac{\frac{P}{\text{dist}(v,w)^\alpha}}{\frac{P}{\text{dist}(u,w)^\alpha} + N} \leq \frac{\frac{P}{(\eta d)^\alpha}}{\frac{P}{((\eta+2)d)^\alpha} + N} < \frac{\frac{P}{(\eta d)^\alpha}}{\frac{P}{((\eta+2)d)^\alpha} + \frac{P}{(\eta d)^\alpha} - \frac{P}{((\eta+2)d)^\alpha}} = \frac{\frac{P}{(\eta d)^\alpha}}{\frac{P}{(\eta d)^\alpha}} = 1 \leq \beta$$

where the first inequality holds since $\text{dist}(v,w) \geq \eta d$ and $\text{dist}(u,w) \leq (\eta + 2)d$ and the strict inequality follows from the selection of η. Hence w cannot receive v's message. Otherwise, if u in S_r the SINR is

$$\frac{\frac{P}{\text{dist}(v,w)^\alpha}}{\frac{P}{\text{dist}(u,w)^\alpha} + N} \leq \frac{\frac{P}{(\eta d)^\alpha}}{\frac{P}{(d)^\alpha} + N} < \frac{1}{\eta^\alpha} < 1 \leq \beta$$

where the first inequality again holds since $\text{dist}(v,w) \geq \eta d$ and $\text{dist}(u,w) \leq d$, the second inequality follows from $0 < N$ and cancelation of $\frac{P}{d^\alpha}$ and the third inequality holds since $\eta^\alpha > 1$. Hence at most one transmission from the left to the right cluster can happen in one time slot. This shows that $\frac{\Delta}{2} \times \frac{\Delta}{2} \in \Omega(\Delta^2)$ time slots are needed to simulate one round of a message-passing algorithm.

We can now prove the main result of this section, which provides a lower bound on the simulation runtime if the global maximum degree is not known to the nodes in the network.

Proposition 1. *Let n be the only knowledge available to the nodes. Then the simulation of a synchronized message-passing algorithm (e.g., $\mathcal{CONGEST}$) that requires τ rounds in the message-passing model cannot be executed in less than $\Omega(D + \tau\Delta^2)$ time slots in the SINR model.*

Proof. According to Lemma 3, $\Omega(\tau\Delta^2)$ is a lower bound for simulating a message-passing algorithm with runtime τ. To show the $\Omega(D)$ lower bound, note that networks with $\Delta = \sqrt{D}$ exist, and hence in those networks at least $\Omega(D)$ time slots are required for each round of the simulation. However, there exist also networks in which $\tau\Delta^2 \notin \Omega(D)$ and still $\Omega(D)$ time slots are required for the simulation. Hence $\Omega(D + \tau\Delta^2)$ is effectively a stronger bound than $\Omega(\tau\Delta^2)$.

Consider two networks. The first is the network depicted in Fig. 4 with $a = \sqrt{n}$, and the second a line network (which is equal to the depicted network without the high-density part on the right, i.e. with $a = 0$). Clearly the line network is a network in which $\tau\Delta^2 \notin \Omega(D)$. For nodes on the left end of both networks the view is exactly the same until at least $\Omega(D)$ time slots have passed and information from the high-density part can reach the left end of the network. Assume for contradiction that there is an algorithm that finishes the simulation on both networks in less than $\Omega(D)$ time slots. This algorithm must compute the number of time slots required for each round of the simulation in order to synchronize the message-passing algorithm. Since the information about the high-density part is not available to nodes on the left end of both networks the algorithm computes the same number of required time slots in the leftmost nodes of both networks. Regardless of the result the algorithm fails to simulate the message-passing algorithm in one of the networks. If the result (i.e., the required number of time slots per simulated round) is in $o(\sqrt{n})$, the algorithm fails in the network depicted in Fig. 4 with $a = \sqrt{n}$, as the network cannot be synchronized. If the result is in $\Omega(\sqrt{n})$ this results in $\Omega(n) = \Omega(D)$ time slots for the simulation, which contradicts the assumption that the algorithm runs in less than $\Omega(D)$ time slots on both graphs. Hence any algorithm that simulates a synchronized message-passing algorithm in the SINR model without the knowledge of Δ requires at least $\Omega(D)$ time slots. $\square$

Note that the proof relies on restrictions on simultaneous transmissions and the synchronization of the $\mathcal{CONGEST}$ algorithm. Hence letting the node know the diameter D or even its position does not circumvent the bound.

4.2   Notes on Location Information

After considering the case that the global maximum degree Δ is unknown, we will now focus on the knowledge of location information. Local broadcasting in $\mathcal{O}(\Delta)$ time slots (after a preprocessing stage of $\mathcal{O}(\Delta \log n)$ time slots) is also possible by allowing nodes to tune their transmission power. Derbel and Talbi describe an algorithm that is based on distributed node coloring with tuned transmission radius in [3] and they achieve a runtime of $\mathcal{O}(\tau\Delta^2 + \Delta\log n)$. However, they state a message size of $\mathcal{O}(s \cdot \log n)$, where s is the original message size. For the simulation of $\mathcal{CONGEST}$ algorithms this results in messages of size $\mathcal{O}(\log^2 n)$ instead of $\mathcal{O}(\log n)$. We claim that messages of size $\mathcal{O}(\log n)$ are possible and hence this additional logarithmic factor is not necessary. The algorithm consists of two parts. In the first part a distributed node coloring is computed. For this only the node ID and the number of the color must be transmitted. Hence messages of size $\mathcal{O}(\log n)$ are sufficient. In the second part the actual simulation takes place. Therefore the original message of size s along with a node ID (in order to identify the receiver) must be transmitted. This requires messages of size $\mathcal{O}(s + \log n)$. For $\mathcal{CONGEST}$ algorithms this results in messages of size $\mathcal{O}(\log n)$, since $s \in \mathcal{O}(\log n)$.

Hence for both cases, using either tuned transmission powers or location information the same runtime of $\mathcal{O}(\tau\Delta^2 + \Delta\log n)$ and messages of size $\mathcal{O}(\log n)$ are sufficient to simulate a $\mathcal{CONGEST}$ algorithm with original runtime τ in the SINR model.

5   Conclusion

In this work we introduced a new algorithm to compute a deterministic local broadcasting schedule of optimal length $\mathcal{O}(\Delta)$ that is feasible in the SINR model of interference. The construction of the schedule requires $\mathcal{O}(\Delta\log n)$ time slots, which is optimal up to the logarithmic factor. The algorithm enables the simulation of algorithms designed for message-passing models in more realistic models of interference such as the SINR model: An algorithm with original runtime of τ rounds in the $\mathcal{CONGEST}$ model can be simulated in $\mathcal{O}(\tau\Delta^2 + \Delta\log n)$ time slots in the SINR model. This is optimal apart from the logarithmic factor. Our algorithm assumes that nodes know their position and the global maximum node degree Δ. We showed a lower bound of $\Omega(D + \tau\Delta^2)$, thus the knowledge of Δ is required in order to achieve an efficient simulation.

Acknowledgments. This work was supported by the German Research Foundation (DFG) within the Research Training Group GRK 1194 "Self-organizing Sensor-Actuator Networks".

References

1. Peleg, D.: Distributed Computing: A Locality-Sensitive Approach. Society for Industrial Mathematics, Philadelphia (2000)
2. Goussevskaia, O., Moscibroda, T., Wattenhofer, R.: Local broadcasting in the physical interference model. In: Proceedings of the 2008 Joint Workshop on Foundations of Mobile Computing (DialM-POMC'08), pp. 35–44. ACM (2008)
3. Derbel, B., Talbi, E.G.: Distributed node coloring in the SINR model. In: Proceedings of the 30th International Conference on Distributed Computing Systems (ICDCS'10), pp. 708–717. IEEE Computer Society (2010)

4. Jurdzinski, T., Kowalski, D.R.: Distributed backbone structure for algorithms in the SINR model of wireless networks. In: Aguilera, M.K. (ed.) DISC 2012. LNCS, vol. 7611, pp. 106–120. Springer, Heidelberg (2012)
5. Moscibroda, T., Wattenhofer, R.: The complexity of connectivity in wireless networks. In: Proceedings of the 25th Annual Joint Conference of the IEEE Computer and Communications Societies (INFOCOM'06), pp. 1–13. IEEE Computer Society Press, April 2006
6. Avin, C., Lotker, Z., Pasquale, F., Pignolet, Y.-A.: A note on uniform power connectivity in the SINR model. In: Dolev, S. (ed.) ALGOSENSORS 2009. LNCS, vol. 5804, pp. 116–127. Springer, Heidelberg (2009)
7. Moscibroda, T., Wattenhofer, R., Zollinger, A.: Topology control meets SINR: the scheduling complexity of arbitrary topologies. In: Proceedings of the 7th ACM International Symposium on Mobile Ad Hoc Networking and Computing (MOBI-HOC'06), pp. 310–321. ACM (2006)
8. Yu, D., Wang, Y., Hua, Q.-S., Lau, F.C.M.: Distributed $(\Delta + 1)$-coloring in the physical model. In: Erlebach, T., Nikoletseas, S., Orponen, P. (eds.) ALGOSENSORS 2011. LNCS, vol. 7111, pp. 145–160. Springer, Heidelberg (2012)
9. Scheideler, C., Richa, A.W., Santi, P.: An O(log n) dominating set protocol for wireless ad-hoc networks under the physical interference model. In: Proceedings of the 9th ACM International Symposium on Mobile Ad Hoc Networking and Computing (MOBIHOC'08), pp. 91–100 (2008)
10. Holzer, S., Wattenhofer, R.: Optimal distributed all pairs shortest paths and applications. In: Proceedings of the 31th ACM Symposium on Principles of Distributed Computing (PODC'12), pp. 355–364. ACM (2012)
11. Derbel, B., Mosbah, M., Zemmari, A.: Fast distributed graph partition and application (extended abstract). In: 20th International Parallel and Distributed Processing Symposium (IPDPS 2006), April 2006
12. Alon, N., Bar-Noy, A., Linial, N., Peleg, D.: On the complexity of radio communication. In: Proceedings of the 21th Annual ACM Symposium on the Theory of Computing (STOC'89), pp. 274–285. ACM (1989)
13. Kuhn, F., Lynch, N., Newport, C.: The abstract MAC layer. Distrib. Comput. 24(3–4), 187–206 (2011)
14. Yu, D., Wang, Y., Hua, Q.S., Lau, F.C.M.: Distributed local broadcasting algorithms in the physical interference model. In: Proceedings of the 2011 International Conference on Distributed Computing in Sensor Systems (DCOSS'11), pp. 1–8. IEEE Computer Society (2011)
15. Yu, D., Hua, Q.S., Wang, Y., Lau, F.C.M.: An O(log n) distributed approximation algorithm for local broadcasting in unstructured wireless networks. In: Proceedings of the 2012 International Conference on Distributed Computing in Sensor Systems (DCOSS'12), pp. 132–139. IEEE Computer Society (2012)
16. Halldórsson, M.M., Mitra, P.: Towards tight bounds for local broadcasting. In: The Eighth ACM International Workshop on Foundations of Mobile Computing (FOMC'12). ACM, July 2012
17. Moscibroda, T., Wattenhofer, M.: Coloring unstructured radio networks. Distrib. Comput. 21(4), 271–284 (2008)
18. Frischknecht, S., Holzer, S., Wattenhofer, R.: Networks cannot compute their diameter in sublinear time. In: Proceedings of the Twenty-Third Annual ACM-SIAM Symposium on Discrete Algorithms, SODA'12, Kyoto, Japan, pp. 1150–1162. SIAM (2012). http://dl.acm.org/citation.cfm?id=2095116.2095207
19. Knuth, D.E.: Fundamental Algorithms. The Art of Computer Programming, vol. 1. Addison-Wesley, Reading (2011)

The Effect of Forgetting on the Performance of a Synchronizer

Matthias Függer[1], Alexander Kößler[1]([✉]), Thomas Nowak[2], Ulrich Schmid[1], and Martin Zeiner[1]

[1] ECS Group, TU Wien, Vienna, Austria
{fuegger,koe,s,mzeiner}@ecs.tuwien.ac.at
[2] Laboratoire d'Informatique, École polytechnique, Palaiseau, France
nowak@lix.polytechnique.fr

Abstract. We study variants of the α-synchronizer by Awerbuch (J. ACM, 1985) within a distributed message passing system with probabilistic message loss. The purpose of synchronizers is to maintain a virtual (discrete) round structure. Their idea essentially is to let processes continuously exchange round numbers and to allow a process to proceed to the next round only after it has witnessed that all processes have already started its own current round.

In this work, we study how four different, naturally chosen, strategies of forgetting affect the performance of these synchronizers. The variants differ in the times when processes discard part of their accumulated knowledge during execution. Such actively forgetting synchronizers have applications, e.g., in sensor fusion where sensor data becomes outdated and thus invalid after a certain amount of time.

We give analytical formulas to quantify the degradation of the synchronizers' performance in an environment with probabilistic message loss. In particular, the formulas allow to explicitly calculate the performance's asymptotic behavior. Interestingly, all considered synchronizer variants behave similarly in systems with low message loss, while one variant shows fundamentally different behavior from the remaining three in systems with high message loss. The theoretical results are backed up by Monte-Carlo simulations.

1   Introduction

A set of sensor nodes collecting in-field data and exchanging it over an ad-hoc wireless network is a common setup for sensor fusion applications [9]. Message loss is typically a non negligible issue within such systems. A common strategy to deal with message loss is to run a synchronizer algorithm, whose purpose is to generate a virtual (discrete) round structure at the application layer such that, at each round step, a process receives all messages from all processes sent in

This work has been partially supported by the Austrian Science Fund (FWF), grant NFN RiSE (S11405).

P. Flocchini et al. (Eds.): ALGOSENSORS 2013, LNCS 8243, pp. 185–200, 2014.
DOI: 10.1007/978-3-642-45346-5_14, © Springer-Verlag Berlin Heidelberg 2014

the current round. In this work we study a retransmission-based variant of the
α-synchronizer, introduced by Awerbuch [1] as the first in a series of synchronizer
algorithms for asynchronous message-passing systems. Its main idea is that each
process continuously broadcasts its current round number together with the
corresponding application data. In systems with high dynamics, this application
data may vary between broadcasts, even within one round. A process starts
the next round when it has received the messages of its current round from all
other processes. Additionally it delivers the most actual data it has received
in its current round to the application layer. The synchronizer then guarantees
a synchronization precision equal to the diameter of the network graph. The
original α-synchronizer by Awerbuch used additional acknowledgment messages,
which we omit. Rather, a message with round number R is treated as an implicit
acknowledge for messages with round numbers less than R.

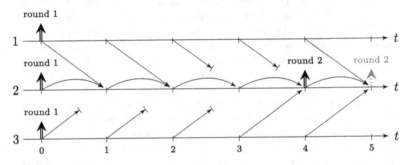

Fig. 1. Messages to process 2 and its resulting round switches without forgetting
(black) and with forgetting (gray).

Figure 1 shows the beginning of an execution of the synchronizer executed in
a system with three processes. For clarity, only messages to process 2 are shown.
Time is assumed to elapse in discrete steps at all processes. We assume the
existence of an underlying mechanism preventing the processes' discrete time
from diverging, i.e., a synchronous system. At each point in time a process
broadcasts its application data, e.g., the current sensor reading. Initially, at time
0, all processes start round 1. By time 4, process 2 has received round 1 messages
from all processes and thus proceeds to round 2. Note, however, that the age of
the round 1 data it hands over to the application layer when switching to round
2 differs significantly per process: while its own data and the data from process
3 is of age 1 (discrete time units), data from process 1 is of age 3. If this data
is time-variant, e.g., the position of a moving object, it is typically represented
by an interval (i.e., a value ± some accuracy) that detoriates with time [9]. A
proper deterioration accounts for the maximum change of the position since the
actual sampling of the data. When merging intervals representing the same data,
from different sources, e.g. using (fault-tolerant) interval-intersection functions
like [7,13], relying on old data obviously yields imprecise results.

A strategy to counteract this problem is to let the synchronizer actively "forget" old data it has received by discarding it. As an extreme, consider a variant of the synchronizer that discards data at each (discrete) time step, resulting in all the data to be of age 1 at each round switch. Clearly, however, this results in a performance loss, i.e., longer times between round switches. The resulting execution is depicted in gray in Fig. 1 with the difference that process 2 then switches to round 2 only at time 5.

In this paper we consider four variants of the α-synchronizer that differ in the conditions of when to forget memory content, that is, reset the variables representing the knowledge to their initial values. While three of the variants, namely the variant that never forgets, the variant in which a process forgets when it makes a round switch, and the variant that forgets at each time step, can be implemented in a distributed manner, the variant which forgets when the last process makes a round switch serves as a theoretical bound only.

We study the impact of forgetting on the performance of the synchronizer variants in an environment where every message transmission succeeds with a certain probability. By giving explicit formulas and simulation results for the performance as well as simulation results for the average age of data when a process makes a round switch, our results can be used to quantify the tradeoff between the different strategies.

Detailed Contribution. We make the following contributions in this paper: (1) We formally introduce the notion of forgetting in the context of a (specific) synchronizer. We consider four different conditions on when processes forget and study the respective degrading effects on the synchronizer's performance in a probabilistic environment. (2) We state explicit formulas for the expected round duration for two of these conditions and give efficient bounds for the other two conditions. These bounds are shown to approximate the true value well if the probability p of successful message transmission is high. (3) We show that for all four conditions, the expected round durations collapse when $p \to 1$: All four expected round durations, as well as its first derivative as a function of p, are equal in $p = 1$. (4) We prove that for $p \to 0$, the expected round duration for three of the conditions has the same order of growth, which we calculate explicitly for all four conditions. (5) We present simulation results of the expected round duration, comparing them to our calculations, and simulation results for the average age of data when a process makes a round switch.

Related Work. Our notion of knowledge is different from that of Fagin et al. [3], who studied the evolution of knowledge in distributed systems with powerful agents; in particular, their agents do not forget. While Mahesh and Varghese [5] use crashing processes and the forgetting during reboot in a destructive way, we use forgetting in a constructive manner. Nowak et al. [10] calculated the expected round duration of a retransmission-based synchronizer when a single transmission arrives with constant probability p, but a message that was retransmitted at least M times is guaranteed to arrive. They did not investigate the impact of forgetting on the synchronizer's performance, and assumed M to be finite, which we do not. Bertsekas and Tsitsiklis [2] proved bounds for the case of constant processing times and exponentially distributed message delays. They did not derive exact performance measures. Rajsbaum [11] presented bounds on the synchronizer rate for the case of exponentially distributed processing times and transmission delays. Rajsbaum and Sidi [12] calculated the rate's exact value in the case of exponentially distributed processing times and negligible transmission delays.

Organization of the Paper. The rest of the paper is organized as follows. Section 2 introduces our system model, the studied synchronizer algorithm, and the four conditions on forgetting we investigate, i.e., when the processes forget the gained knowledge. In Sect. 3 we give the performance measure and we derive explicit formulas for two of the four conditions in Sect. 4. Section 5 uses a Markov chain model to compute the expected round duration of the remaining two conditions on forgetting and states results on the asymptotic behavior of the expected round duration. It also presents analytical lower bounds that facilitate estimations of the expected round duration. In Sect. 6 we compare the performance as well as the average age of data achieved by the different conditions on forgetting against each other.

2   System Model and Algorithm

In this paper we study the performance of variants of the α-synchronizer [1] running in a fully-connected message passing system with processes $1, 2, \ldots, N$. Processes take steps simultaneously at all integral times $t \geqslant 0$, but messages may be lost. Messages that do arrive have a transmission delay of 1, i.e., a message sent at time t arrives at time $t + 1$, or not at all. A step consists in (a) receiving messages from other processes, (b) performing local computations, and (c) broadcasting a message to the other processes.

The synchronizer variants have two local variables, specified for every process i at time t: The *local round number* $R_i(t)$ and the *knowledge vector* $\big(K_{i,1}(t), K_{i,2}(t), \ldots, K_{i,N}(t)\big)$. Processes continuously broadcast their local round number. The knowledge vector contains information on other processes' local round numbers, accumulated via received messages. A process increments its local round number, and thereby starts the next round, after it has gained knowledge that all other processes have already started the current round. The round increment rule assures a precision of 1, i.e., $|R_i(t) - R_j(t)| \leqslant 1$ for all t. We write $R_G(t) = \min_i R_i(t)$ and call it the *global round number* at time t.

After updating its local round number, a process may *forget*, i.e., lose its knowledge about other processes' local round numbers. We are considering four different conditions COND, describing the times when process i forgets:

 I. Never, i.e., COND := *false*.
 II. At every *local round switch*, i.e., COND := $\big[R_i(t) = R_i(t-1) + 1\big]$.
 III. At every *global round switch*, i.e., COND := $\big[R_G(t) = R_G(t-1) + 1\big]$.
 IV. Always, i.e., COND := *true*.

Formally, we write $\mathcal{M}_{i,j}(t) = 0$ if process j's message to process i sent at time t was lost, and $\mathcal{M}_{i,j}(t) = 1$ if it arrives (at time $t+1$). Process i's computation in its step at time t consists of the following:

1. *Update knowledge according to received messages:*
   $K_{i,j}(t) \leftarrow R_j(t-1)$ if $\mathcal{M}_{i,j}(t-1) = 1$, and $K_{i,j}(t) \leftarrow K_{i,j}(t-1)$ otherwise.
2. *Increment round number if possible:* $R_i(t) \leftarrow R_i(t-1)+1$ if $K_{i,j}(t) \geqslant R_i(t-1)$
   for all j, and $R_i(t) \leftarrow R_i(t-1)$ otherwise.
3. *Conditional forget:* $K_{i,j}(t) \leftarrow 0$ if COND is true.

Initially, $K_{i,j}(0) = 0$, and no messages are received at time 0. In particular, $R_i(0) = 1$. In the remainder of this paper, when we refer to $K_{i,j}(t)$, we mean its value after step 3.

We assume that the $\mathcal{M}_{i,j}(t)$ are pairwise independent random variables with

$$\mathbb{P}\big(\mathcal{M}_{i,j}(t) = 1\big) = p \text{ if } i \neq j \text{ and } \mathbb{P}\big(\mathcal{M}_{i,i}(t) = 1\big) = 1 . \tag{1}$$

We call the parameter p the *probability of successful transmission*.

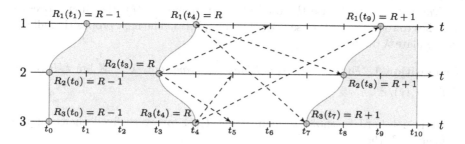

Fig. 2. An execution of the synchronizer

Figure 2 shows part of an execution for condition I on forgetting. Times are labeled t_0 to t_{10}. Processes 1 and 3 start their local round R at time t_4 while process 2 has already started its local round R at time t_3. The arrows in the figure indicate the time until the first successful reception of a message sent in round R: The tail of the arrow is located at time t a process i starts round R and thus broadcasts R for the first time. The head of the arrow marks the smallest time after t at which a process j receives a message from i. Messages from processes to themselves are always received at the next time step and thus are not explicitly shown in the figure. For example, processes 1 and 3 start round R at time t_4 sending R for the first time. While process 2 receives the message from 3 in the next step, it needs an overall amount of 4 time steps and consecutive retransmissions to receive a message from process 1 at time t_8.

3   Performance Measure

For a system with N processes and probability p of successful transmission, we define the *expected round duration* of process i by $\lambda_i(N, p) = \mathbb{E} \lim_{t\to\infty} t/R_i(t)$.

Since our synchronization algorithm guarantees precision 1, it directly follows that $\lambda_i(N, p) = \lambda_j(N, p)$ for any two processes i and j. We will henceforth refer to this common value as $\lambda(N, p)$, or simply λ if the choice of parameters N and p is clear from the context. To distinguish the four proposed conditions on forgetting, I to IV, we will write λ^I, λ^{II}, λ^{III}, and λ^{IV}, respectively.

In the rest of the paper, we study the expected round duration λ for the four conditions on forgetting. Note that the condition in case III cannot be detected locally and thus does not allow for a distributed implementation. We rather use λ^{III} as a bound (cf. Eq. 2). For case IV, where processes always forget, and for case III, where processes forget on global round switches, λ can be calculated efficiently with explicit formulas, which we give in Sect. 4 in Theorems 1 and 2. For the remaining cases, I and II, we could compute $\lambda(N, p)$ by means of a steady state analysis of a finite Markov chain with time complexity exponential in N. We show how to do this in Sect. 5.1. The Markov chain model is also useful to study the behavior of λ, for all four conditions on forgetting, when $p \to 1$ and $p \to 0$. We do this in Sects. 5.2 and 5.3, respectively. We derive explicit lower bounds on λ^I and λ^{II} in Sect. 5.4.

We will repeatedly use the dual of $R_i(t)$, namely $T_i(r)$, the time process i switches to round r. Further set $T_G(r) = \max_i T_i(r)$. The next proposition allows to calculate λ dually by:

Proposition 1. *For all four conditions on forgetting,* $\lambda = \mathbb{E} \lim_{t \to \infty} t/R_i(t) = \mathbb{E} \lim_{r \to \infty} T_i(r)/r.$

It is not hard to show, by comparing $T_i(r)$ for every fixed choice of the sequence $\mathcal{M}$, that

$$\lambda^I \leqslant \lambda^{II} \leqslant \lambda^{III} \leqslant \lambda^{IV} . \qquad (2)$$

4   Explicit Formulas for λ^{III} and λ^{IV}

In this section, by elementary probability theory and calculations, we derive explicit formulas for λ^{III} and λ^{IV} in Theorems 1 and 2, respectively. Both use a formula for the expected maximum of geometrically distributed random variables (Proposition 2). For that purpose define for pairwise independent with parameter p geometrically distributed random variables G_i

$$\Lambda(M, p) = \mathbb{E} \max_{1 \leqslant i \leqslant M} G_i .$$

We will make use of the following well-known proposition [6,14].

Proposition 2. $\Lambda(M, p) = \displaystyle\sum_{i=1}^{M} \binom{M}{i} (-1)^i \frac{1}{(1-p)^i - 1}$

Consider case III, i.e., processes forget on global round switches. Initially, all processes i are in round $R_i(0) = 1$, and their knowledge is $K_{i,j}(0) = 0$.

Observe that processes switch to round 2 as messages are received. At the time t at which the last process switches to round 2, it holds that (i) all processes i have $R_i(t) = 2$, (ii) all processes have knowledge $K_{i,j}(t) \geqslant 1$ for all j before forgetting, and (iii) all processes forget, since a global round switch occurred, ultimately resulting in $K_{i,j}(t) = 0$. The only difference between the initial state and the state at time t is the constant round number offset $R_i(t) = R_i(0) + 1$. By repeated application of the above arguments we obtain that the system is reset to the initial state modulo a constant offset in round numbers R_i, each time a global round switch occurs. This allows to determine the expected average round duration by analyzing the expected time until the first round switch.

We will now state explicit formulas for the expected round duration in cases III and IV. We will use these formulas in particular in Sect. 5.3 when studying the behavior of λ for $p \to 0$.

Theorem 1. $\lambda^{\mathrm{III}}(N, p) = \Lambda\big(N(N-1), p\big) = \displaystyle\sum_{i=1}^{N(N-1)} \binom{N(N-1)}{i} \frac{(-1)^i}{(1-p)^i - 1}$

Proof. Recall that the events that i receives a message from j at time t are pairwise independent for all i, j and times t. Thus the smallest time t, at which i receives a message from j is geometrically distributed with parameter p. Noting that the first global round switch occurs at time $T_G(2) = \max_i(T_i(2))$, we obtain

$$\lambda(N, p) = \mathbb{E} \lim_{r \to \infty} T_G(r)/r = \mathbb{E} T_G(2) = \mathbb{E} \max_{1 \leqslant i \leqslant N(N-1)} G_i$$

where the G_i are geometrically distributed with parameter p. The theorem now follows from Proposition 2.   □

Theorem 2. $\lambda^{\mathrm{IV}}(N, p) = \Lambda\big(N, p^{N-1}\big) = \displaystyle\sum_{i=1}^{N} \binom{N}{i}(-1)^i \frac{1}{(1 - p^{N-1})^i - 1}$

Proof. Observe that the first global round switch occurs at the minimum time t by which each of the processes has received messages from all processes simultaneously; and that $R_i(t) = 2$ as well as $K_{i,j}(t) = 0$ holds at this time. Again the state at time t is identical to the initial state with all round numbers incremented by 1. Repeated application of the above arguments allows to calculate the expected round duration by $\lambda(N, p) = \mathbb{E} T_G(2)$. The first time i receives a message from all processes simultaneously is geometrically distributed with parameter p^{N-1}. Since we have N nodes, we take the maximum over N such geometrically distributed random variables. The theorem now follows from Proposition 2.   □

5   Markovian Analysis

Determining λ^{I} and λ^{II}, the expected round duration in the cases that processes never forget or forget at local round switches, is more involved. In the following, we will calculate λ by modeling the system as a finite Markov chain and analyzing its steady state distribution. Additionally, we derive the asymptotic behaviors for $p \to 1$ and for $p \to 0$ from the Markov chain model. As the computation of the chain's steady state distribution is computationally very expensive, we will give analytical lower bounds in Sect. 5.4.

Let $A(t)$ be the sequence of matrices with $A_{i,i}(t) = R_i(t)$ and $A_{i,j}(t) = K_{i,j}(t)$ for $i \neq j$. It is easy to see that $A(t)$ is a Markov chain, i.e., the distribution of $A(t+1)$ depends only on $A(t)$. Since both $R_i(t)$ and $K_{i,j}(t)$ are unbounded, the state space of Markov chain $A(t)$ is infinite.

We therefore introduce the sequence of *normalized* states $a(t)$, defined by $A(t) - \min_k A_{k,k}(t)$ cropping negative entries to -1, i.e., $a_{i,j}(t) = \max \{A_{i,j}(t) - \min_k A_{k,k}(t) , -1\}$. Normalized states belong to the finite set $\{-1, 0, 1\}^{N \times N}$.

The sequence of normalized states $a(t)$ is a Markov chain: The probability that $A(t+1) = Y$, given that $A(t) = X$, is equal to the probability that $A(t+1) = Y + c$, given that $A(t) = X + c$. We may thus restrict ourselves without loss of generality to considering the system being in state $X - \min_i(X_{i,i})$ at time t. Further, by the algorithm and the fact that the precision is 1, cropping the entries of $X - \min_i(X_{i,i})$ at -1 does not lead to different transition probabilities: the probability that $A(t+1) = Y$ given that $A(t) = X - \min_i(X_{i,i})$ is equal to the probability that $A(t+1) = Y$ given that $A(t)$ is $X - \min_i(X_{i,i})$ cropped at -1. It follows that $a(t)$ is a finite Markov chain, for the algorithm with any of the four conditions on forgetting.

We will repeatedly need to distinguish whether there is a global round switch at time t or not. Let $\hat{a}(t)$ be the Markov chain obtained from $a(t)$ by adding to each state a an additional flag Step such that $\text{Step}(\hat{a}(t)) = 1$ if there is a global round switch at time t, and 0 otherwise.

5.1   Using the Steady State to Calculate λ

Call a Markov chain *good* if it is aperiodic, irreducible, Harris recurrent, and has a unique steady state distribution. It is not difficult to see that $\hat{a}(t)$ is good for all four conditions on forgetting.

Theorem 3. *Let $X(r)$ be good Markov chain with state space $\mathcal{X}$ and steady state distribution π. Further, let $g : \mathcal{X} \to \mathbb{R}$ be a function such that $\sum_{X \in \mathcal{X}} |g(X)| \cdot \pi(X) < \infty$. Then, $\lim_{r \to \infty} \frac{1}{r} \sum_{k=1}^{r} g(X(k)) = \sum_{X \in \mathcal{X}} g(X) \cdot \pi(X)$ with probability 1 for every initial distribution.*

Proof. [8, Theorem 17.0.1(i)] $\qquad \square$

A standard method, given the chain's transition matrix P, to compute the steady state distribution π is by matrix inversion:

$$\pi = e \cdot \left(P^{(n \to 1)} - I^{(n \to 0)}\right)^{-1} \tag{3}$$

where $M^{(k \to x)}$ denotes matrix M with its kth column set to x, I is the identity matrix, and $e = (1, 1, \dots, 1)$.

We call a processes i a *1-process* in state $\hat{a}$ if $\hat{a}_{i,i} = 1$. Likewise, we call i a *0-process* in $\hat{a}$ if $\hat{a}_{i,i} = 0$. Denote by $\#_{-1}(\hat{a})$ the number of -1 entries in rows of matrix $\hat{a}$ that correspond to 0-processes in $\hat{a}$.

Proposition 3. *For all conditions of forgetting, $R_i(t)/t \to 1/\lambda$ with probability 1 as $t \to \infty$. Furthermore, $\lambda = 1/\left(\sum_{\hat{a}} p^{\#_{-1}(\hat{a})} \cdot \pi(\hat{a})\right)$.*

Proof. It holds that $R_G(t) = \sum_{k=1}^{t} \text{Step}(\hat{a}(k))$. By Theorem 3, with probability 1 it holds that:

$$\lim_{t \to \infty} R_i(t)/t = \lim_{t \to \infty} R_G(t)/t = \lim_{t \to \infty} \frac{1}{t} \sum_{k=1}^{t} \text{Step}(\hat{a}(k)) = \sum_{\hat{a}} \text{Step}(\hat{a}) \cdot \pi(\hat{a}) .$$

Since $\hat{a}(t)$ is a finite Markov chain, the last sum is finite. It follows that $R_i(t)/t$ converges to a constant, say c, with probability 1. Thus $t/R_i(t)$ converges to $1/c$ with probability 1. By definition of λ, it follows that $\lambda = 1/c$. This shows the first part of the proposition.

The second part of the proposition is proved by the following calculation:

$$1/\lambda = \mathbb{E} \lim_{t \to \infty} R_i(t)/t = \mathbb{E} \lim_{t \to \infty} R_G(t)/t = \mathbb{E} \lim_{t \to \infty} \frac{1}{t} \sum_{k=1}^{t} \text{Step}\left(\hat{a}(k)\right)$$

$$= \sum_{\hat{a}} \lim_{t \to \infty} \frac{1}{t} \sum_{k=1}^{t} \mathbb{P}\left(\hat{a}(k-1) = \hat{a}\right) \cdot \mathbb{E}\left(\text{Step}(\hat{a}(k)) \mid \hat{a}(k-1) = \hat{a}\right)$$

$$= \sum_{\hat{a}} p^{\#_{-1}(\hat{a})} \lim_{t \to \infty} \frac{1}{t} \sum_{k=1}^{t} \mathbb{P}\left(\hat{a}(k-1) = \hat{a}\right) = \sum_{\hat{a}} p^{\#_{-1}(\hat{a})} \cdot \pi(\hat{a}) . \qquad \square$$

5.2 Behavior of λ for $p \to 1$

The next theorem provides means to approximate the expected round duration for all conditions on forgetting when messages are successfully received with high probability. Since this is typically the case for real-world systems, it allows to characterize their expected round duration very efficiently.

Theorem 4. *For all four conditions on forgetting, $\dfrac{d}{dp}\lambda(N, p)\big|_{p=1} = -N(N-1)$.*

Proof. Let $p \in (0, 1)$. Let $\pi_{N,p}(\hat{a})$ be the steady state probability of state $\hat{a}$ of Markov chain $\hat{a}(t)$. From Proposition 3, $1/\lambda(N, p) = \sum_{\hat{a}} p^{\#_{-1}(\hat{a})} \cdot \pi_{N,p}(\hat{a})$. Then

$$\frac{d}{dp} 1/\lambda(N, p) = \sum_{\hat{a}} \#_{-1}(\hat{a}) \cdot p^{\#_{-1}(\hat{a})-1} \cdot \pi_{N,p}(\hat{a}) + \sum_{\hat{a}} p^{\#_{-1}(\hat{a})} \cdot \frac{d}{dp} \pi_{N,p}(\hat{a}) .$$

Evaluation of the derivative at $p = 1$ leads to

$$\frac{d}{dp} 1/\lambda(N, p)\bigg|_{p=1} = \sum_{\hat{a}} \#_{-1}(\hat{a}) \cdot \pi_{N,1}(\hat{a}) + \sum_{\hat{a}} \frac{d}{dp} \pi_{N,p}(\hat{a})\bigg|_{p=1} .$$

Observe that as p goes to 1, $\pi_{N,p}(\hat{a})$ goes to 0 for all states $\hat{a}$, except for $\hat{a}_0$, the state with 0 in the diagonal, -1 everywhere else, and $\mathrm{Step}(\hat{a}) = 1$. It is $\#_{-1}(\hat{a}_0) = N(N-1)$. Moreover, as p goes to 1, $\pi_{N,p}(\hat{a}_0)$ approaches 1. Hence,

$$= N(N-1) + \frac{d}{dp}\left(\sum_{\hat{a}} \pi_{N,p}(\hat{a})\right)\Bigg|_{p=1} = N(N-1) + 0 \ ,$$

as the sum of the steady state probabilities over all states a equals 1. The theorem follows from $\frac{d}{dp}\lambda(N,p)\big|_{p=1} = -\frac{d}{dp}1/\lambda(N,p)\big|_{p=1} \cdot \lambda^2(N,1)$ and $\lambda(N,1) = 1$. $\qquad\square$

5.3   Behavior of λ for $p \to 0$

In systems with unreliable communication, in which Theorem 4 is not valuable, the following theorem on the asymptotic behavior of the expected round duration for all our conditions on forgetting, is useful. It turns out that λ^{I}, λ^{II}, and λ^{III} have the same order of growth for $p \to 0$, namely p^{-1}, while λ^{IV} has a higher order of growth.

Theorem 5. *For $p \to 0$, $\lambda^{\mathrm{I}}(N,p)$, $\lambda^{\mathrm{II}}(N,p)$ and $\lambda^{\mathrm{III}}(N,p)$ are in $\Theta(p^{-1})$, and $\lambda^{\mathrm{IV}}(N,p)$ is in $\Theta(p^{-(N-1)})$.*

Proof. We first show the statement for λ^{III}. It is $(1-p)^i - 1 = \sum_{j=1}^{i}\binom{i}{j}(-p)^j = \Omega(p)$ for $p \to 0$. Hence by Theorem 1, $\lambda^{\mathrm{III}}(N,p) = O(p^{-1})$ for $p \to 0$.

For all conditions on forgetting, all transition probabilities of the Markov chain $\hat{a}(t)$ are polynomials in p. Hence by Eq. 3, all steady state probabilities $\pi(\hat{a})$ are rational functions in p. Proposition 3 then in particular implies that $\lambda^{\mathrm{I}}(N,p)$ is also rational in p. Clearly, $\lambda^{\mathrm{I}}(N,p) \to \infty$ as $p \to 0$. Hence $\lambda^{\mathrm{I}}(N,p)$ has a pole at $p = 0$ of order at least 1. This implies $\lambda^{\mathrm{I}}(N,p) = \Omega(p^{-1})$. From the inequalities $\lambda^{\mathrm{I}} \leqslant \lambda^{\mathrm{II}} \leqslant \lambda^{\mathrm{III}}$, the first part of the theorem follows.

To show the asymptotic behavior of $\lambda^{\mathrm{IV}}(N,p)$, observe that by $(1-p)^i - 1 = -p\sum_{j=1}^{i}\binom{i}{j}(-p)^{j-1} \sim -p \cdot i$ for $p \to 0$ and by Proposition 2, we have

$$p \cdot \Lambda(M,p) \sim \sum_{i=1}^{M}\binom{M}{i}(-1)^{i+1}\frac{1}{i} \ .$$

As shown in the textbook by Graham et al. [4, (6.72) and (6.73)] this sum equals H_M, denoting the Mth harmonic number. This concludes the proof. $\qquad\square$

5.4   Lower Bounds on λ^{I} and λ^{II}

Determining the expected round duration for cases I and II by means of the Markov chain $a(t)$ is computationally intensive, even for small system sizes N. We can, however, compute efficient lower and upper bounds on $\lambda(N,p)$: For both, case I and II, $\lambda^{\mathrm{III}}(N,p)$ is an upper bound. We will next derive computationally feasible lower bounds for $\lambda^{\mathrm{I}}(N,p)$ and $\lambda^{\mathrm{II}}(N,p)$.

From Proposition 1 and Theorem 3 follows, by considering the conditional expectation of T_G:

$$\lambda = \frac{1}{\sum_{\hat{a}} \mathrm{Step}(\hat{a}) \cdot \pi(\hat{a})} \sum_{\hat{a}} \mathrm{Step}(\hat{a}) \cdot \pi(\hat{a}) \cdot \mathbb{E}(T_G(2) \mid \hat{a}(0) = \hat{a}) \ ,$$

where $\mathbb{E}(T_G(2) \mid \hat{a}(0) = \hat{a})$ is the expected time until the first global round switch, given that the system initially is in state $\hat{a}$. It holds that $\mathbb{E}(T_G(2) \mid \hat{a}(0) = \hat{a}) = \Lambda(\#_{-1}(\hat{a}), p)$.

Let $[n]$ denote the set of states $\hat{a}$ with $\#_{-1}(\hat{a}) = n$ and $\mathrm{Step}(\hat{a}) = 1$, and denote by $\bigcup[n]$ the union of all $[n]$ for $0 \leqslant n \leqslant N(N-1)$. Further let $\hat{\pi}(n) = \sum_{\hat{a} \in [n]} \pi(\hat{a})/(\sum_{\hat{a}} \mathrm{Step}(\hat{a}) \cdot \pi(\hat{a}))$. It follows that $\hat{\pi}(n) = 0$ for $n < 2N-2$ in case II and $\hat{\pi}(n) = 0$ for $n < N-1$ in case I.

The basic idea of the bounds on λ is to bound $\hat{\pi}(n)$. Let $\mathbb{P}(\hat{a} \leadsto [n])$ be the probability that, given the system is in state $\hat{a}$ at some time t, for the minimum time $t' > t$ at which a global round switch occurs, $\hat{a}(t') \in [n]$. We obtain for $\hat{\pi}(n)$:

$$\hat{\pi}(n) = \sum_{\hat{a}} \mathrm{Step}(\hat{a}) \cdot \hat{\pi}(\hat{a}) \cdot \mathbb{P}(\hat{a} \leadsto [n]) = \sum_{\hat{a} \in \bigcup[n]} \hat{\pi}(\hat{a}) \cdot \mathbb{P}(\hat{a} \leadsto [n])$$

$$= \sum_{\hat{a} \in [n]} \hat{\pi}(\hat{a}) \cdot \mathbb{P}(\hat{a} \leadsto [n]) + \sum_{\hat{a} \in \bigcup[n] \setminus [n]} \hat{\pi}(\hat{a}) \cdot \mathbb{P}(\hat{a} \leadsto [n])$$

$$\geqslant \hat{\pi}(n) \min_{\hat{a} \in [n]} \mathbb{P}(\hat{a} \leadsto [n]) + (1 - \hat{\pi}(n)) \min_{\hat{a} \in \bigcup[n] \setminus [n]} \mathbb{P}(\hat{a} \leadsto [n])$$

$$\geqslant \hat{\pi}(n) c_n + (1 - \hat{\pi}(n)) d_n$$

for c_n, d_n suitably chosen. One can derive valid choices for both parameters for cases I and II by excessive case inspection of transition probabilities for all state equivalence classes $[k]$, $k \geqslant 0$. We provide only a proof for case II in Sect. 5.5, as the proof for case I is by analogous arguments.

Partitioning the above sum into a one term from states in $[n]$ to states in $[n]$, and one remaining term, allows us to finally state inequality

$$\hat{\pi}(n) \geqslant \frac{d_n}{1 + d_n - c_n} =: \pi_n \ . \tag{4}$$

The resulting lower bounds on $\hat{\pi}(n)$, denoted by π_n^{I} and π_n^{II} for cases I and II respectively, finally yield lower bounds on λ. Since Λ is nondecreasing in its first argument, we can bound $\lambda(N,p)$ by

$$\left(1 - \sum_{n=N}^{N(N-1)} \pi_n^{\mathrm{I}}\right) \Lambda(N-1, p) + \sum_{n=N}^{N(N-1)} \pi_n^{\mathrm{I}} \Lambda(n, p) \leqslant \lambda^{\mathrm{I}}(N, p) \tag{5}$$

in case I. For case II we obtain

$$\left(1 - \sum_{n=2N-1}^{N(N-1)} \pi_n^{\mathrm{II}}\right) \Lambda(2N-2,p) + \sum_{n=2N-1}^{N(N-1)} \pi_n^{\mathrm{II}} \Lambda(n,p) \leqslant \lambda^{\mathrm{II}}(N,p) \ . \qquad (6)$$

5.5   Lower Bound on Parameters for λ^{II}

We next show how to derive bounds on parameters c_n and d_n, in the following denoted by d_n^{II} and c_n^{II}. From these we obtain bounds on $\pi_{N(N-1)}$ from (4).

We start our analysis with determining $\pi_{N(N-1)}$. Since $\mathbb{P}(\hat{a} \rightsquigarrow [N(N-1)])$ is greater than the probability that $\hat{a}(t+1) \in [N(N-1)]$, given that $\hat{a}(t) = \hat{a}$, for arbitrary t, we have $\mathbb{P}(\hat{a} \rightsquigarrow [N(N-1)]) \geqslant p^{\#_{-1}(\hat{a})}$. Thus we may choose $c_{N(N-1)}^{\mathrm{II}} = p^{N(N-1)}$, $d_{N(N-1)}^{\mathrm{II}} = p^{N(N-1)-1}$ and obtain

$$\pi_{N(N-1)} = \frac{p^{N(N-1)-1}}{1 + p^{N(N-1)-1}(1-p)} \ .$$

Next we turn to the analysis of $\pi_{N(N-1)-1}$. Since it is not possible to make a direct transition from a state $\hat{a} \in \bigcup[n]$ to a state in $[N(N-1)-1]$, we consider bounds on the probability that the system is in a state within $[N(N-1)-1]$ at time $t+2$, given that $\hat{a}(t) = \hat{a}$. Fix in $\hat{a}$ one column j whose all non-diagonal entries equal -1. Clearly such a column must exist, since $\mathrm{Step}(\hat{a}) = 1$. Given that $\hat{a}(t) = \hat{a}$, assume that at time $t+1$, all messages from processes $i \neq j$ to all processes i' with $K_{i',i}(t) = -1$, and one message from process j to some fixed $j' \neq j$, are received. That is, $N(N-2) + 2 - \#_0(\hat{a})$ messages are received. Moreover, at time $t+1$, k (up to $N-3$) of the remaining $N-2$ message sent by j are received. By construction, $k+2$ of the processes are 1-processes at time $t+1$. For $\hat{a}(t+1) \in [N(N-1)-1]$ to hold, it is sufficient that: For all 0-processes i with $\hat{a}_{i,j}(t+1) = -1$, process i must receive a message from j at time $t+2$; exactly one of the messages from a 1-process to a 1-process is received. Since at time $t+1$ there are $(k+2)(k+1)$ messages from 1-processes to 1-processes, we obtain: For all $\hat{a} \in \bigcup[n]$,

$$\mathbb{P}(\hat{a} \rightsquigarrow [N(N-1)-1]) \geqslant$$

$$\geqslant \sum_{k=0}^{N-3} \binom{N-2}{k} p^{N(N-2)+2-\#_0(\hat{a})+k} (1-p)^{N-2-k} \cdot$$

$$\cdot p^{N-2-k} \cdot p \cdot (1-p)^{(k+2)(k+1)-1} \cdot ((k+2)(k+1))$$

$$= p^{N(N-1)-\#_0(\hat{a})+1} \cdot$$

$$\cdot \sum_{k=0}^{N-3} \binom{N-2}{k} ((k+2)(k+1))(1-p)^{N+k^2+2k-1}$$

$$=: \beta(\#_0(\hat{a})) \ .$$

So we choose $c_{N(N-1)-1}^{\mathrm{II}} = \beta(1)$ and $d_{N(N-1)-1}^{\mathrm{II}} = \beta(0)$.

Finally we turn to the analysis of π_n for $n = 2(N - 1) + x$, where $0 \leqslant x \leqslant (N - 2)(N - 1) - 2$. Again we bound $\mathbb{P}(\hat{a} \rightsquigarrow [2(N - 1) + x])$, for $\hat{a} \in \bigcup[n]$, by analyzing the probability that $\hat{a}(t + 2) \in [2(N - 1) + x]$, given that $\hat{a}(t) = \hat{a}$. Fix a row j of $\hat{a}$ with T non-diagonal entries equal to 0. Given that $\hat{a}(t) = \hat{a}$, assume that at time $t + 1$, all messages to processes $i \neq j$ from all processes i' with $K_{i,i'}(t) \neq 0$ are received. That is, $(N - 1)(N - 1) - \#_0(\hat{a}) + T$ messages are received. Moreover, at time $t + 1$, k (up to $N - T - 2$) of the remaining $N - T - 1$ messages to j are received. Hence, all processes different from j are 1-processes at time $t + 1$. At time $t + 2$ all remaining messages to process j are received. From the $(N - 2)(N - 1)$ messages sent by 1-processes to 1-processes exactly x are not allowed to be received for $\hat{a}(t + 2) \in [2(N - 1) + x]$ to hold. Thus, for fixed row j and $\hat{a} \in \bigcup[n]$,

$$\mathbb{P}(\hat{a} \rightsquigarrow [2(N - 1) + x] \mid \text{row } j) \geqslant$$

$$\geqslant \sum_{k=0}^{N-2-T} \binom{N - T - 1}{k} p^{k + (N-1)^2 - \#_0(\hat{a}) + T}$$

$$\cdot (1 - p)^{N-1-k-T} p^{N-1-k-T} p^{(N-2)(N-1)-x} (1 - p)^x$$

$$\cdot \binom{(N - 1)(N - 2)}{x}$$

$$= \binom{(N - 1)(N - 2)}{x} (1 - p)^x p^{N(N-1) - \#_0(\hat{a}) + (N-2)(N-1) - x}$$

$$\cdot ((2 - p)^{N-1-T} - 1)$$

$$=: \gamma(\#_0(\hat{a}), T, x) .$$

Note that γ is nonincreasing in its second and third argument. Every state $\hat{a}$ has at least one row with $T = 0$ non-diagonal entries equal to 0. All other rows must have $T \leqslant N - 2$ non-diagonal entries equal to 0, since a row must have at least one entry equal to -1. Thus, we have

$$\mathbb{P}(\hat{a} \rightsquigarrow [2(N - 1) + x]) \geqslant$$
$$\gamma(\#_0(\hat{a}), 0, x) + (N - 1) \cdot \gamma(\#_0(\hat{a}), N - 2, x) =: \tilde{\gamma}(\#_0(\hat{a}), x).$$

We thus choose $c^{II}_{2(N-1)+x} = \tilde{\gamma}((N - 1)(N - 2) - x, x)$ and $d^{II}_{2(N-1)+x} = \tilde{\gamma}(0, x)$. The lower bound on λ^{II} follows from (4) and (6).

6   Discussion of Results

In this section we present the results obtained by calculating the expected round duration λ for the four conditions on forgetting that we consider. Additionally, we used Monte-Carlo simulations to estimate λ and the average age of data when a process performs a round switch.

Figure 3 shows, with varying probability p, the exact value of the expected round duration for conditions on forgetting I–IV in a system with $N = 3$

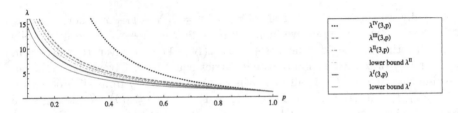

Fig. 3. Expected round durations for $N = 3$ and lower bounds for cases I and II.

processes. As stated in Sect. 5.3, the figure shows the gap between the cases I, II, and III, having an asymptotic growth in $\Theta(1/p)$ when p approaches 0, and the case IV, which has an asymptotic growth in $\Theta(1/p^{N-1})$. Furthermore, as depicted in Sect. 5.2, all the plots have the same slope in the point $p = 1$ resulting in a good approximation for the hard to calculate cases I and II in a system with reliable communication.

In settings with unreliable communication, for which the approximation result on the derivative of λ at $p = 1$ is not valuable, cases I and II can be approximated by their analytical lower bounds (Sect. 5.4), and bounded from above by the λ for case III (Theorem 1). A comparison between the lower bounds and the actual systems is illustrated in Fig. 3.

As the calculations of the exact values for the expected round duration using the Markov chain model are computationally very expensive, we used Monte-Carlo simulations to compare them with our calculations. To this end, we simulated systems with $2 \leqslant N \leqslant 12$ processes for 100 000 steps and averaged over 30 runs. The simulations were done using three different values for p. Figures 4 and 5 show the obtained average round durations with the calculated lower bound and with case III as upper bound. The average round durations for case I (where processes never forget) is shown in Fig. 4(a) – (c) and the case II (where processes forget after a local round switch) is shown in Fig. 5(a) – (c). Figure 6(a) – (c) depict the calculated expected round duration for case IV, i.e., the synchronizer variant that forgets at each time step. Note that it is significantly higher than all the other variants when message loss is considerable.

Figure 7 shows Monte-Carlo simulation results of the average age of data when a process performs a round switch, for cases I and II, both of which can be implemented in a distributed manner. Case IV, for which the same holds, by definition has an average age of data of 1. One immediately observes that while the average age of both cases I and II is significantly higher than in case IV, forgetting at each processes' round switch only has a marginal effect on the average age compared to not forgetting at all.

7 Conclusion

We studied the effect of actively discarding memory content on a variant of the α-synchronizer. For practically relevant applications, e.g., in sensor fusion,

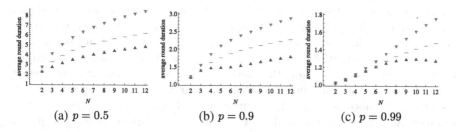

(a) $p = 0.5$ (b) $p = 0.9$ (c) $p = 0.99$

Fig. 4. Monte-Carlo simulation results for case I compared against the calculated lower bound and the calculated expected round duration of case III serving as an upper bound.

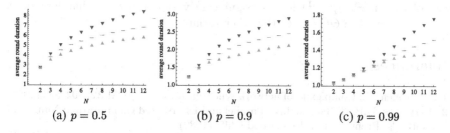

(a) $p = 0.5$ (b) $p = 0.9$ (c) $p = 0.99$

Fig. 5. Monte-Carlo simulation results for case II compared against the calculated lower bound and the calculated expected round duration of case III serving as an upper bound.

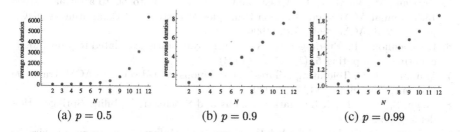

(a) $p = 0.5$ (b) $p = 0.9$ (c) $p = 0.99$

Fig. 6. Calculated expected round duration of case IV.

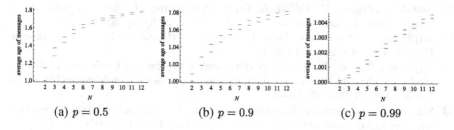

(a) $p = 0.5$ (b) $p = 0.9$ (c) $p = 0.99$

Fig. 7. Average age Monte-Carlo simulation results for case I (blue, upper) and II (green, lower) (Colour figure online).

forgetting turns out to be a simple strategy to decrease the average age of data handed over to the application layer when a process makes a round switch. In case the accuracy of data degrades over time, e.g., data samples taken from a timed process, a decreased average age of the samples results in an increased accuracy of the merged data. To assess the inevitable drawback of forgetting strategies, namely degraded performance, we analyzed four naturally chosen strategies of forgetting. We obtained analytic formulas for the behavior of the expected round duration $\lambda(N, p)$ as the probability of successful transmission $p \to 0$ and $p \to 1$, as well as means to calculate $\lambda(N, p)$ for arbitrary N and p, allowing to assess whether the resulting loss of performance is acceptable for a specific application. Interestingly, it turned out that the behavior of all four variants is similar for $p \to 1$. For $p \to 0$ only two asymptotic behaviors of the expected round duration were observed: $\Theta(1/p^{N-1})$ for the significantly slower variant that forgets at each time step, and $\Theta(1/p)$ for the other variants.

References

1. Awerbuch, B.: Complexity of network synchronization. J. ACM **32**, 804–823 (1985)
2. Bertsekas, D.P., Tsitsiklis, J.N.: Parallel and Distributed Computation: Numerical Methods. Prentice Hall, Englewood Cliffs (1989)
3. Fagin, R., Halpern, J.Y., Moses, Y., Vardi, M.Y.: Reasoning About Knowledge. MIT Press, Cambridge (1995)
4. Graham, R.L., Knuth, D.E., Patashnik, O.: Concrete Mathematics. Addison-Wesley, Reading (1989)
5. Jayaram, M., Varghese, G.: Crash failures can drive protocols to arbitrary states. 15th Annual ACM Symposium on Principles of Distributed Computing (PODC), pp. 247–256. ACM, New York (1996)
6. Kirschenhofer, P., Prodinger, H.: A result in order statistics related to probabilistic counting. Computing **51**(1), 15–27 (1993)
7. Marzullo, K.A.: Tolerating failures of continuous-valued sensors. ACM Trans. on Comput. Syst. **8**(4), 284–304 (1990)
8. Meyn, S., Tweedie, R.L.: Markov Chains and Stochastic Stability. Springer, Heidelberg (1993)
9. Nakamura, E.F., Loureiro, A.A.F., Frery, A.C.: Information fusion for wireless sensor networks: methods, models, and classifications. ACM Comput. Surv. **39**(3), 9 (2007). doi:10.1145/1267070.1267073
10. Nowak, T., Függer, M., Kößler, A.: On the Performance of a Retransmission-Based Synchronizer. Theor. Comput. Sci. (2012). doi:10.1016/j.tcs.2012.04.035
11. Rajsbaum, S.: Upper and lower bounds for stochastic marked graphs. Inform. Process. Lett. **49**, 291–295 (1994)
12. Rajsbaum, S., Sidi, M.: On the performance of synchronized programs in distributed networks with random processing times and transmission delays. IEEE Trans. Parall. Distr. **5**, 939–950 (1994)
13. Schmid, U., Schossmaier, K.: How to reconcile fault-tolerant interval intersection with the Lipschitz condition. Distrib Comp. **14**(2), 101–111 (2001)
14. Szpankowski, W., Rego, V.: Yet another application of a binomial recurrence order statistics. Computing **43**(4), 401–410 (1990)

On the Complexity of Barrier Resilience for Fat Regions

Matias Korman[1]([⊠]), Maarten Löffler[2],
Rodrigo I. Silveira[1,3], and Darren Strash[4]

[1] Departmento de Matemàtica Aplicada II, Universitat Politècnica de Catalunya,
Barcelona, Spain
{matias.korman,rodrigo.silveira}@upc.edu
[2] Department of Computing and Information Sciences, Utrecht University,
Utrecht, The Netherlands
m.loffler@uu.nl
[3] Departmento de Matemática, Universidade de Aveiro, Aveiro, Portugal
rodrigo.silveira@ua.pt
[4] Computer Science Department, University of California, Irvine, USA
dstrash@uci.edu

Abstract. In the *barrier resilience* problem (introduced by Kumar *et al.*, Wireless Networks 2007), we are given a collection of regions of the plane, acting as obstacles, and we would like to remove the minimum number of regions so that two fixed points can be connected without crossing any region. In this paper, we show that the problem is NP-hard when the regions are arbitrarily fat regions (even when they are axis-aligned rectangles of aspect ratio $1 : (1 + \varepsilon)$). We also show that the problem is fixed-parameter tractable (FPT) for such regions. Using our FPT algorithm, we show that if the regions are β-fat and their arrangement has bounded ply Δ, there is a $(1 + \varepsilon)$-approximation that runs in $O(2^{f(\Delta,\varepsilon,\beta)} n^7)$ time, where $f \in O(\frac{\Delta^2 \beta^6}{\varepsilon^4} \log(\beta \Delta/\varepsilon))$.

1 Introduction

The *barrier resilience* problem asks for the minimum number of spatial regions from a collection $\mathcal{D}$ that need to be removed, such that two given points p and q are in the same connected component of the complement of the union of the remaining regions. This problem was posed originally in 2005 by Kumar et al [11,12], motivated from sensor networks. In their formulation, the regions are unit disks (sensors) in some rectangular strip $B \subset \mathbb{R}^2$, where each sensor is able to detect movement inside its disk. The question is then how many sensors need to fail before an entity can move undetected from one side of the strip to the other (that is, how *resilient* to failure the sensor system is). Kumar *et al.* present a polynomial time algorithm to compute the resilience in this case. They also consider the case where the regions are disks in an annulus, but their approach cannot be used in that setting.

P. Flocchini et al. (Eds.): ALGOSENSORS 2013, LNCS 8243, pp. 201–216, 2014.
DOI: 10.1007/978-3-642-45346-5_15, © Springer-Verlag Berlin Heidelberg 2014

1.1   Related Work

Despite the seemingly small change from a rectangular strip to an annulus, the second problem still remains open, even for the case in which regions are unit disks in $\mathbb{R}^2$. There has been partial progress towards settling the question: Bereg and Kirkpatrick [2] present a factor 5/3-approximation algorithm for the unit disk case. Afterwards, Alt *et al.* [1] and Tseng and Kirkpatrick [16] independently showed that if the regions are line segments in $\mathbb{R}^2$, the problem is NP-hard. We also note that Tseng and Kirkpatrick [16] also sketched how to extend their proof for the case in which the input consists of (translated and rotated) copies of a fixed square or ellipse, but no formal proof was given.

The problem of covering barriers with sensors is very current and has received a lot of attention in the sensor network community (e.g. [3,4,8]). In the algorithms community, closely related problems involving region intersection graphs have also became quite popular. Gibson *et al.* [7] study the opposite problem of ours: compute the maximum number of disks one can remove such that p and q are still separated.

1.2   Results

We present constructive algorithms for two natural restricted variants of the problem. In Sect. 3 we show that the problem is fixed-parameter tractable on the resilience when the regions satisfy an upper bound on the *fatness* [6] (intuitively speaking, the regions must have some resemblance to a unit disk). In Sect. 4 we also show that if the collection of regions has bounded *ply* [14] (that is, sensors are more or less evenly distributed in the plane), the FPT result can be used to obtain an approximation scheme. In particular, the constructive results apply to the original unit disk coverage setting (formal definitions of fatness and ply are given in the corresponding sections).

As a complement to these algorithms, in Sect. 5 we show that the problem is NP-hard even when the input is a collection of arbitrary fat regions in $\mathbb{R}^2$. The result holds even if regions consist of axis-aligned rectangles of aspect ratio $1 : 1 + \varepsilon$ and $1 + \varepsilon : 1$.

Our results rely on tools and techniques from both computational geometry and graph theory.

Due to lack of space, several proofs and extensions have been omitted. A full version of the paper containing all the omitted details can be found in [10].

2   Preliminaries

We denote with p and q the points that need to be connected, and with $\mathcal{D}$ the set of regions that represent the sensors. To simplify the presentation of our results, we make the following general position assumption: all intersections between boundaries of regions in $\mathcal{D}$ consist of isolated points. We say that a collection of objects in the plane are *pseudodisks* if the boundaries of any two of them intersect at most twice.

We formally define the concepts of *resilience* and *thickness* introduced in [2]. The *resilience of a path* π between two points p and q, denoted $r(\pi)$, is the number of regions of $\mathcal{D}$ intersected by π. Given two points p and q, the *resilience of p and q*, denoted $r(p,q)$, is the minimum resilience over all paths connecting p and q. In other words, the resilience between p and q is the minimum number of regions of $\mathcal{D}$ that need to be removed to have a path between p and q that does not intersect any region of $\mathcal{D}$. From now on, we assume that neither p nor q are contained in any region of $\mathcal{D}$. Note that such regions must always be counted in the minimum resilience paths, hence we can ignore them (and update the resilience we obtain accordingly).

Often it will be useful to refer to the arrangement (i.e., the subdivision of the plane into faces, see e.g. [5] for a formal definition) induced by the regions of $\mathcal{D}$, which we denote by $\mathcal{A}(\mathcal{D})$. Based on this arrangement we define a weighted dual graph $G_{\mathcal{A}(\mathcal{D})}$ as follows. There is one vertex for each cell (i.e., face) of $\mathcal{A}(\mathcal{D})$. Each pair of neighboring cells A, B is connected in $G_{\mathcal{A}(\mathcal{D})}$ by two directed edges, (A, B) and (B, A). The weight of an edge is 1 if, when traversing from the starting cell to the destination one, we enter a region of $\mathcal{D}$ (or 0 if we leave a region[1]).

The *thickness* of a path π between p and q, denoted $t(\pi)$, equals the number of sensor region intersections of π (possibly counting the same region multiple times). Given two points p and q, the *thickness of p and q*, denoted $t(p,q)$, is the value $|\mathfrak{p}_{G_{\mathcal{A}(\mathcal{D})}}(p,q)| + \Delta(p)$, where $\mathfrak{p}_{G_{\mathcal{A}(\mathcal{D})}}(p,q)$ is a shortest path in $G_{\mathcal{A}(\mathcal{D})}$ from the cell of p to the cell of q, and $\Delta(p)$ equals the number of regions that contain p. Also note that the resilience (or thickness) between two points only depends on the cells to which the points belong to. Hence, we can naturally extend the definitions of thickness to encompass two cells of $\mathcal{A}(\mathcal{D})$, or a cell and a point.

Note that thickness and resilience can be different (since entering the same region several times has no impact on the resilience, but is counted every time for the thickness). In fact, the thickness between two points can be efficiently computed in polynomial time using any shortest path algorithm for weighted graphs (for example, using Dijkstra's algorithm). However, as we will see later, the thickness (and the associated shortest path) will help us find a path of low resilience.

Throughout the paper we often use the following fundamental property of disks, already observed in [2]. In the statement below, "well-separated" is in the sense used in [2]—i.e., the distance between p and q is at least $2\sqrt{3}$.[2]

Lemma 1 ([2], Lemma 1). *Let $\mathcal{D}$ be a set of unit disks, and let π^* be a path from p to q of minimum resilience. If p, q are well-separated, then π^* encounters no disk of $\mathcal{D}$ more than twice.*

Corollary 1 ([2]). *When the regions of $\mathcal{D}$ are unit disks, the thickness between two well-separated points is at most twice their resilience.*

---

[1] Note that no other option is possible under our general position assumption.

[2] Note that the well-separatedness of p and q is used to prove a factor 2 instead of 3. Everything still works for ill-separated points, at a slight increase of the constants. Our most general statements for β-fat regions do not make this requirement.

3   Fixed-Parameter Tractability

In this section we introduce a single-exponential fixed-parameter tractable (FPT) algorithm, where the parameter is the length of the optimal solution. Thus, our aim is to obtain an algorithm that given a problem instance, determines whether or not there is a path of resilience r between p and q, and runs in $O(2^{f(r)}n^c)$ time for some constant c and some polynomial function f.

For clarity we first explain the algorithm for the special case of unit disks. Afterwards, in Sect. 3.2, we show how to adapt the solution to the case in which $\mathcal{D}$ is a collection of β-fat objects. Note that for treating the case of unit disk regions we assume that p and q are well-separated, so we can apply Lemma 1. This requirement is afterwards removed in Sect. 3.2.

First we give a quick overview of the method of Kumar *et al.*[11] for open belt regions. Their idea consists in considering the intersection graph of $\mathcal{D}$ together with two additional artificial vertices s,t with some predefined adjacencies. There is a path from the bottom side to the top side of the belt if and only if there is no path between s and t in the graph. Hence, computing the resilience of the network is equal to finding a minimum vertex cut between s and t.

Our approach is to find a low thickness path that passes through p and q, cut open through it, and transform the problem instance into one with something similar to an open belt region. We then follow the approach of Kumar *et al.* taking into account that the right and left boundaries of our region correspond to the same point. Hence, instead of using a regular vertex cut, we will use a vertex multicut [17].

Consider the shortest path τ between the cells containing p and q in $G_{\mathcal{A}(\mathcal{D})}$, let t be the number of traversed disks (recall that we assumed that p is not contained in any region, hence this number is exactly the *thickness* of p and q). We observe that cells with high thickness to p or q can be ignored when we look for low resilience paths.

Lemma 2. *The minimum resilience path between p and q cannot traverse cells whose thickness to p or q is larger than $1.5t$.*

Proof. We argue about thickness to p; the argument with respect to q is analogous. Let ρ be a path of minimum resilience between p and q, and let r be the resilience of ρ. Recall that ρ does not enter a disk more than twice, hence the thickness of ρ is at most $2r \leq 2t$. Assume, for the sake of contradiction, that the thickness of some cell C traversed by ρ is greater than $1.5t$. Let ρ_C be the portion of ρ from C to q. By the triangle inequality, the thickness of ρ_C is less than $0.5t$. However, by concatenating τ and ρ_C we would obtain a path that connects p with C whose thickness is less than $1.5t$, giving a contradiction. $\boxtimes$

Let R be the union of the cells of the arrangement that have thickness from p at most $1.5t$; we call R the *domain* of the problem. Observe that R is connected, but need not be simple (see Fig. 1(a)). By the previous result, cells that do not belong to R can be discarded, since they will never belong to a path of resilience r. Note that the number of cells remaining in R might still be quadratic, hence

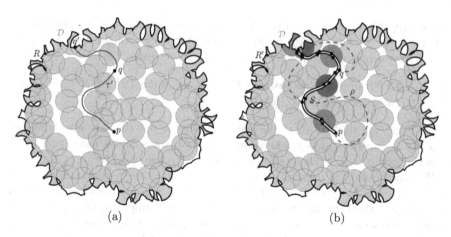

Fig. 1. (a) The reduced domain R, and a path τ' from p to q' via q. (b) After cutting along τ', we get the domain R'. We add a set S of extra vertices on the boundary of R', and we now have two copies of q. A crossing pattern, consisting of a topological path ρ (defined by the sequence of points of S it passes) and a binary assignment to the disks of $\mathcal{D}$ intersected by τ', is also shown.

asymptotically speaking the instance size has not decreased (the purpose of this pruning will become clear later). We extend the minimum resilience path τ from q until a point q' on the boundary of the domain. Let τ' denote the extended path (Fig. 1(a)).

Lemma 3. *There exists a path τ' of minimum resilience from p to a point q' on the boundary of R via q, whose thickness is at most $1.5t$.*

Proof. Consider any shortest path tree from p in the dual graph of the reduced domain R, defined as the corresponding subgraph of $G_{\mathcal{A}(\mathcal{D})}$. All leaves of the tree correspond to cells on the boundary of R, which by definition have thickness at most $1.5t$ from p. Therefore all other cells in the tree lie on a path from p to a boundary cell that has length exactly $1.5t$, including q.   ⊠

We "cut open" through τ', removing the cut region from our domain. Note that cells that are traversed by τ' are split by two copies of the same Jordan curve (Fig. 1(b)). After this cut we have two paths from p to q. We arbitrarily call them the left and right paths. Consider now a minimum resilience path denoted ρ; let $r = r(\rho)$ denote its resilience. This path can cross τ' several times, and it can even coincide with τ' in some parts (shared subpaths). Although we do not know how and where these crossings occur, we can *guess* (i.e., try all possibilities) the topology of ρ with respect to τ'. For each disk that τ' passes through, we either remove it (at a cost of 1) or we make it an obstacle. That way we explicitly know which of the regions traversed by τ' could be traversed by ρ. Additionally, we guess how many times ρ and τ' share part of their paths (either for a single crossing in one cell, or for a longer shared subpath). For each shared subpath, we guess from which cell ρ arrives and leaves (and if the entry or exit was from the left or right path). We call each such configuration a *crossing pattern* between τ' and ρ. Figure 1(b) illustrates a crossing pattern.

Lemma 4. *For any problem instance $\mathcal{D}$, there are at most $2^{4r \log r + o(r \log r)}$ crossing patterns between τ' and ρ, where $r = r(\rho)$.*

Proof. First, for all disks in τ', we guess whether or not they are also traversed by ρ. By Lemma 3, τ' has thickness at most $1.5t$, there are at most such many disks (hence up to $2^{1.5t}$ choices for which disks are traversed by ρ).

We now bound the number of (maximal) shared subpaths between ρ and τ': recall that ρ passes through exactly $r = r(\rho)$ disks, and visits each disk at most twice. Hence, there cannot be more than $2r$ shared subpaths. Observe that τ' cannot traverse many cells of $\mathcal{A}(\mathcal{D})$: when moving from a cell to an adjacent one, we either enter or leave a disk of $\mathcal{D}$. Since we cannot leave a disk we have not entered and τ' has thickness at most $1.5t$, we conclude that at most $3t$ cells will be traversed by τ' (other than the starting and ending cells).

For each shared subpath we must pick two of the cells traversed in τ' (as candidates for first and last cell in the subpath). By the previous observation there are at most $3t$ candidates for first and last cell (since that is the number of cells traversed by τ'). Additionally, for each shared subpath we must determine from which side ρ entered and left the subpath (four options in total). Since these choices are independent, in total we have at most $2r \times (3t \times 3t \times 4)^{2r} = 2r \cdot 36^{2r} \cdot t^{4r}$ options for the number of crossing patterns. Combining both bounds and using the fact that $t \leq 2r$, we obtain:

$$2^{1.5t} \cdot 2r \cdot 36^{2r} \cdot t^{4r} \leq 2^{5r} \cdot 2r \cdot 36^{2r} \cdot (2r)^{4r}$$
$$= 2^{9r+1+\log r + 2r \log 36 + 4r \log r}$$
$$= 2^{4r \log r + o(r \log r)}$$

$\boxtimes$

Note that the bound is very loose, since most of the choices will lead to an invalid crossing pattern. However, the importance of the lemma is in the fact that the total number of crossing patterns only depends on r.

Our FPT algorithm consists in considering all possible crossing patterns, finding the optimal solution for a fixed crossing pattern, and returning the solution of smallest resilience. From now on, we assume that a given pattern has been fixed, and we want to obtain the path of smallest resilience that satisfies the given pattern. If no path exists, we simply discard it and associate infinite resilience to it.

3.1 Solving the Problem for a Fixed Crossing Pattern

Recall that the crossing pattern gives us information on how to deal with the disks traversed by τ'. Thus, we remove all cells of the arrangement that contain

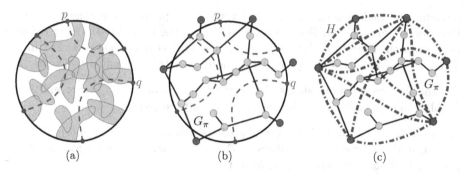

Fig. 2. (a) We may schematically represent W as a circle, since the geometry no longer plays a role. Partial paths are shown dashed. (b) The intersection graph of the regions after adding extra vertices for boundary pieces between points of $S \cup \{p, q\}$, shown green (Colour figure online). (c) The secondary graph H, representing the forbidden pairs.

one or more disks that are forbidden to ρ. Similarly, we remove from $\mathcal{D}$ the disks that ρ must cross. After this removal, several cells of our domain may be merged.

Since we do not use the geometry, we may represent our domain by a disk W (possibly with holes). After the transformation, each remaining region of $\mathcal{D}$ becomes a pseudodisk, and ρ becomes a collection of disjoint partial paths, each of which has its endpoints on the boundary of W (see Fig. 2(a)). To solve the subproblem associated with the crossing pattern we must remove the minimum number of disks so that all partial paths are feasible.

We consider the intersection graph G_I between the remaining regions of $\mathcal{D}$. That is, each vertex represents a region of $\mathcal{D}$, and two vertices are adjacent if and only if their corresponding regions intersect. Similarly to [11], we must augment the graph with boundary vertices. The partial paths split the boundary of R into several components. We add a vertex for each component (these vertices are called *boundary vertices*). We connect each such vertex to vertices corresponding to pseudodisks that are adjacent to that piece of boundary (Fig. 2(b)). Let $G_{\mathcal{X}} = (V_{\mathcal{X}}, E_{\mathcal{X}})$ be the resulting graph associated to crossing pattern $\mathcal{X}$. Note that no two boundary vertices are adjacent.

We now create a secondary graph H as follows: the vertices of H are the boundary vertices of $G_{\mathcal{X}}$. We add an edge between two vertices if there is a partial path that separates the vertices in $G_{\mathcal{X}}$ (Fig. 2(c)). Two vertices connected by an edge of H are said to form a *forbidden pair* (each partial path that would create the edge is called a *witness* partial path). We first give a bound on the number of forbidden pairs that H can have.

Lemma 5. *Any crossing pattern has at most $2r^2 + r$ forbidden pairs.*

Proof. Observe that $G_{\mathcal{X}}$ only adds edges between boundary vertices. Thus, it suffices to show that $G_{\mathcal{X}}$ has at most $2r + 1$ boundary vertices. Since partial paths cannot cross, each such path creates a single cut of the domain. This cut introduces a single additional boundary vertex (except the first partial path that introduces two vertices). Recall that we can map the partial paths to crossings between paths τ' and ρ and, by Lemma 4, these paths can cross at most $2r$ times. Thus, we conclude that there cannot be more than $2r + 1$ boundary vertices. ☒

The following lemma shows the relationship between the vertex multicut problem and the minimum resilience path for a fixed pattern.

Lemma 6. *There are k vertices of $G_{\mathcal{X}}$ whose removal disconnects all forbidden pairs if and only if there are k disks in $\mathcal{D}$ whose removal creates a path between p and q that obeys the crossing pattern $\mathcal{X}$.*

Proof. Let $\mathcal{A}'$ be the regions of $\mathcal{A}(\mathcal{D})$ inside R that are not covered by any disk after the k disks have been removed and let R' be their union. By definition, there is a path between p and q with the fixed crossing pattern if all partial paths are feasible (i.e., there exists a path connecting the two endpoints that is totally within R'). The reasoning for each partial path is analogous to the one used by Kumar *et al.* [11]. If all partial paths are possible, then no forbidden pair can remain connected in $G_{\mathcal{X}}$, since—by definition—each forbidden pair disconnects at least one partial path (the witness path). On the other hand, as soon as one forbidden pair remains connected, there must exist at least one partial path (the witness path) that crosses the forbidden pair. Thus if a forbidden path is not disconnected, there can be no path connecting p and q for that crossing pattern.   ☒

That is, thanks to Lemma 6, we can transform the barrier resilience problem to the following one: given two graphs $G = (V, E)$, and $H = (V, E')$ on the same vertex set, find a set $D \subset V$ of minimum size so that no pair $(u, v) \in E'$ is connected in $G \setminus D$. This problem is known as the (vertex) *multicut* problem [17]. Although the problem is known to be NP-hard if $|E'| > 2$ [9], there exist several FPT algorithms on the size of the cut and on the size of the set E' [13, 17]. Among others, we distinguish the method of Xiao ([17], Theorem 5) that solves the vertex multicut problem in roughly $O((2k)^{k+\ell/2} n^3)$ time, where k is the number of vertices to delete, $\ell = |E'|$, and n is the number of vertices of G.

Theorem 1. *Let $\mathcal{D}$ be a collection of unit disks in $\mathbb{R}^2$, and let p and q be two well-separated points. There exists an algorithm to test whether $r(p, q) \leq r$, for any value r, and if so, to compute a path with that resilience, in $O(2^{f(r)} n^3)$ time, where $f(r) = r^2 \log r + o(r^2 \log r)$.*

Proof. Recall that our algorithm considers all possible crossings between ρ and τ'. For any fixed crossing pattern $\mathcal{X}$, our algorithm computes $G_{\mathcal{X}}$, and all associated forbidden pairs. We then execute Xiao's FPT algorithm [17] for solving the vertex multicut problem. By Lemma 6, the number of removed vertices (plus the number of disks that were forced to be deleted by $\mathcal{X}$) will give the minimum resilience associated with $\mathcal{X}$.

Regarding the running time, the most expensive part of the algorithm is running an instance of the vertex multicut problem for each possible crossing pattern. Observe that the parameters k and ℓ of the vertex multicut problem are bounded by functions of r as follows: $k \leq r$ and $\ell \leq 2r^2 + r$ (the first claim is direct from the definition of resilience, and the second one follows from Lemma 5). Hence, a single instance of the vertex multicut problem will need $O((2r)^{r+(2r^2+r)/2}n^3) = O(2^{(1+\log r)(r^2+1.5r)}n^3) = O(2^{r^2 \log r + o(r^2 \log r)}n^3)$ time. By Lemma 4 the number of crossing patterns is bounded by $2^{4r \log r + o(r \log r)}$. Thus, by multiplying both expressions we obtain the bound on the running time, and the theorem is shown.   ⊠

We remark that the importance of this result lies in the fact that an FPT algorithm exists. Hence, although the dependency on r is high, we emphasize that the bounds are rather loose. We also note that both the minimum resilience path and the disks to be deleted can be reported.

3.2   Extension to Fat Regions

We now generalize the algorithm to similarly-sized β-fat regions. A region D is β-fat if there exist two concentric disks C and C' whose radii differ by at most a factor β, such that $C \subseteq D \subseteq C'$ (whenever the constant β is not important, the region D is simply called *fat*). Since we need the regions to be of similar size, we assume without loss of generality that the radius of C is 1 and the radius of C' is β; in this case we will call D a β-*fat unit region*. For the purpose, we must extend Lemmas 1, 2, 3, 4 and 5 to consider β-fat unit regions. The dependency of β in most of the Lemmas is quadratic (see details in the full version [10]), but the rest of the algorithm remains unchanged: the only property of unit disks that is still used is the fact that they are connected, to be able to phrase the problem as a vertex cut in the region intersection graph.

Theorem 2. *Let $\mathcal{D}$ be a collection of n connected β-fat unit regions in $\mathbb{R}^2$, and let p and q be two points. Let r be a parameter. There exists an algorithm to test whether $r(p,q) \leq r$, and if so, to compute a path with that resilience, in $O(2^{f(\beta,r)}n^3)$ time, where $f(\beta,r) \in O(\beta^4 r^2 \log(\beta r))$.*

4 $(1 + \varepsilon)$-Approximation

The arrangement formed by a collection of regions $\mathcal{D}$ is said to have bounded ply Δ if no point $p \in \mathbb{R}^2$ is contained in more than Δ elements of $\mathcal{D}$. In this section we present an efficient polynomial-time approximation scheme (EPTAS) for computing the resilience of an arrangement of disks of bounded ply Δ.

The general idea of the algorithm is very simple: first, we compute all pairs of regions that can be reached by removing at most k disks, for $k = \lceil 4\Delta/\varepsilon^2 \rceil$. Then, we compute a shortest path in the dual graph of the arrangement of regions, augmented with extra edges. We prove that the length of the resulting path is a $(1 + \varepsilon)$-approximation of the resilience.

As in the previous section, we first consider the case in which $\mathcal{D}$ is a set of n unit disks in $\mathbb{R}^2$ of ply Δ. Let $\mathcal{A}(\mathcal{D})$ be the arrangement induced by the regions of $\mathcal{D}$, and let $G_{\mathcal{A}(\mathcal{D})}$ be the dual graph of $\mathcal{A}(\mathcal{D})$. Recall that $G_{\mathcal{A}(\mathcal{D})}$ has a vertex for every cell of $\mathcal{A}(\mathcal{D})$, and a directed edge between all pairs of adjacent cells of cost 1 when entering a disk, and cost 0 when leaving a disk. For any given k, let G_k be the graph obtained from $G_{\mathcal{A}(\mathcal{D})}$ by adding, for each pair of cells $A, B \in \mathcal{A}(\mathcal{D})$ with resilience at most k, a *shortcut edge* $\overrightarrow{AB}$ of cost $r(A, B)$.

For a pair of cells of $\mathcal{A}(\mathcal{D})$, we can test whether $r(A, B)$ is smaller than k, and if it is, compute it, in $O(2^{f(k)}n^3)$ time (where $f(k) = r^2 \log r + o(r^2 \log r)$) by applying Theorem 1 to a point $p \in A$ and a point $q \in B$. Since there are $O(n^2)$ cells in $\mathcal{A}(\mathcal{D})$, we can compute G_k by doing this $O(n^4)$ times, leading to a total running time of $O(2^{f(k)}n^7)$. Observe that this running time is polynomial in n, and exponential in k. In particular, it is an EPTAS since $k = 4\Delta/\varepsilon^2$. Again, we emphasize that the bounds are loose, and that our objective is to show the existence of an EPTAS to the resilience problem.

4.1   Analysis

Lemma 7. *Let $D \in \mathcal{D}$, where $\mathcal{A}(\mathcal{D})$ has ply Δ, and let s, t be any two points inside D. Then the resilience between s and t in D is at most Δ.*

Proof. Let c be the number of disks containing either s or t ($c \geq 1$, since D contains both points). These c disks clearly must be removed. Now we analyze what other disks, not containing neither s nor t, may need to be removed too. For each other disk D_1 (not containing both s and s) that needs to be removed in an optimal solution, there must be another disk D_2 that intersects D_1 and, together, separate s and t inside D. We call such a pair of disks a *separating pair*.

Thus if the resilience is $(c + c')$, there must be at least c' *disjoint* separating pairs intersecting D. Moreover, since disks have unit-size, if two disks form a separating pair, at least one of them must intersect the center of D. Figure 3 illustrates the argument.

Since the ply of $\mathcal{A}(\mathcal{D})$ is Δ, this implies that there can be at most $\Delta - c$ separating pairs, and thus the resilience is at most Δ.   ⊠

The previous lemma implies that in an optimal resilience path, if a disk appears twice, its two occurrences cannot be more than 2Δ apart (when counting the cells traversed by the path between the two occurrences of the disk).

To prove our result it will be convenient to focus on the sequence of disks encountered by a path when going from p to q. It turns out that such problem is essentially a string problem, where each symbol represents a disk encountered by the path. In that context, the thickness will be equivalent to the number of symbols of the string (recall that we assume that p is not contained in any disk), and the resilience to the number of distinct symbols.

Let $S = \langle s_1 \dots s_n \rangle$ be a string of n symbols from some alphabet $\mathfrak{A}$, such that no symbol appears more than twice. Let T be a substring of S. We define $\ell(T)$ to be the length of T, and $d(T)$ to be the number of distinct symbols in T. Clearly,

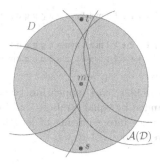

Fig. 3. The point m is contained in at least $c' + 1$ disks of $\mathcal{A}(\mathcal{D})$. Therefore, one of the neighboring cells has ply at least $c' + c + 1$

$\frac{1}{2}\ell(T) \leq d(T) \leq \ell(T)$. Let σ and k be two fixed integers such that $\sigma < k$. We define the *cost* of a substring T of S to be:

$$\psi(T) = \begin{cases} \sigma & \text{if } T = \langle \mathfrak{a}\lambda\mathfrak{a}\rangle \text{ for some } \mathfrak{a} \in \mathfrak{A}, \text{ string } \lambda \text{ s. t.} \mathfrak{a} \notin \lambda, \text{ and } \ell(T) > \sigma \\ d(T) & \text{if } \ell(T) \leq k \\ \ell(T) & \text{otherwise} \end{cases}$$

Note that, in the string context, d acts as the resilience, ℓ as the thickness, and ψ is the approximation we compute. Intuitively, if T is short (i.e., length at most k) we can compute the exact value $d(T)$. If T has a symbol whose two appearances are far away we will use a "shortcut" and pay σ (i.e., for unit disk regions, by Lemma 7, we will take $\sigma = \Delta$). Otherwise, we will approximate d by ℓ. Given a long string, we wish to subdivide S into a *segmentation* $\mathcal{T}$, composed of m disjoint segments (i.e. substrings of S) $T_1, \ldots, T_m$, that minimize the total cost $\psi(\mathcal{T}) = \sum_i \psi(T_i)$. Clearly, $\psi(\mathcal{T}) \leq \ell(S)$.

Lemma 8. *Let S be a sequence. There exists a segmentation $\mathcal{T}$ such that $\psi(\mathcal{T}) \leq (1 + \varepsilon)d(S)$, where $\varepsilon = 2\sqrt{\sigma/k}$.*

Proof. Let λ be an integer such that $\sigma < \lambda < k$, of exact value to be specified later. First, we consider all pairs of equal symbols in S that are more than λ apart. We would like to take all of these pairs as separate segments; however, we cannot take segments that are not disjoint. So, we greedily take the leftmost symbol $\mathfrak{s}$ whose partner is more than λ further to the right, and mark this as a segment. We recurse on the substring remaining to the right of the rightmost $\mathfrak{s}$.[3] Finally, we segment the remaining pieces greedily into pieces of length k. Figure 4 illustrates the resulting segmentation.

Now, we prove that the resulting segmentation has a cost of at most $(1 + \varepsilon)d(S)$. First, consider a symbol to be *counted* if it appears in only one short

---

[3] In fact, we could choose any disjoint collection such that after their removal there are no more segments of this type longer than λ.

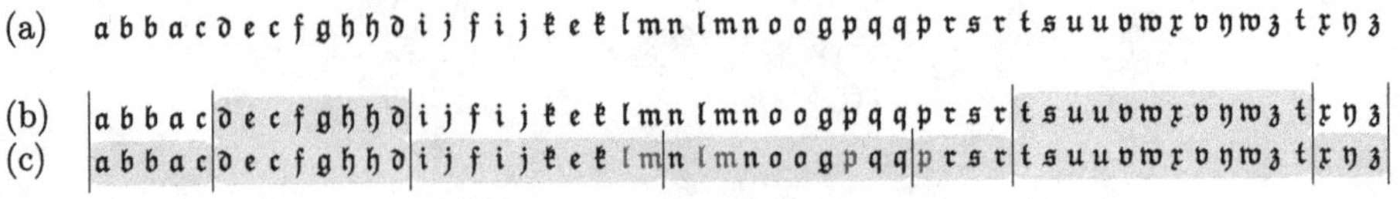

(a) abbacɔecfgɦɦɔijfijʇeʇɭmnɭmnoogpqqprsrtsuuʋꭥɤʋɳꭥɜtꭓɳɜ

(b) |abbac|ɔecfgɦɦɔ|ijfijʇeʇɭmnɭmnoogpqqprsr|tsuuʋꭥɤʋɳꭥt|ꭓɳɜ|
(c) |abbac|ɔecfgɦɦɔ|ijfijʇeʇɭm|nɭmnoogpqq|prsr|tsuuʋꭥɤʋɳꭥt|ꭓɳɜ|

Fig. 4. (a) A string of 52 symbols, each appearing twice. (b) First, we identify a maximal set of segments bounded by equal symbols, and longer than $\lambda = 4$. (c) Then, we segment the remaining pieces into segments of length $k = 10$. Red symbols are double-counted (Colour figure online).

(blue) segment, and to be *double-counted* if it appears in two different short segments. Suppose s is double-counted. Then the distance between its two occurrences must be smaller than λ, otherwise it would have formed a long (red) segment. Therefore, it must appear in two adjacent short segments. The leftmost of these two segments has length exactly k, but only λ of these can have a partner in the next segment. So, at most a fraction λ/k symbols are double-counted.

Second, we need to analyze the cost of the long (red) segments. In the worst case, all symbols in the segment also appear in another place, where they were already counted. In this case, the true cost would be 0, and we pay σ too much. However, we can assign this cost to the at least λ symbols in the segment; since each symbol appears only twice they can be charged at most once. So, we charge at most σ/λ to each symbol. The total cost is then bounded by $(1 + \lambda/k + \sigma/\lambda)d(S)$. To optimize the approximation factor, we choose λ such that $\lambda/k = \sigma/\lambda$; more precisely we take $\lambda = \lceil \sqrt{k\sigma} \rceil$. Recall that we initially set $k = \lceil 4\sigma/\varepsilon^2 \rceil$.   □

4.2 Application to Resilience Approximation

We now show that the shortest path between any p, q in G_k is a $(1 + \varepsilon)$-approximation of their resilience. Let π be a path from p to q in $\mathbb{R}^2$, and let $S(\pi)$ be the sequence that records every disk of $\mathcal{D}$ we enter along π, plus the disks that contain the start point of π, added at the beginning of the sequence, in any order. Then we have $|S(\pi)| = t(\pi)$.

Lemma 9. *For every path π from p to q and every segmentation $\mathcal{T}$ of $S(\pi)$, there exists a path from p to q in G_k of cost at most $\psi(\mathcal{T})$.*

Lemma 10. *For any $p, q \in \mathbb{R}^2$, it holds $r(\wp_{G_k}(p,q)) \leq (1+\varepsilon)r(p,q)$.*

Theorem 3. *Let $\mathcal{D}$ be a set of unit disks of ply Δ in $\mathbb{R}^2$. We can compute a path π between any two given points $p, q \in \mathbb{R}^2$ whose resilience is at most $(1+\varepsilon)r(p,q)$ in $O(2^{f(\Delta,\varepsilon)}n^7)$ time, where $f(\Delta,\varepsilon) = 16\frac{\Delta^2 \log(\Delta/\varepsilon)}{\varepsilon^4} + o(\frac{\Delta^2 \log(\Delta/\varepsilon)}{\varepsilon^4})$.*

4.3 Extension to Fat Regions

As in Sect. 3.2, we now generalize the result to arbitrary β-fat unit regions.

Lemma 11. *Let $D \in \mathcal{D}$, where $\mathcal{A}(\mathcal{D})$ has ply Δ, and let p,q be any two points inside D. Then the resilience between p and q in $\mathcal{D}$ is at most $(2\beta + 1)^2 \Delta$.*

As before, the rest of the arguments do not rely on the geometry of the regions anymore, and we can proceed as in the disk case. The only difference is that the value σ of doing a shortcut has increased to $(2\beta + 1)^2 \Delta$.

Theorem 4. *Let $\mathcal{D}$ be a set of unit disks of ply Δ in $\mathbb{R}^2$. We can compute a path π between any two points $p, q \in \mathbb{R}^2$ whose resilience is at most $(1+\varepsilon)r(p,q)$ in $O(2^{f(\Delta,\beta,\varepsilon)}n^7)$ time, where $f \in O(\frac{\Delta^2 \beta^6}{\varepsilon^4} \log(\beta\Delta/\varepsilon))$.*

5 NP-Hardness

In this section we show that computing the resilience of certain types of fat regions is NP-hard. We recall that NP-hardness was shown in [1] and [16] for the case in which regions are line segments in $\mathbb{R}^2$. In this section we show hardness extends for the case in which ranges have bounded fatness (i.e., ranges are not skinny). We note that Tseng [15] sketched how to extend the proof given in [16] for the case in which $\mathcal{D}$ is a collection of (translated and rotated) copies of a fixed square or ellipse. Although the spirit of the construction is clear, no details and no formal proof of correctness were given.

In addition to providing completeness to the rest of our results, our construction is of independent interest, since it is completely different from those given in [1] and [16]. Moreover, our proof has the advantage of being very easy to extend to other shapes. We also note that the construction of Tseng uses several rotations of a fixed shape (i.e., 3 for a square, 4 for an ellipse), whereas our construction only needs two different rotations of the same shape.

First we show NP-hardness for general connected regions, and later we extend it to axis-aligned rectangles of aspect ratio $1 : 1 + \varepsilon$ and $1 + \varepsilon : 1$. We start the section establishing some useful graph-theoretical results.

Let G be a graph, and let p be a point in the plane. Let Γ be an embedding of G into the plane, which behaves properly (vertices go to distinct points, edges are curves that don't meet vertices other than their endpoints and don't triple cross), and such that p is not on a vertex or edge of the embedding. We say Γ is an *odd* embedding around p if it has the following property: every cycle of G has odd length if and only if the winding number of the corresponding closed curve in the plane in Γ around p is odd. We say a graph G is *oddly embeddable* if there exists an odd embedding Γ for it. We begin by proving that vertex cover is still NP-hard for this constrained class of graphs. (Omitted proofs can be found in the full version [10].)

Lemma 12. *Minimum vertex cover on oddly embeddable graphs of maximum degree 3 is NP-hard.*

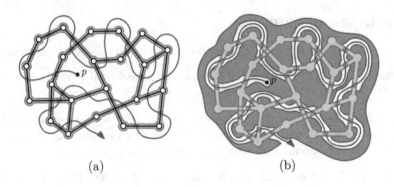

(a)   (b)

Fig. 5. Creating regions to follow Γ and T.

Given an embedded graph Γ, we say that a curve in the plane is an *odd Euler path* if it does not go through any vertex of Γ and it crosses every edge of Γ an odd number of times.

Lemma 13. *Let p be a point in the plane, and Γ an oddly embedded graph around p. Then there exists an odd Euler path for Γ that starts at p and ends in the outer face. Moreover, such path can be computed in polynomial time.*

Lemma 14. *Let p be a given point in the plane, and Γ an oddly embedded graph (not necessarily planar) around p. Furthermore, let T be a curve that forms an odd Euler path from p to the outer face. Then we can construct a set $\mathcal{D}$ of connected regions such that a minimum set of regions from $\mathcal{D}$ to remove corresponds exactly to a minimum vertex cover in Γ.*

Proof. If T is self-intersecting, then we can rearrange the pieces between self-intersections to remove all self-intersections. Thus we assume that T is a simple path.

If T crosses any edge of Γ more than once, we insert an even number of extra vertices on that edge such that afterwards, every edge is crossed exactly once. Let Γ' be the resulting graph. Since we inserted an even number of vertices on every edge, finding a minimum vertex cover in Γ' will give us a minimum vertex cover in Γ.

Now, for each vertex v in Γ', we create one region D_v in $\mathcal{D}$. This region consists of the point where v is embedded, and the pieces of the edges adjacent to v up to the point where they cross T. Figure 5(a) shows an example (the regions have been dilated by a small amount for visibility; if the embedding Γ has enough room this does not interfere with the construction). Note that all regions are simply connected.

Finally, we create one more special region W in $\mathcal{D}$ that forms a corridor for T. Then W is duplicated at least n times to ensure that crossing this "wall" will always be more expensive than any other solution. Figure 5(b) shows this.

Now, in order to escape, anyone starting at p must roughly follow T in order to not cross the wall. This means that for every edge of Γ' that T passes, one of the regions blocking the path (one of the vertices incident to the edge) must be disabled. The smallest number of regions to disable to achieve this corresponds to a minimum vertex cover in Γ'.   ⊠

Combining this result with Lemma 12, we obtain our first hardness result for the barrier resilience problem. As mentioned before, our construction can be modified so that it works for a much more restricted class of regions: axis-aligned rectangles of sizes $1 \times (1 + \varepsilon)$ and $(1 + \varepsilon) \times 1$ for any $\varepsilon > 0$ (as long as ε depends polynomially on n). Details on the necessary changes can be found in the full version [10].

Theorem 5. *The barrier resilience problem for regions that are axis-aligned rectangles of aspect ratio $1 : (1 + \varepsilon)$ is NP-hard.*

A similar approach can likely be used to show NP-hardness for other specific shapes of regions. However, it seems that a vital property is that they need to be able to completely cross each other: that is, the regions in $\mathcal{D}$ should not be pseudodisks.[4] Thus, if one were to prove that vertex cover for oddly embeddable graphs of bounded degree is NP-hard would also imply that the barrier resilience problem for unit disks is also NP-hard.

Acknowledgements. M.K was partially supported by the Secretary for Universities and Research of the Ministry of Economy and Knowledge of the Government of Catalonia and the European Union. M.L. was supported by the Netherlands Organisation for Scientific Research (NWO) under grant 639.021.123. R.S. was partially supported by FP7 Marie Curie Actions Individual Fellowship PIEF-GA-2009-251235 and by FCT through grant SFRH/BPD/88455/2012. M.K and R.S. were also supported by projects MINECO MTM2012-30951 and Gen. Cat. DGR2009SGR1040 and by ESF EURO-CORES program EuroGIGA-ComPoSe IP04-MICINN project EUI-EURC-2011-4306.

References

1. Alt, H., Cabello, S., Giannopoulos, P., Knauer, C.: On some connection problems in straight-line segment arrangements. In: Proceedings of the EuroCG, pp. 27–30 (2011)
2. Bereg, S., Kirkpatrick, D.: Approximating barrier resilience in wireless sensor networks. In: Dolev, S. (ed.) ALGOSENSORS 2009. LNCS, vol. 5804, pp. 29–40. Springer, Heidelberg (2009)
3. Chang, C.-Y.: The k-barrier coverage mechanism in wireless visual sensor networks. In: Proceedings of the WCNC, pp. 2318–2322 (2012)
4. Chen, D.Z., Gu, Y., Li, J., Wang, H.: Algorithms on minimizing the maximum sensor movement for barrier coverage of a linear domain. In: Fomin, F.V., Kaski, P. (eds.) SWAT 2012. LNCS, vol. 7357, pp. 177–188. Springer, Heidelberg (2012). http://arxiv.org/abs/1207.6409

---

[4] A similar fact was also observed in [16].

5. de Berg, M., Cheong, O., van Kreveld, M., Overmars, M.: Computational Geometry: Algorithms and Applications, 3rd edn. Springer, Heidelberg (2008)
6. de Berg, M., Katz, M.J., van der Stappen, A.F., Vleugels, J.: Realistic input models for geometric algorithms. In: Proceedings of the SoCG, pp. 294–303 (1997)
7. Gibson, M., Kanade, G., Varadarajan, K.: On isolating points using disks. In: Demetrescu, C., Halldórsson, M. (eds.) ESA 2011. LNCS, vol. 6942, pp. 61–69. Springer, Heidelberg (2011)
8. He, S., Chen, J., Li, X., Shen, X., Sun, Y.: Cost-effective barrier coverage by mobile sensor networks. In: Proceedings of the INFOCOM, pp. 819–827 (2012)
9. Hu, T.C.: Multi-commodity network flows. Oper. Res. 11(3), 344–360 (1963)
10. Korman, M., Löffler, M., Silveira, R.I., Strash, D.: On the complexity of barrier resilience for fat regions. CoRR, abs/1302.4707 (2013)
11. Kumar, S., Lai, T.-H., Arora, A.: Barrier coverage with wireless sensors. In: Proceedings of the MOBICOM, pp. 284–298 (2005)
12. Kumar, S., Lai, T.-H., Arora, A.: Barrier coverage with wireless sensors. Wirel. Netw. 13(6), 817–834 (2007)
13. Marx, D.: Parameterized graph separation problems. Theor. Comput. Sci. 351(3), 394–406 (2006)
14. Miller, G., Teng, S., Thurston, W., Vavasis, S.: Separators for sphere-packings and nearest neighbor graphs. J. ACM 44(1), 1–29 (1992)
15. Tseng, K.-C.R.: Resilience of wireless sensor networks. Master's thesis, University of British Columbia (2011)
16. Tseng, K.-C.R., Kirkpatrick, D.: On barrier resilience of sensor networks. In: Erlebach, T., Nikoletseas, S., Orponen, P. (eds.) ALGOSENSORS 2011. LNCS, vol. 7111, pp. 130–144. Springer, Heidelberg (2012)
17. Xiao, M.: Simple and improved parameterized algorithms for multiterminal cuts. Theor. Comput. Syst. 46(4), 723–736 (2010)

A Distributed Approximation Algorithm for Strongly Connected Dominating-Absorbent Sets in Asymmetric Wireless Ad-Hoc Networks

Christine Markarian[1]([✉]), Friedhelm Meyer auf der Heide[2],
and Michael Schubert[3]

[1] Heinz Nixdorf Institute and Computer Science Department,
Fellow of the International Graduate School, Dynamic Intelligent Systems,
University of Paderborn, 33102 Paderborn, Germany
[2] Heinz Nixdorf Institute and Computer Science Department,
University of Paderborn, 33102 Paderborn, Germany
[3] Mathematics Department, Fellow of the International Graduate School,
Dynamic Intelligent Systems, University of Paderborn, 33102 Paderborn, Germany
{christine.markarian,fmadh,mischub}@upb.de
http://www.uni-paderborn.de

Abstract. Dominating set based virtual backbones are used for routing in wireless ad-hoc networks. Such backbones receive and transmit messages from/to every node in the network. Existing distributed algorithms only consider undirected graphs, which model symmetric networks with uniform transmission ranges. We are particularly interested in the well-established disk graphs, which model asymmetric networks with non-uniform transmission ranges. The corresponding graph theoretic problem seeks a strongly connected dominating-absorbent set of minimum cardinality in a digraph. A subset of nodes in a digraph is a strongly connected dominating-absorbent set if the subgraph induced by these nodes is strongly connected and each node in the graph is either in the set or has both an in-neighbor and an out-neighbor in it. We introduce the first distributed algorithm for this problem in disk graphs. The algorithm gives an $O(k^4)$-approximation ratio and has a runtime bound of $O(Diam)$ where $Diam$ is the diameter of the graph and k denotes the transmission ratio r_{max}/r_{min} with r_{max} and r_{min} being the maximum and minimum transmission range, respectively. Moreover, we apply our algorithm on the subgraph of disk graphs consisting of only bidirectional edges. Our algorithm gives an $O(\ln k)$-approximation and a runtime bound of $O(k^8 \log^* n)$, which, for bounded k, is an optimal approximation for the problem, following Lenzen and Wattenhofer's $\Omega(\log^* n)$ runtime lower bound for distributed constant approximation in disk graphs.

Keywords: Dominating set · Absorbent set · Distributed Algorithms · Ad-hoc networks · Virtual backbone

This work was partially supported by the German Research Foundation (DFG) within the Collaborative Research Center "On-The-Fly Computing" (SFB 901).

P. Flocchini et al. (Eds.): ALGOSENSORS 2013, LNCS 8243, pp. 217–227, 2014.
DOI: 10.1007/978-3-642-45346-5_16, © Springer-Verlag Berlin Heidelberg 2014

1   Introduction

Unlike cellular networks, wireless ad-hoc networks have no fixed routing infrastructure. Communication in these networks relies on a shared medium through
either single hops or multihops, often resulting in inevitable collision. Virtual
backbones are often used as a potential solution where only the nodes in the
virtual backbone become responsible to handle communication among all nodes
in the network. Clearly, a small virtual backbone is desirable to reduce routing
overhead. In addition, the nodes in the backbone must induce a (strongly) connected subgraph that can be reached by all nodes. Thus, in case of undirected
graphs, we speak of a connected dominating set based virtual backbone. A subset of nodes in an undirected graph is a *connected dominating set*, henceforth
CDS, if each node in the network is either in the set or has a neighbor in it and
the subgraph induced by these nodes is connected.

Previous work on dominating set based virtual backbones propose distributed algorithms for undirected graphs, especially for unit disk graphs, which
model symmetric networks in which all nodes have the same transmission range
[2–9]. In practice, however, wireless ad-hoc networks are asymmetric, because
the transmission ranges of nodes need not be the same, due to differences in
power and functionality. Therefore, a disk graph, which considers nodes with
different transmission ranges, better models a wireless ad-hoc network. In a *disk
graph* G, each node $v \in V$ is fixed on the Euclidean plane and has a transmission
range $r_v \in [r_{min}, r_{max}]$, where r_{min} and r_{max} denotes the minimum and maximum transmission range, respectively. For two disjoint nodes $u, v \in V$, there
is a directed edge (u, v) if and only if $d_{u,v} \leq r_u$ where $d_{u,v}$ is the *Euclidean
distance* between u and v. An edge $(u, v) \in E$ is *unidirectional* if $(v, u) \notin E$ and
bidirectional if $(v, u) \in E$.

We consider the corresponding graph theoretic problem which seeks a minimum *strongly connected dominating-absorbent set* ($SCDAS$) in a directed disk
graph, namely DG. A subset of nodes in a directed graph is an $SCDAS$ if the
subgraph induced by these nodes is strongly connected and each node in the
graph is either in the set or has both an in-neighbor and an out-neighbor in it.

On the other hand, many routing protocols in wireless ad-hoc networks use
acknowledgments for the data packets exchanged between the nodes. This motivates the restriction to only employ the bidirectional edges of the graph for
routing. Therefore, we also consider disk graphs from which all unidirectional
edges are dropped. These were first studied in [19] and are called disk graphs
with bidirectional edges, namely DGBs. The corresponding problem asks for a
CDS in an undirected graph.

The two problems, minimum $SCDAS$ in a DG and minimum CDS in a
DGB, are NP-hard by a simple reduction from minimum CDS in unit disk
graphs [1]. Previous work on minimum $SCDAS$ in a DG includes only sequential approximation algorithms based on constructing spanning trees, such as in
[13,14]. For minimum CDS in a DGB, there are a number of distributed approximation algorithms such as in [19–22], which also use spanning trees. But since
constructing a spanning tree distributively cannot be done in better than linear

time (in the order of the graph) [2], the techniques used in these algorithms cannot construct time-efficient distributed algorithms.

We introduce the first distributed algorithm for minimum $SCDAS$ in a DG. The algorithm gives an $O(k^4)$-approximation ratio and has a runtime bound of $O(Diam(G))$ where $Diam(G)$ is the diameter of the graph, and k denotes the transmission ratio r_{max}/r_{min} with r_{max} and r_{min} being the maximum and minimum transmission range, respectively.

Rather than constructing a spanning tree, our algorithm first constructs a maximal independent set (MIS), which forms an absorbent-dominating set then strongly connects the MIS constructed. To find a maximal independent set, our algorithm uses the $O(\log^* n)$ time algorithm for undirected growth-bounded graphs in [9]. This is done by finding an MIS in the subgraph $\bar{G} = (V, \bar{E})$ resulting from $G = (V, E)$ by deleting all unidirectional edges from G.

Moreover, when applied to minimum CDS in a DGB, our algorithm gives an $O(\ln k)$-approximation in $O(k^8 \log^* n)$ time, which, for bounded k, is an optimal approximation for the problem, following Lenzen and Wattenhofer's $\Omega(\log^* n)$ runtime lower bound for distributed constant approximation in unit disk graphs [18].

The rest of this paper is organized as follows. In Sect. 2, we give some preliminaries. We present the recent work in Sect. 3. Then, we describe the algorithm and give its analysis (running time and approximation ratio) in Sects. 4 and 5, respectively. We extend the algorithm to $DGBs$ in Sect. 6 and finally present some possible future research directions in Sect. 7.

2 Preliminaries

In this paper, basic graph theoretic notations such as *degree* of a node, *maximum degree* Δ in a graph, and diameter $Diam(G)$ of a graph G, are adopted.

In a *disk graph*, $G = (V, E)$, each node $v \in V$ is fixed on the Euclidean plane and has a transmission range $r_v \in [r_{min}, r_{max}]$, where r_{min} and r_{max} denote the minimum and maximum transmission range, respectively. For two disjoint nodes $u, v \in V$, there is a directed edge (u, v) if and only if $d_{u,v} \leq r_u$ where $d_{u,v}$ is the *Euclidean distance* between u and v. An edge $(u, v) \in E$ is called *unidirectional* if $(v, u) \notin E$ and *bidirectional* if $(v, u) \in E$. A *disk graph* with only bidirectional edges, i.e., undirected graph, is a DGB.

We say G is (strongly) connected if for any two nodes $u, v \in V$, there exists a (directed) path from u to v. An *independent set* (IS) in G is a subset S of V such that there is no bidirectional edge between any two nodes of S. We then say S does not violate independence. A subset of V is a *maximal independent set* (MIS) if it is an IS to which no node can be added without violating independence.

Let (u, v) be a unidirectional edge. Then, u is *absorbed* by v and v is *dominated* by u. We say a graph is *growth-bounded* if there is a polynomial function $f(r)$ such that every r-neighborhood in the graph contains at most $f(r)$ *independent* nodes.

Communication Model. In this paper, communication among nodes is done in synchronous rounds. In each round, each node sends a message of size $O(\log n)$ bits to its neighbors.

3   Recent Work

Previous work on dominating set based virtual backbones propose distributed algorithms mainly for unit disk graphs [3–9]. For general undirected graphs, the authors in [2] give a polylogarithmic-time randomized distributed approximation algorithm. Their algorithm starts with a dominating set constructed by a randomized distributed algorithm for dominating sets and then it destroys cycles to preserve connectivity while at the same time adding as few more nodes as possible. Their approximation ratio of $O(\log \Delta)$ is the best possible for general graphs unless $NP \subseteq D_{TIME}[N^{O(\log \log n)}]$. Some other works such as [26] and [27] also propose distributed approximation algorithms but only for dominating sets which do not require connectivity.

Thai et al. [19] introduced the first distributed algorithm for minimum CDS in $DGBs$. They gave a constant approximation algorithm for bounded transmission ratio, with $O(n^2)$ time and message complexity. Later, Raei et al. [20] improved the two bounds. They gave an $O(n)$ time complexity and $O(n \log n)$ message complexity, which was further improved to $O(n)$ in [21]. Lately, the authors in [22] presented a timer based energy-aware algorithm with similar results.

Algorithms for minimum $SCDAS$ in directed DGs have been proposed in [10–12,25]. These are either heuristic based where no approximation bound is given or sequential. Du et al. [23] introduced the minimum Strongly Connected Dominating Set problem and presented a centralized constant approximation algorithm for bounded transmission ratio in disk graphs. Later Thai et al. [13] added the 'absorbent' property and introduced the minimum $SCDAS$ problem. They presented a centralized constant approximation algorithm for the problem, based on constructing two spanning trees, and bounded their approximation ratio using the geometric properties of disk graphs. A similar result was given in [24] for minimum $SCDAS$, based on using a Steiner tree with minimum number of Steiner nodes.

For general directed graphs, Li et al. [14] gave a polynomial time logarithmic approximation algorithm based on constructing directed rooted trees.

4   Approximation Algorithm

This section has two parts. In the first part, we present our distributed algorithm, which we call DistributedSCDAS, and in the second part, we prove its correctness.

Algorithm Description

Given a strongly connected directed graph $G = (V, E)$, algorithm Distributed-SCDAS first constructs $\bar{G} = (V, \bar{E})$ by deleting all unidirectional edges from

G. Then, it constructs an $MISI$ in $\bar{G}$, for an illustration see Fig. 1. Note that, since $\bar{G}$ has only bidirectional edges, we may apply any distributed algorithm for computing the MIS in undirected graphs, e.g. Luby's algorithm in [17]. We will later (Sect. 4, Subsection Runtime) present a faster possibility in case of disk graphs. Clearly, I forms a dominating-absorbent set, DAS, in $G = (V, E)$. To strongly connect I in G, the algorithm constructs $G' = (I, E')$ from G, which contains a directed edge from u to v if there is a path from u to v of length at most three, whose (at most two) inner nodes are in $V \setminus I$. This construction may lead to multiple edges from u to v. In this case, we remove all but one of them. The nodes in I along with the inner nodes in G corresponding to the remaining edges of G' form an $SCDAS$ (See Fig. 2).

Algorithm 1. DistributedSCDAS

Input: A strongly connected directed graph $G = (V, E)$
Output: An $SCDAS$ $S \subset V$
-*Step 1:* (Construct a DAS) Construct $\bar{G} = (V, \bar{E})$ by deleting all unidirectional edges from G and then find an MIS in $\bar{G}$. Let I be such an MIS.
-*Step 2:* (Strongly connect DAS)
2.1: Construct $G' = (I, E')$ from $G = (V, E)$ as follows. For each pair u, v of nodes from I, E' contains a directed edge from u to v if there is a path from u to v of length at most three, whose (at most two) inner nodes are in $V \setminus I$. This construction may lead to multiple edges from u to v.
2.2: For each pair (u, v) of nodes from I, for which G' contains multiple edges from u to v, remove all but one of these edges.
-Output $I \cup C$, where C is the set of the inner nodes in G corresponding to the remaining edges of G'.

Distributed Implementation

All nodes are assigned IDs. The algorithm above can be implemented distributedly as follows. At the end of the algorithm, each node knows whether it belongs to the $SCDAS$ or not.

Step 1: At each round, each node runs a distributed MIS algorithm and decides whether it belongs to an $MISI$ or not.

2.1: Each MIS node in I sends an *edgeRequest* packet, which is a packet requesting to form an edge. The edgeRequest packet includes a *source* which contains the ID of the node requesting to form an edge. If a non-MIS node in $V \setminus I$ receives such a request, it adds its ID to the packet and forwards the packet. Note that:

- a node in I ignores an edgeRequest packet it receives, if itself is the source of it.
- a node in $V \setminus I$ ignores an edgeRequest packet it receives, if it has already received a packet from the same source.

2.2: A node in I may receive multiple edgeRequests with the same source. For each source, it ignores all but the first edgeRequest. Thus with each source, it forms an edge consisting of the nodes with the IDs on the packet.

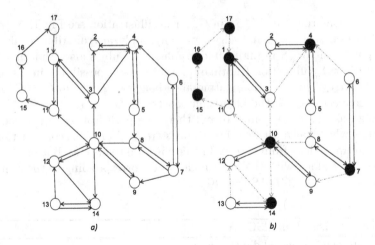

Fig. 1. a) The original graph $G = (V, E)$, b) $\bar{G} = (V, \bar{E})$ where the dotted edges are the unidirectional edges that were removed from G and the black nodes form the MIS nodes.

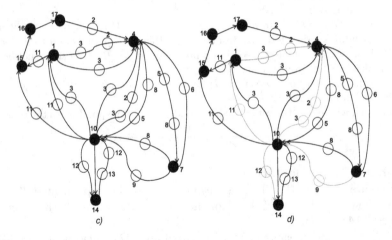

Fig. 2. c) $G' = (I, E')$ where the node set I is in black. d) $G' = (I, E')$, the gray edges are those which were removed in *Step 2.2*, and the nodes on the black edges form the set C.

- Each node in I informs the source nodes about its edge formations with them, which in turn inform the participating nodes, i.e., nodes on the selected packets(edges).

In the next subsection, we prove the correctness of the algorithm. Then, in Sect. 5, we bound its approximation ratio and runtime.

Correctness

Note that the following theorem also holds for a DGB since a DG is a generalization of a DGB.

Theorem 1. *Given a strongly connected directed graph $G = (V, E)$, algorithm DistributedSCDAS constructs an SCDAS.*

Proof. DistributedSCDAS outputs $I \cup C$, where I is an MIS in $\bar{G}$ and thus in G, and consequently a DAS. Thus, it remains to show that $I \cup C$ is strongly connected. The next lemma gives the main insight.

Lemma 1. *If G is strongly connected, then G' is strongly connected.*

Proof of Lemma 1. Let G be a strongly connected graph and I be an MIS in G. Then, the following two properties hold.

i) If there is a path $P = (u, v, w)$ with $u, v, w \in V \setminus I$ then v must be dominated and absorbed by a node $y \in I$.
ii) For each pair of nodes (u, w) there exists a walk $W = (u = v_1, ...v_n = w)$ (v_i not necessarily distinct) such that each node in $I \cap W$ is followed by at most two nodes of $V \setminus I$.

Proof of i). Assume i) is not true. Then $I \cup \{v\}$ is an independent set. Thus I is not maximal. $\square$

Proof of ii). Assume ii) is not true. Let A be the set including all walks W from u to w in G. For a walk $W \in A$ let m_W be the maximum length of a subpath given by consecutive nodes of $V \setminus I$ and $c(m_W)$ be the number of those subpaths with length m_W in W. We choose a walk $W' = (u = v_1, ...v_n = w) \in A$ such that $m_{W'}$ and $c(m_{W'})$ is minimum. Let v_i, v_{i+1}, v_{i+2} ($i = 1..n-2$) be three nodes of a longest subpath with nodes in $V \setminus I$. Due to i), v_{i+1} is dominated and absorbed by a node $y \in I$. Thus, $W = (v_1, ..., v_i, v_{i+1}, y, v_{i+1}, v_{i+2}, ...v_n)$ is a walk from u to w. If $c(m_{W'}) = 1$ then $m_W = m_{W'} - 1$ and if $c(m_{W'}) > 1$ then $c(m_W) = c(m_{W'}) - 1$, contradicting the choice of W'. $\square$

Let I be the underlying MIS of G and $u, w \in I$. It follows from ii) that we can find a walk W, such that each node in $I \cap W$ is followed by at most two nodes of $V \setminus I$. Thus, W corresponds to a path P in G' from u to w. $\square$

Proof of Theorem 1. Lemma 1 implies that the graph G' constructed in *Step 2.1* is strongly connected. Then, since *Step 2.2* only removes edges from E' whose deletion does not affect the strong connectivity of G', the resulting $I \cup C$ is strongly connected in G. $\square$

5   Analysis (Approximation Ratio and Running Time)

In this section, we prove that Algorithm DistributedSCDAS has a constant approximation ratio and completes in $O(Diam(G))$ time in directed disk graphs, DGs.

Theorem 2 [(*Main Theorem-DG*)]. *Given a directed disk graph, $G = (V, E)$ with transmission ratio $k = r_{max}/r_{min}$. Algorithm DistributedSCDAS constructs an $O(k^4)$-approximation for minimum SCDAS in $O(Diam(G))$ time.*

Proof. To prove the $O(k^4)$-approximation ratio, we need the following three lemmas.

Lemma 2. *For any two independent nodes u, v in G, the distance between u and v is greater than r_{min}.*

Proof of Lemma 2. If the distance between u and v is at most r_{min}, then, u and v are connected in both directions.   □

Lemma 3. *At the end of Step 2.2, the degree of each node in G' is upper bounded by $\lfloor 49k^2 - 1 \rfloor$.*

Proof of Lemma 3. Due to Lemma 2, for each pair of independent nodes (x, y) in G, their distance $d_{x,y}$ is greater than r_{min}. At the end of *Step 2.2*, each node in G' has at most one out-going edge for each neighbor. Therefore, the degree of each node in G' is bounded by the number of independent nodes in u's 3-hop neighborhood in G. Thus, for each edge (u, v) in G', the distance $d_{u,v}$ is bounded by $r_{min} < d_{u,v} < 3r_{max}$. The maximum area which may be covered by the disks of the independent nodes in u's 3-hop neighborhood in G is given by the difference of the areas between two disks with radii $3.5r_{max}$ and $\frac{r_{min}}{2}$. Moreover, the minimum area of a disk is $\pi(\frac{r_{min}}{2})^2$. It follows that u has at most $\left\lfloor \frac{(3.5r_{max})^2 - (\frac{r_{min}}{2})^2}{(\frac{r_{min}}{2})^2} \right\rfloor = \lfloor 49k^2 - 1 \rfloor$ neighbors in G'.   □
The following lemma is proven in [13].

Lemma 4. *([13]) The size of any independent set in G is bounded from above by $2.4(k + \frac{1}{2})^2 \cdot |SCDAS_{opt}| + 3.7(k + \frac{1}{2})^2$, where $SCDAS_{opt}$ denotes an optimal SCDAS.*

Now we can conclude the bound on the approximation ratio as follows:
The size of the $SCDAS$ constructed by the algorithm is bounded by

$$|SCDAS| \le |IS|\,(1 + 2\Delta(G')), \qquad (1)$$

because *Step 2* of the algorithm adds for each node in I at most $2\Delta(G')$ nodes from $V \setminus I$ in order to strongly connect I.

Plugging in the bounds from lemmas 3 and 4 yields:

$$|SCDAS| \le \left(2.4(k + \frac{1}{2})^2 \cdot |SCDAS_{opt}| + 3.7(k + \frac{1}{2})^2\right) \cdot \left(1 + 2\lfloor 49k^2 - 1 \rfloor\right) \qquad (2)$$

This implies:

$$|SCDAS| = O(k^4) \cdot |SCDAS_{opt}| \qquad (3)$$

□

We now show that DistributedSCDAS requires $O(Diam(G))$ time on *DGs*.

- For computing the MIS in *Step 1*, we might use Luby's $O(\log n)$ time randomized algorithm to construct an MIS. However, this is not the best we can do when it comes to disk graphs. In fact, there is a deterministic $O(\log^* n)$ time algorithm for finding an MIS in bounded-growth graphs [9]. Moreover, when the transmission ratio $k = r_{max}/r_{min}$ is bounded, disk graphs become bounded-growth graphs with $f(r) = O(r^2 k^2)$. Therefore, for disk graphs, rather than using Luby's $O(\log n)$ time randomized algorithm for finding an MIS in *Step 1*, we use the deterministic $O(\log^* n)$ time algorithm for bounded-growth graphs in [9]. When applied to disk graphs, the algorithm takes $O(k^8 \log^* n)$.
- In *Step 2* constructing $G' = (I, E')$ from $G = (V, E)$ needs only three broadcasts because the edgeRequest packets stop after at most two inner nodes. Once 'connecting' nodes are selected, it remains to inform them.
- Informing the participating nodes takes $O(Diam(G))$ time where each source node informs other source nodes about the selected edges and consequently the selected nodes. Note that the number of nodes to be informed are at most $2k^2$, thus bounding the number of propagated messages by $O(k^2)$.

Therefore, DistributedSCDAS requires $O(Diam(G))$ time in total.   □

6   Undirected Disk Graphs, $DGBs$

In this section, we modify DistributedSCDAS and apply it on a DGB. Then, we show its approximation ratio and running time.

Given a connected undirected graph $G = (V, E)$, our modified Distributed-SCDAS constructs an MIS I in G. Clearly, I forms a dominating set in G. To connect I in G, each node in I ignores all edgeRequest packets from all source nodes except one, i.e., the first one it receives. Since the graph is undirected, it is enough for each node in I to connect to one other node in I.

Theorem 3 *[(**Main Theorem-DGB**)]. Given an undirected disk graph, $G = (V, E)$ with transmission ratio $k = r_{max}/r_{min}$. Algorithm DistributedSCDAS constructs a $O(\ln k)$-approximation for minimum CDS in $O(k^8 \log^* n)$ time.*

Proof. To show the $O(\ln k)$-approximation ratio, we need the following lemma.

Lemma 5. *([19]) The size of any independent set in G is at most $O(\ln k)$.*

Since each node in the MIS adds at most two non-MIS nodes to connect, the approximation ratio holds.

We now show that DistributedSCDAS requires $O(k^8 \log^* n)$ time on $DGBs$.

For computing an MIS in G, we use the same deterministic $O(\log^* n)$ time algorithm for finding an MIS in bounded-growth graphs [9], which when applied to disk graphs, takes $O(k^8 \log^* n)$ time. Moreover, since the graph is undirected, $O(1)$ time is needed to connect the constructed MIS.   □

7 Conclusion

This paper presents the first distributed algorithm, DistributedSCDAS, for the minimum Strongly Connected Dominating-Absorbent Set problem in directed disk graphs. Not only does DistributedSCDAS give a constant approximation ratio when the transmission ratio is bounded, but it also terminates in $O(Diam(G))$ time. Moreover, when applied to the minimum Connected Dominating Set problem in undirected disk graphs with bounded transmission ratio, our algorithm gives an optimal approximation. The main question which remains open is whether we can achieve good approximation ratios not only for disk graphs but other classes of graphs, such as general digraphs and bounded-degree digraphs, as well. Moreover, nodes in the virtual backbone are often subject to failure. Thus, fault tolerance cannot be avoided. Symmetric networks have been extensively studied within the fault tolerant consideration, whereas only few heuristics, [15,16], have studied the problem in asymmetric networks. Motivated by this, we aim to give distributed approximation algorithms for the underlying fault-tolerant virtual backbone problem in asymmetric networks.

References

1. Clark, B., Colbourn, C., Johnson, D.: Unit disk graphs. Discrete Math. **86**, 165–177 (1990)
2. Dubhashi, D., Mei, A., Panconesi, A., Radhakrishnan, J., Srinivasan, A.: Fast distributed algorithms for (weakly) connected dominating sets and linear-size skeletons. In: Symposium on Discrete Algorithms (SODA), pp. 717–724 (2003)
3. Wan, P., Alzoubi, K.M., Frieder, O.: Distributed construction of connected dominating set in wireless ad hoc networks. In: IEEE International Conference on Computer Communications (Infocom) (2002)
4. Wu, J., Li, H.: A dominating-set-based routing scheme in ad hoc wireless networks. Telecommun. Syst. J. **3**, 63–84 (2001)
5. Alzoubi, K.M., Wan, P.J., Frieder, O.: New distributed algorithm for connected dominating set in wireless ad hoc networks. In: Hawaii International Conference System Sciences (2002)
6. Funke, S., Kesselman, A., Meyer, U., Segal, M.: A simple improved distributed algorithm for minimum CDS in unit disk graphs. In: Wireless and Mobile Computing, Networking and Communications (WiMob) (2005)
7. Czyzowicz, J., Dobrev, S., Fevens, T., González-Aguilar, H., An, H.-C., Opatrny, J., Urrutia, J.: Local algorithms for dominating and connected dominating sets of unit disk graphs with location aware nodes. In: Laber, E.S., Bornstein, C., Nogueira, L.T., Faria, L. (eds.) LATIN 2008. LNCS, vol. 4957, pp. 158–169. Springer, Heidelberg (2008)
8. Czygrinow, A., Hańćkowiak, M.: Distributed approximation algorithms in unit-disk graphs. In: Dolev, S. (ed.) DISC 2006. LNCS, vol. 4167, pp. 385–398. Springer, Heidelberg (2006)
9. Schneider, J., Wattenhofer, R.: A log* distributed maximal independent set algorithm for growth-bounded graphs. In: Principles of Distributed Computing (PODC) (2008)

10. Markarian, C., Abu-Khzam, F.: A degree-based heuristic for strongly connected dominating-absorbent sets in wireless ad-hoc networks. In: Innovations in Information Technology (IIT) (2012)
11. Wu, J.: An extended dominating-set-based routing in ad hoc wireless networks with unidirectional links. IEEE Trans. Parallel Distrib. Syst. 13(9), 866–881 (2002)
12. Dai, F., Wu, J.: An extended localized algorithm for connected dominating set formation in ad hoc wireless networks. IEEE Trans. Parallel Distrib. Syst. 15(10), 902–920 (2004)
13. Park, M.A., Willson, J., Wang, C., Thai, M., Wu, W., Farago, A.: A dominating and absorbent set in a wireless adhoc network with different transmission ranges. In: Mobile Ad hoc Networking and Computing (MobiHoc), pp. 22–31 (2007)
14. Li, D., Duc, H., Wan, P.: Construction of strongly connected dominating sets in asymmetric multihop wireless networks. Theor. Comput. Sci. 410, 661–669 (2009)
15. Tiwari, R., Thai, M.T.: On enhancing fault tolerance of virtual backbone in a wireless sensor network with unidirectional links sensors. In: Boginski, V.L., Commander, C.W., Pardalos, P.M., Ye, Y. (eds.) Theory, Algorithms, and Applications. Springer, New York (2011)
16. Tiwari, R., Mishra, T., Li, Y.: k-strongly connected dominating and absorbing set in wireless ad hoc networks with unidirectional links. In: Wireless Algorithm, Systems and Applicaitons (WASA) , pp. 103–112 (2007)
17. Luby, M.: A simple parallel algorithm for the maximal independent set problem. SIAM J. Comput. 15(4), 1036–1053 (1986)
18. Lenzen, C., Wattenhofer, R.: Leveraging Linials locality limit. Distrib. Comput. 5218, 394407 (2008)
19. Thai, M., Wang, F., Liu, D., Zhu, S., Du, D.: Connected dominating sets in wireless networks with different transmission ranges. Mobile Comput. 6(7), 721–730 (2007)
20. Raei, H., Sarram, M., Adibniya, F.: Distributed algorithm for connected dominating sets in wireless sensor networks with different transmission ranges. In: Telecommunications, pp. 337–342 (2008)
21. Raei, H., Fathi, M., Akhhlaghi, A., Ahmadipoor, B.: A new distributed algorithm for virtual backbone in wireless sensor networks with different transmission ranges. Computer Systems and Applications, pp. 983–988 (2009)
22. Raei, H., Sarram, M., Salimi, B., Adibnya, F.: Energy-aware distributed algorithm for virtual backbone in wireless sensor networks. In: Innovations in Information Technology, pp. 435–439 (2008)
23. Du, D.-Z., Thai, M.T., Li, Y., Liu, D., Zhu, S.: Strongly connected dominating sets in wireless sensor networks with unidirectional links. In: Zhou, X., Li, J., Shen, H.T., Kitsuregawa, M., Zhang, Y. (eds.) APWeb 2006. LNCS, vol. 3841, pp. 13–24. Springer, Heidelberg (2006)
24. Thai, M., Tiwari, R., Du, D.: On construction of virtual backbone in wireless adhoc networks with unidirectional links. Mobile Comput. 7(9), 1098–1109 (2008)
25. Kassaei, H., Narayanan, L.: A new algorithm for backbone formation in ad hoc wireless networks of nodes with different transmission ranges. In: Wireless and Mobile Computing(WiMob), pp. 83–90 (2010)
26. Jia, L., Rajaraman, R., Suel, T.: An efficient distributed algorithm for constructing small dominating sets. In: Principles of Distributed Computing, pp. 3342 (2001)
27. Rajagopalan, S., Vazirani, V.: Primal-dual RNC approximation algorithms for (multi)set (multi)cover and covering integer programs. SIAM J. Comput. 28(2), 525540 (1998)

Uniform Dispersal of Asynchronous Finite-State Mobile Robots in Presence of Holes

Eduardo Mesa Barrameda[1], Shantanu Das[2]([✉]), and Nicola Santoro[1]

[1] School of Computer Science, Carleton University, Ottawa, Canada
[2] LIF, Aix-Marseille University and CNRS, Marseille, France
`shantanu.das@lif.univ-mrs.fr`

Abstract. We consider the problem of uniformly dispersing mobile robots in an unknown, connected, and closed space, so as to cover it completely. The robots are autonomous and identical, they enter the space from a single point, and move in coordination with other robots, relying only on sensed local information within a restricted radius. The existing solutions for the problem require either the robots to be synchronous or the space to be without holes and obstacles. In this paper we allow the robots to be fully asynchronous and the space to contain holes. We show how, even in this case, the robots can uniformly fill the unknown space, avoiding any collisions, when endowed with only $O(1)$ bits of persistent memory and $O(1)$ visibility radius. Our protocols are asymptotically optimal in terms of visibility and memory requirements, and these results can be achieved without any direct means of communication among the robots.

1 Introduction

Unlike their static counterparts, mobile sensors and robots can self-deploy within a target space S to "cover" it so to satisfy some optimization criteria. To achieve such a goal without the help of any central coordination or external control is a rather complex task, and designing localized algorithms for efficient and effective deployment of these mobile entities is a challenging research issue. Such a task has been studied by several authors and continues to be the subject of extensive research. Most of the work is focused on the (uniform) *self-deployment* problem; that is, how to achieve uniform deployment in S (usually assumed to be polygonal) starting from an initial random placement of the sensors in S (e.g., [2,6–10,12–14,16,17]).

The problem has recently been studied within the context of weak robots: the mobile robots rely only on sensed local information within a restricted range, called the *visibility radius*; usually they have no explicit means of communication or they can communicate only within a very limited range, called the *communication radius*. Localized solution algorithms for such weak robots have been developed for special spaces such as a *line* (e.g., a rectilinear corridor), a *ring* (e.g., the boundary of a convex region), a *grid*, etc. (e.g., [1,3,4]; see [5] for a recent survey).

Research partially supported by NSERC Canada.

P. Flocchini et al. (Eds.): ALGOSENSORS 2013, LNCS 8243, pp. 228–243, 2014.
DOI: 10.1007/978-3-642-45346-5_17, © Springer-Verlag Berlin Heidelberg 2014

We are interested in a specific instance of the self-deployment problem, called the *Uniform Dispersal* (or *Filling*) problem, where the robots have to completely cover an unknown space S entering through a designated entry point. In the process, the robots must avoid colliding with each other, and must terminate (i.e., reach a quiescent state) within finite time [9,11,15]. The space S may be of arbitrary shape but we assume it to be an orthogonal grid of unit cells, where each cell can contain exactly one robot (see Fig. 1). Such orthogonal spaces are of particular interest because they can be used to model indoor and urban environment (e.g., floorplans, city maps, etc.). The robots enter the space through one of the cells in S, called the *door*, and they must eventually fill up all the cells of S while ensuring that there is never a collision (two or more robots in the same cell).

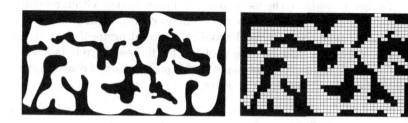

Fig. 1. An arbitrary closed space (Left) represented as an orthogonal grid of cells (Right). Each cell can contain a single robot.

Our focus is on the minimum capabilities required by the sensors in order to effectively complete the uniform dispersal task. There are some intrinsic limitations on the amount of memory and the minimum visibility/communication range needed to solve the uniform dispersal problem for any connected orthogonal space whose shape is a priori unknown. In particular, it is known that the robots need to have some *persistent* memory of the past; the problem is in fact unsolvable by *oblivious* robots even if the system is fully synchronous and the robots have unlimited visibility and an unlimited communication range [15]. It is similarly unsolvable by robots that cannot see nor communicate, even if the system is fully synchronous and the robots have unlimited persistent memory [15].

Since oblivious robots cannot deterministically solve the problem, the question is whether the problem is solvable by *finite-state* robots, that is robots with a constant number of bits of non-volatile memory. More specifically, since robots that cannot see nor communicate cannot deterministically solve the problem, the main research question is whether the problem is solvable by finite-state robots that have just a constant visibility radius (and constant communication range).

Two existing results hints that, under very restrictive conditions, the answer to those questions is positive. In fact, it has been proven that finite-state robots with constant visibility and communication radius are able to solve the problem if the system is *fully-synchronous* (FSYNC), which allows perfect coordination between the robots [11]. A similar positive result has been established for *asynchronous* systems (ASYNC) if the space S contains *no holes* (i.e., obstacles completely within S) [15].

In this paper we lift these two restrictions: we allow the robots to be fully asynchronous and the space to contain holes; and we show that the answer is indeed positive in all cases. We constructively prove that robots endowed with only $O(1)$ bits of persistent memory and $O(1)$ visibility radius can always uniformly fill the unknown space, avoiding any collisions; this results can be achieved without any direct means of communication.

In particular, we present a solution protocol that, without using any direct communication, solves the problem for robots having only a few states and a visibility radius $v = 6$ units (i.e. a robot can see up to 6 consecutive cells in each direction). We then investigate the use of direct communication within constant range and show how to use it to decrease the visibility radius, without increasing the memory requirement. Namely, we introduce a second protocol that, using communication radius $c = 1$, solves the problem for finite-state robots having visibility radius $v = 1$; transmitted messages in this protocol are of constant size. A summary of these results is shown in Table 1.

Table 1. Summary of contributions.

Algorithm	Memory of robots	Visibility radius	Communication radius	Asynchronous	Holes allowed
MUTE	$O(1)$	6	No communication	YES	YES
TALK	$O(1)$	1	1	YES	YES

2   Model and Definitions

The system is composed by a set R of mobile entities, called *robots*, whose task is to completely cover a space S that they enter sequentially from the same place. The space S is a connected finite region of the plane possibly with holes; the shape of S is arbitrary (see Fig. 1). Connected means that it is possible to reach any point of S from any other point of S passing only through points in S. A *hole* is a region of the plane which is not part of S and is surrounded completely by points of S; the boundaries of S and holes are called *obstacles*. The region S and its holes are assumed to be partitioned into square cells; each cell can be covered completely by one robot, and each hole is composed of an integral number of cells. Let $|S| = n$ be the number of cells of S. We assume $|R| \geq |S|$.

The robots in R are simple computational entities with sensory and locomotion capabilities. The sensory devices on the robot allows it to have a vision of its immediate surroundings up to a fixed distance called visibility radius[1] v.

---

[1] A visibility radius of one means that the robot sees all eight neighboring cells.

The robots may have or may not have an explicit means of communicating; if available, this ability is also restricted to a fixed distance called communication range[2] c, and each transmission is restricted to a constant number of bits. Both the visibility radius and the communication range remain constant in time. The robots have local sense of orientation, that is a consistent notion of *up-down* and *right-left*; however, they do not have any global positioning mechanism, and have no knowledge of the shape of S other than that it is connected orthogonal grid. The robots are *anonymous*, in the sense that there are no distinct ids and they are externally identical, autonomous, and they all follow the same protocol. Each robot has a constant number of bits of non-volatile working memory; thus the robots are *finite-state* machines with $O(1)$ distinct states.

On entering S, a robot operates in continuous *active-inactive* cycles. When *active*, a robot performs a *Look-Compute-Move* sequence of operations: It first takes a snapshot of its surrounding inside the visibility radius to know which cells are occupied, empty or obstacle (*Look*). Using the information provided by the snapshot and its local state, the robot executes the protocol to determine a new state and a destination cell, which is either the same where it currently resides or one of the four adjacent cells (*Compute*). Finally, it moves to the computed destination (*Move*). It then becomes *inactive*. In case the robots have communication capabilities, a robot may send messages to any robot within its communication range, during the *Compute* stage. The message is immediately received and causes the receiving robot to change its state. Each cycle of activity is assumed to be non-interruptible, in the sense that once they are started they will be completed. However, the robots are *asynchronous*: there is no global synchronization among the cycles of different robots, and the time elapsed between two consecutive operations by the same robot, as well as between two consecutive activations, is finite but arbitrary.

The robots enter S through a special cell called *door*. This cell could be located anywhere and it is indistinguishable from other cells; that is, a robot cannot distinguish the door from other cells using its sensory vision. However, unlike other cells, the door is never empty: whenever a robot leaves the door cell, a new robot appears it. If a robot is in a cell, it completely *covers* it. If two or more robots are in the same cell at the same time then there is a *collision*. Two cells are called neighbors if a robot can move from one to the other in one step. The *distance* between two cells is the smallest number of steps a robot needs to move to reach one cell from the other. The *successor* of a robot r is the robot that entered S just after r and its predecessor is the robot that entered just before r.

The problem to be solved called *uniform dispersal* (or *filling*), requires that within finite time, the entire space is completely *filled*, i.e., every cell of the space is occupied by exactly one robot. Furthermore the system configuration at that time must be *quiescent*, i.e., no robot moves thereafter.

---

[2] A communication range of one means that the robot can communicate directly to the robots located in the eight neighboring cells.

3   Algorithm for Filling without Communication

In this section, we consider robots that do not have any explicit means of communication. The robot can still see other robots (if they are within the visibility range). So the only way for robots to coordinate with each other is by means of their vision and their movements. We present an algorithm (**MUTE**) that succeeds in filling the entire space with such robots, without any collisions, even under the restriction that both the amount of memory available to a robot and its visibility range are constants (independent of the size of the space and the number of robots).

The strategy of *Follow the Leader* introduced in [11], solves the filling problem by moving the robots in a single file, with one leading robot. However, when there are holes, it is possible there are cycles in the path of the leader, which would lead to either a deadlock or a collision. To avoid forming any cycles, the algorithm **MUTE** uses the simple trick of putting up a wall to block any secondary access to the path. In other word, whenever the Leader reaches a cell from which it can move in more than one direction (henceforth such a cell is called a *bifurcation* cell), the leadership is passed to the next robot and the old leader moves to one of the neighbouring cell to block this access until the current path is completely filled.

During the algorithm **MUTE**, the robots can be in one of the following four states:

- **Leader:** There is at most one robot in this state at any time during the execution of the algorithm[3].
- **Follower:** A robot in this state is always in the path followed by the current leader.
- **Blocking:** A robot in this state was a Leader that reached a bifurcation cell and moved to block one of the access to the path. A robot in this state remains stationary until all other possible directions[4] become completely filled. At this moment, if it is possible for this robot to move, then it reassumes the leadership and starts moving again.
- **Stopped:** Any robot in this state was a Leader that reached a dead-end. This is a terminating state so robots that enter this state never move again.

The transition from one state to another, during any execution of algorithm **MUTE**, is defined by the diagram shown in Fig. 2.

Note that the robots need to coordinate with each other to pass the leadership, exchange state information, avoid collisions and so on. Since the robots lack any means of explicit communication, we need to use the visual capabilities of the robots to implement some sort of implicit communication. In algorithm MUTE, this is achieved by assuming a visibility range of six cells (i.e. the robots can see up to a distance of six cells in each direction). This visibility range is

---

[3] Except during the process of transferring the leadership as explained later.

[4] the blocking robot only takes into account the direction that were still open at the moment it enters in the Blocking state.

Fig. 2. States transition diagram of algorithm MUTE where *None* is the state of the robot when it appears at the door.

enough to permit each robot to see the preceding and succeeding robots, and still maintain an appropriate distance between them so that the robots can move in special patterns as a signalling mechanism to exchange information whenever necessary. In order to communicate implicitly using vision, both robots, the sender and the receiver of the information, effectuate some special movements. This is reminiscent of the way some species of animals or insects communicate among themselves (e.g. the dancing of the bees when they find a source of food).

Thus, during the algorithm the robots move along a path maintaining a fixed pattern (i.e. a fixed distance between successive robots). This pattern is broken when a robot (typically the leader) needs to communicate with its successor. The successor realizes the break in the pattern and thus, it knows that a communication process has started. The receiver acknowledges that it has received the information by making a special movement (specified later). Once the information exchange is completed, the two robots continue moving along the path, maintaining the fixed pattern as before. Algorithm 1 presents the rules for movement of a robot r when it is not is the process of communicating with another robot. Algorithm 2 defines the rules for two consecutive robots in the path that are in the process of communicating something.

Algorithm 1. (MUTE): Rules for movement

A robot r, that is not in the process of communicating, moves forward if and only if the following conditions hold:

1. Robot r is not is state **Blocking** or **Stopped**.
2. The cell to which r wants to move is not **occupied**
3. r does not have a predecessor **or** the predecessor is at distance four.
4. r does not have a successor **or** the successor is at distance three **or** the successor is at the door, at a distance less than three.

During the algorithm, the system of robots is always in a *correct configuration* as defined below.

Definition 1. *During algorithm* **MUTE**, *at any time t, the system is said to be in a correct configuration if and only if the following conditions hold:*

- *There is at most one Leader, or, the leadership is being passed from one robot to another (by means of a special communication process).*
- *Let Π be the path followed by the Leader from the door to its current location (If the leadership is being passed we consider the old leader), then:*
  - *Except the Leader, any robot located on Π is in state Follower.*
  - *Any robot in state Follower is located in Π.*
  - *Every cell $p' \notin \Pi$ that has a neighboring cell $p \in \Pi$ is occupied by a robot either in state Blocking or in state Stopped.*
- *Let $R = \{r_i : 0 \le i \le l\}$, be the ordered sequence of robots in Π where l is the number of robot in state Follower, r_l is the leader and r_0 is the robot at the door. And let r_i and r_j, $i, j > 0$, $j = i + 1$, be two consecutive robots in R*
  - *If neither r_i nor r_j is in the middle of a communication process then the distance in Π between them is either three or four.*
  - *If r_i and r_j are at distance less than three, then they are in the process of communicating with each-other.*

Note that the robot r_0 which is at the door, is not required to satisfy the last condition of a correct configuration. This is because, whenever the door becomes empty, a new robot appears at the door. Thus, the pattern of distances cannot be maintained near the door. However, as we will show this does not affect the behavior of the rest of the robots.

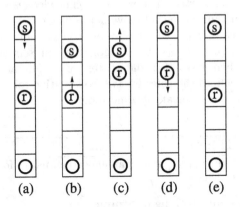

Fig. 3. Behavior of sender robot s and receiver robot r, during a communication process.

We shall now explain in more details when and how communications take place in algorithm **MUTE**. During any execution of the algorithm there are four situations in which a robot needs to communicate some information to its successor. We describe below these four types of communications called **Blocking**, **Stopping**, **Bifurcation** and **Reopening**.

Algorithm 2. (MUTE): Rules for implicit communication

Robot s needs to notify something to its successor r which is at distance three (see Figure 3).

1. s waits until r's successor becomes visible. This is to ensure that r's successor knows that there is a communication process in progress ahead so it does not move until the communication is completed.
2. s moves one step back to notify a communication of either a **Bifurcation** or **Reopening** (Figure 3 (b)). As explained later, the visibility range of the robots and the pattern maintained by the robots in the path, allows robot r to realize which kind of communication is taking place. After the move, robot s waits for an acknowledgement from robot r.
3. To acknowledge the receipt of the communication, robot r now moves one step ahead and waits for a notification from robot s (Figure 3 (c)).
4. s signals the end of the communication process by moving one step ahead; it then waits for an acknowledgement again (Figure 3 (d)).
5. Robot r acknowledges by moving one step back. At this time, the communication has ended and both robots are back in their original positions (Figure 3 (e)).

---

1. **Blocking:** This situation occurs when the leader reaches a cell with more than one possible direction of movement (see Fig. 4). In this case the leader (the dark-gray robot in the figure) does not need to take any special action. The leader waits until its successor (the white robot) is at distance three; it then moves to one of the possible directions and changes its state to Blocking (the light-gray robot in the figure). When the successor robot was at distance four (Fig. 4(a)), it realizes, before its next move, that its predecessor has more than one possible movement. So after the move, the successor becomes the new leader (the dark-gray robot in Fig. 4(b)), while the predecessor changes to state Blocking. As a result, the leadership was passed from one robot to another and the communication process ends here.

2. **Stopping:** This situation occurs when the leader r_l is moving to the end of the path and it has only one more move to perform (Fig. 5(b)). In this case, the Leader's successor r_s, cannot distinguish whether its predecessor is in the middle of a communication process with some robot ahead (Fig. 3) or, the predecessor is the leader going to a dead-end (Fig. 5(b) and (e)). Thus r_s will wait at distance four from r_l. The Leader r_l moves one step ahead to communicate to the successor r_s that the leader has reached a dead end and the leadership should be passed to another robot. Note that if r_l was at a bifurcation cell (before the move) then it will now become **Stopped** (Fig. 5(c)). In this case, the next **Blocking** robot will notify r_s ·which is the new direction. Otherwise, if r_l was not at a bifurcation cell, it will become **Blocking** (Fig. 5(f)), and then communicate to r_s that the path is going to a dead-end. This communication is similar to the **Reopening** communication (described below) when a new direction of movement is communicated

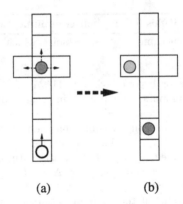

(a)   (b)

Fig. 4. (a) The leader (colored dark-grey) reaches a cell with more than one possible moves (b) It blocks one direction and passes leadership to the successor.

(except that the new direction is actually a dead-end in this case, and the robot r_s will realize that).

3. **Bifurcation:** This situation occurs when a robot, either a Follower or a Leader, is in a bifurcation cell with only one possible movement (Fig. 6), so the robot must communicate to its successor the information that this is at a bifurcation cell.

4. **Reopening:** A *Blocking* robot b must reopen the blocked direction because all other directions that were open before, are now completely filled (Fig. 5(c) and (e)). It is possible that are other *Blocking* robots in neighboring cells, but if those directions were blocked before b became *Blocking*, then robot b has the higher priority to reopen its blocked direction.

In the two last types of communication (*Bifurcation* and *Reopening*) robots must take some special action in order to communicate. The behavior of the two robots during these two types of communication is essentially the same and follows the rules of Algorithm 2. Notice that whenever two robots start such a communication process they do nothing else until the communication has terminated. In this sense, the communication process can be thought of as an atomic action.

There are a few complications that can occur during the communication process of Algorithm 2, due to the special condition for robots at the door. First consider what happens if at the start of the communication process, the successor of the receiver (r_s) is at distance less than three from the receiver (r). There are two cases when this situation could occur. Either when r_s is at door or when r_s itself is communicating with r. In the first case r_s will not move since its predecessor is at distance less than three. And, in the second case, r_s will move back just one step returning to distance three from r. Notice that r will not move until r_s moves back since they are in a communication process. After that, r can start communicating with its predecessor s. Thus, both communications would be successful. Now, let us consider the case when the receiver robot r is

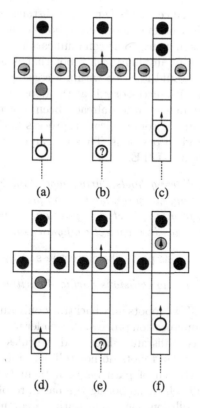

Fig. 5. The leader (colored dark-grey) is moving to the end of the path. The white robot is a Follower which is next robot in the path. Black robots are in state Stopped, while Grey robots with an arrow are in state Blocking (i.e. blocking the exit in the direction of the arrow).

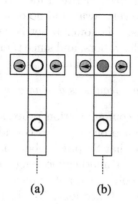

Fig. 6. Either a follower (colored white) or the leader (colored grey) is in a bifurcation cell.

at the door. In this case, since robots can see at distance six and all the robots enter the space through a unique door, any robot at the door can distinguish among the four scenarios in which some communication is needed. Thus it is not actually necessary to notify anything to a robot that is at the door. Moreover since the sender robot s would itself be within a constant distance of the door, it can remember whether its successor r is at the door or not. Thus the robot s would act as if the communication has already been sent to r.

Finally notice that after a communication process both, the sender and the receiver, end at their original positions. Thus we can ensure the following properties about the algorithm **MUTE**.

Proposition 1. *The following holds during algorithm* **MUTE***: (1) Starting from a correct configuration, every communication process ends in a correct configuration. (2) Starting from a correct configuration, every movement following the rules of Algorithm 1 ends in a correct configuration.*

The following propositions prove the correctness of the algorithm.

Proposition 2. *There are no collisions during Algorithm* **MUTE***.*

Proof. In algorithm **MUTE**, robots may backtrack only during a communication process. During the communication process, the sender moves back only when the receiver (its successor) is at distance three, and the rules of the communication process ensure that these two robots do not collide. Moreover these two robots do not go beyond the stretch of path between them (when they started the communication), and all cells neighboring the path are blocked, in any correct configuration. Thus no collisions can occur during a communication process.

Let us now assume that no communication takes place and suppose a collision occur the normal movement of the robots. Since robots never move to an occupied cell, such a collision must involve either Follower or Leader robots that move to the same cell simultaneously. Now, the movements of a Follower is restricted to the part of the path between its current cell and its predecessor which is always at least three cells away and no other robot can enter this path as all exits are blocked. Thus, a Follower robot is never involved in a collision. This leaves us with only Leader robots, and since there is at most one such robot at any time, there cannot be any collisions.

Proposition 3. *There are no deadlocks during any execution of the algorithm.*

Proof. First notice that any communication process ends within a finite time. Outside of the communication process, let us consider the cases when a robot waits for another robot. Note that Stopped robots have already terminated and a blocking robot never waits for another blocking robot, so any deadlock must involve at least a Leader or Follower robot. The only case when a leader waits is when its successor s is at distance four. But in this case successor s would move unless the successor of s is also at a distance four from s. This implies that every robot in the path has its successor at distance four. Otherwise the robot that is closer than distance four from its successor, would move, contradicting the existence of deadlock. However, since the robot at the door does not have any successor, it could move and this contradicts the existence of a deadlock. Thus no leader or follower could be involved in a deadlock and thus there are no deadlocks during the algorithm.

Proposition 4. *Algorithm* **MUTE** *always terminates within a finite time.*

Proof. We have already seen that the algorithm ensures there are no cycles in the path and thus no robot visits the same cell twice except during a communication process. A communication process always ends in finite time, and at that both the receiver and the sender are back in their original positions. Moreover, according to the rules of the algorithm no pair of robots communicates more than twice from the same cells. So, within a finite time the robots involved in a communication, must either moves to a new cell or stop moving. Since there are only a finite number of new cells, all robots must stop after a finite time.

Proposition 5. *Algorithm* **MUTE** *always completely fills the space S.*

Proof. Suppose for the sake of contradiction that after the algorithm terminates there are some empty cells and let us consider an empty cell p that is adjacent to an occupied cell q (such a cell must exist as S is connected). Since the algorithm has terminated, the robot r in cell q is in state *Stopped*. This implies that when r became Stopped, either cell p was occupied or it was the previous location of robot r. In either case cell p had been occupied by some robot in the past. Consider the last robot r_l to visit p. When robot r_l moved out of cell p, it must have a successor and thus the successor must have visited cell p too. This contradicts the fact r_l was the last robot to visit cell p. Thus, there cannot be any empty cells after the termination of the algorithm.

The above results prove the correctness of the algorithm. Recall that the robots can be in only five states and in each state a robot needs to remember only a constant amount of additional information. To summarize:

Theorem 1. *Algorithm* **MUTE** *solves the filling problem for any connected space of n cells, without collisions, and using exactly n robots, each of which has $O(1)$ memory and a visibility radius of six.*

4   Algorithm for Filling with Explicit Communication

The algorithm in the previous section used robots having a visibility radius of six to solve the Filling problem. We now show that if explicit communication is allowed between robots (even within short ranges), then the visibility radius of the robots could be reduced one. Note that this is the minimum visibility radius necessary to solve Filling without any collisions [15].

In this section we present an algorithm (**TALK**) which solves the filling problem when both the visibility and communication ranges are limited to distance one. The algorithm still requires only a constant amount of memory for each robot, and fills the complete space without any collisions. Algorithm **TALK** also uses the basic strategy of follow-the-leader, but no robot moves until its previous position is occupied by the successor. This ensures that the distance between consecutive robots in the path is never more than two (at most one cell between consecutive robots). If the leader robot r is about to move to a cell x that was visited before (e.g. when the path has a cycle), robot r can determine whether cell x has been already visited, by asking the robots in the cells neighboring x (Robot r can see at least two such cells and one of them would contain a robot). Thus the algorithm can avoid collisions. During the algorithm, the leader always moves to an unvisited cell, while the followers always move to the previous position of their predecessor. If the leader does not have any unvisited neighboring cell, it stops moving and passes the leadership to its successor.

The robots could be in one of the following states while executing this algorithm:

1. **Leader:** At any time, there is at most one robot in this state. This robot always moves to an unvisited cell. If there is no possible movement it passes the leadership to its successor and changes to state **Stopped**.
2. **Follower:** A robot in the state is always on the path from the door to the current **Leader**. A Follower robot always moves to the previous position of its predecessor unless it receive the leadership (in this case it changes to state **Leader**).
3. **Stopped:** A Robot in this state has terminated and does not ever move again during the execution of the algorithm.

A new robot appearing at the door changes to state *Follower* or *Leader* depending on whether or not it sees any robots in the neighboring cells.

The following variables and functions are used by a robot r to store and share information about past and future moves:

- $r.Entry$ is the cell from which r entered its current position. This variable is NULL if r is at the door.
- $r.Exit$ is the cell to which r should move in the next step. This variable is NULL for the Leader and the Stopped robots.
- *Find-Next-Move(NewCell)* is a function that returns true or false depending on whether or not there is a valid next move (i.e. an unvisited neighboring cell). The possible destination is returned as parameter *NewCell*.

The communication between robots uses the following types of messages:

1. **Leadership:** This message is sent by the Leader to transfer the leadership to its successor.
2. **Next-Pos:** This message is sent by either a Leader or a Follower, just before moving. This informs their successor about the next position in the path.

When a robot r receives a *Next-Pos* message from the cell x, the variable $r.Exit$ is set to cell x. The rules of the algorithm are presented in Algorithm 3. The implementation of function *Find-Next-Move()* is straightforward and it is omitted.

Algorithm 3. (TALK) : Rules followed by robot r

---

if ($r.State = None$) **then**
  if (all neighboring cells are **empty**) **then**
    $r.State \leftarrow Leader$
  else
    $r.State \leftarrow Follower$
  end if
else if ($r.State = Follower$) **then**
  if (($r.Exit \neq$ **NULL**) $\wedge$ ($r.Entry$ is **occupied**)) **then**
    $NewPos \leftarrow r.Exit$
    $r.Exit \leftarrow$ **NULL**
    Send message $Next\text{-}Pos$ to $r.Entry$.
    $r.Entry \leftarrow$ Current cell.
    Move to $NewPos$
  end if
else if ($r.State = Leader$) **then**
  if ($r.Entry$ is **occupied**) **then**
    if ($Find\text{-}Next\text{-}Move(NewCell) = $ **true**) **then**
      Send message $Next\text{-}Pos$ to $r.Entry$.
      $r.Entry \leftarrow$ Current cell.
      Move to $NewCell$
    else
      Send message $Leadership$ to $r.Entry$.
      $r.State \leftarrow Stopped$
    end if
  end if
end if

We can prove the following properties for the algorithm.

Proposition 6. *There are no collisions during Algorithm* **TALK**.

Proof. As in the previous algorithm, a robot never moves to an occupied cell. Furthermore, according to the rules of the algorithm, no robot backtracks and a Follower robot can move only to the last position of its predecessor, when this cell becomes empty. Since only the Leader can move to unvisited cells, no collisions are possible. Therefore, algorithm TALK is collision free.

Proposition 7. *Algorithm* **TALK** *terminates in finite time.*

Proof. According to the rules of the algorithm, the Follower robots always move on the path of Leader while the Leader only moves to unvisited cells and passes the leadership to its successor whenever it cannot find any empty cell to move to. Since the number of cells are finite, any Leader must eventually stop and pass the leadership. Every new leader will also stop in finite time and, thus every robot on the path will stop. Therefore the algorithm terminates in a finite time when every robot has stopped.

Proposition 8. *Algorithm* **TALK** *completely fills the space* S.

Proof. Suppose that after termination of the algorithm, at time t, there are some cells that remain empty. Let p be such a cell. Notice that if this cell was visited before then there must be a *Follower* robot r in one of its neighboring cells and the next time this robot executes, it will move to cell p, contradicting the fact of the termination of the algorithm. Thus, every empty cell must be an unvisited cell.

However since the space is connected, there must be an empty (and hence unvisited) cell adjacent to an occupied cell q. Consider the last robot that visited cell q. This robot is now Stopped (since the algorithm has terminated). But according to the rules of the algorithm, a robot cannot become *Stopped* if there is an unvisited cell adjacent. This contradiction proves that there cannot be any empty cells after the termination of the algorithm. Thus, the algorithm completely fills S.

We summarize the above results as follows:

Theorem 2. *Algorithm* **TALK** *solves the filling problem for any connected space of size n without collisions and using only n robots each having a constant amount of memory and a visibility and communication radius of one.*

5   Conclusions

We showed how to solve the filling (or, uniform dispersal) problem for unknown spaces of arbitrary shapes possibly containing holes, using asynchronous robots having $O(1)$ visibility and $O(1)$ memory. While our solutions are asymptotically optimal in the memory and visibility requirements, it would be interesting to determine the exact bounds on v, the visibility radius required for solving the problem. Our solution gives an upper bound of $v = 6$ when there is no explicit communication, while only a trivial lower bound of $v = 1$ is known. Another possible research direction is to investigate fault-tolerant algorithms for the problem (e.g. one can consider self-stabilizing algorithms to tolerate corruptions in the memory of the robots).

References

1. Barriere, L., Flocchini, P., Mesa Barrameda, E., Santoro, N.: Uniform scattering of autonomous mobile robots in a grid. Int. J. Found. Comput. Sci. **22**(3), 679–697 (2011)
2. Cheng, T.M., Savkin, A.V.: A distributed self-deployment algorithm for the coverage of mobile wireless sensor networks. IEEE Commun. Lett. **13**(11), 877–879 (2009)
3. Cohen, R., Peleg, D.: Local algorithms for autonomous robot systems. In: Flocchini, P., Gąsieniec, L. (eds.) SIROCCO 2006. LNCS, vol. 4056, pp. 29–43. Springer, Heidelberg (2006)

4. Flocchini, P., Prencipe, G., Santoro, N.: Self-deployment of mobile sensors on a ring. Theor. Comput. Sci. **402**(1), 67–80 (2008)
5. Flocchini, P., Prencipe, G., Santoro, N.: Computing by mobile robotic sensors. In: Nikoletseas, S., Rolim, J. (eds.) Theoretical Aspects of Distributed Computing in Sensor Networks. Springer, Berlin (2011)
6. Ganguli, A., Cortes, J., Bullo, F.: Visibility-based multi-agent deployment in orthogonal environments. In: Proceedings American Control Conference, pp. 3426–3431 (2007)
7. Heo, N., Varshney, P.K.: A distributed self spreading algorithm for mobile wireless sensor networks. In: Proceedings IEEE Wireless Communication and Networking Conference, vol. 3, pp. 1597–1602 (2003)
8. Heo, N., Varshney, P.K.: Energy-efficient deployment of intelligent mobile sensor networks. IEEE Trans. Syst. Man Cybern. Part A **35**(1), 78–92 (2005)
9. Howard, A., Mataric, M.J., Sukahatme, G.S.: An incremental self-deployment algorithm for mobile sensor networks. IEEE Trans. Robot. Autom. **13**(2), 113–126 (2002)
10. Howard, A., Mataric, M.J., Sukhatme, G.S.: Mobile sensor network deployment using potential fields: a distributed, scalable solution to the area coverage problem. In: Proceedings 6th International Symposium on Distributed Autonomous Robotics Systems (DARS), pp. 299–308 (2002)
11. Hsiang, T.R., Arkin, E., Bender, M.A., Fekete, S., Mitchell, J.: Algorithms for rapidly dispersing robot swarms in unknown environment. In: Proceedings of the 5th Workshop on Algorithmic Foundations of Robotics (WAFR), pp. 77–94 (2002)
12. Li, X., Frey, H., Santoro, N., Stojmenovic, I.: Strictly localized sensor self-deployment for optimal focused coverage. IEEE Trans. Mob. Comput. **10**(11), 1520–1533 (2011)
13. Loo, L., Lin, E., Kam, M., Varshney, P.: Cooperative multi-agent constellation formation under sensing and communication constraints. In: Cooperative Control and Optimization, pp. 143–170. Kluwer Academic, Dordrecht (2002)
14. Martinson, E., Payton, D.: Lattice formation in mobile autonomous sensor arrays. In: Proceedings of International Workshop on Swarm Robotics (SAB'04), pp. 98–111 (2004)
15. Mesa Barrameda, E., Das, S., Santoro, N.: Deployment of asynchronous robotic sensors in unknown orthogonal environments. In: Fekete, S. (ed.) ALGOSENSORS 2008. LNCS, vol. 5389, pp. 125–140. Springer, Heidelberg (2008)
16. Poduri, S., Sukhatme, G.S.: Constrained coverage for mobile sensor networks. In: Proceedings of the IEEE International Conference on Robotic and Automation, pp. 165–173 (2004)
17. Wang, G., Cao, G., La Porta, T.: Movement-assisted sensor deployment. In: Proceedings of the IEEE INFOCOM, vol. 4, pp. 2469–2479 (2004)

On the Complexity of Fixed-Schedule Neighbourhood Learning in Wireless Ad Hoc Radio Networks

Avery Miller[✉]

Department of Computer Science, University of Toronto, Toronto, Canada
a4miller@cs.toronto.edu

Abstract. Consider a synchronous static radio network of n nodes represented by an undirected graph with maximum degree Δ. Suppose that each node has a unique ID from $\{1, \ldots, N\}$, where $N \gg n$. In the *complete neighbourhood learning task*, each node p must produce a set L_p of IDs such that ID $i \in L_p$ if and only if p has a neighbour with ID i. We study the complexity of this task when it is assumed that each node fixes its entire transmission schedule at the start of the algorithm. We prove a $\Omega(\frac{\Delta^2}{\log \Delta} \log N)$-slot lower bound on schedule length that holds in very general models, e.g., when nodes possess collision detectors, messages can be of arbitrary size, and nodes know the schedules being followed by all other nodes. We also prove a similar result for the SINR model of radio networks. To prove these results, we introduce a new generalization of cover-free families of sets, which may be of independent interest. We also show a separation between the class of fixed-schedule algorithms and the class of algorithms where nodes can choose to leave out some transmissions from their schedule.

1 Introduction

Neighbourhood learning is an important step in wireless network initialization and in algorithms for tasks such as routing, medium-access control, topology control, and gossiping. Further, in the study of local computation in distributed computing, it is assumed that information about neighbouring nodes has already been collected, which is non-trivial in the case of wireless radio networks. If it is not known how to collect neighbourhood information in an efficient way, or, if we are able to prove a strong lower bound for neighbourhood learning, then the actual running time of a solution that depends on this information can be significantly worse than its running-time analysis suggests. In many of these applications, it is important that each node learns their *entire* neighbourhood before proceeding to other tasks, so we focus our attention on deterministic algorithms that guarantee full neighbourhood discovery within a bounded amount of time that is known in advance.

The main challenge that is encountered when designing algorithms for wireless radio networks is the possibility of radio interference. When several nodes

P. Flocchini et al. (Eds.): ALGOSENSORS 2013, LNCS 8243, pp. 244–259, 2014.
DOI: 10.1007/978-3-642-45346-5_18, © Springer-Verlag Berlin Heidelberg 2014

transmit during a single time slot, the signals from all transmissions may prevent a nearby listening node from receiving any message. This is known as a *transmission collision*. Unless otherwise specified, we consider the Unit-Disk Graph (UDG) radio network model in this paper: a collision occurs at p if two or more neighbours of p transmit in the same time slot, or, if a neighbour of p transmits during the same time slot that p does. A node p receives a message from a neighbour q if q transmits and no collision occurs at p. In the basic model, p cannot tell the difference between a collision and the case where no neighbour transmits. With a *weak* collision detector, p can make this distinction, as long as p is not transmitting. With a *strong* collision detector, p can always distinguish between a collision and silence.

Our eventual goal is to determine the complexity of the neighbourhood learning in both static and mobile networks. Most of the results in the literature concentrate on determining upper bounds for this task in various models. However, good lower bounds are missing, even if we restrict attention to the class of algorithms where nodes follow a fixed schedule. Specifically, a t-slot *fixed-schedule algorithm* for the complete neighbourhood learning task is a deterministic algorithm run by each node p such that: p knows its entire t-slot schedule at time slot 0; p transmits a message during each slot for which the entry in its schedule is 1, and otherwise stays silent; and, at the end of slot t, p outputs a list of all of its neighbours. We can classify existing fixed-schedule algorithms as one of three types:

1. **Collision-Free Algorithms:** In every execution, there is never a time slot during which a transmission collision occurs.
2. **Local Broadcast Algorithms:** For each node p in the network, there exists a time slot during which p transmits and no transmission collisions occur at p's neighbours.
3. **Direct-Discovery Algorithms:** For each pair of nodes (p, q) in the network, there exists a time slot during which p transmits and no transmission collision occurs at q.

The best known fixed-schedule solution is a $O(\Delta^2 \log N)$-slot direct-discovery algorithm based on cover-free families (equivalently, strongly-selective families), which we describe in Sect. 3.2. However, it is not known if this is the optimal fixed-schedule solution. Of course, this might depend on the particular choice of model: direct-discovery algorithms work in models where nodes do not possess collision detectors, have no knowledge of the schedules used by other nodes, and can only send their own ID in every message. Perhaps, without these restrictions, there are algorithms that can do better. For example, in multi-hop networks, nodes can forward messages that they have received in previous slots, which can help other nodes infer who their neighbours are. As another example, with strong collision detection and knowledge of every node's schedule, a node can learn from silence: all nodes scheduled to transmit in a slot where no message was received and no collision was detected can be eliminated as possible neighbours. Even if it is difficult to imagine how to use such additional information to devise algorithms that do significantly better than direct discovery, it is important to formally verify whether or not it is possible, and to understand the reasons why.

Our main contribution in this paper is a $\Omega(\frac{\Delta^2}{\log \Delta} \log N)$ lower bound for general fixed-schedule neighbourhood learning algorithms for UDG networks models, even when nodes possess strong collision detectors, have knowledge of every node's schedule, and can send arbitrary messages (Sect. 4). Nodes with these features will be referred to as *strong* nodes. The fact that the lower bound holds for networks of strong nodes strengthens our result, since the lower bound automatically applies to models where nodes do not have these features. For networks of strong nodes, we also prove a separation between fixed-schedule algorithms and the class of algorithms where nodes can choose to leave out transmissions from their schedule (Sect. 5). Our lower bound for fixed-schedule neighbourhood learning depends on size bounds for a new generalization of cover-free families that we call *thick* cover-free families (Sects. 2.4 and 6). Further, we use our results about these combinatorial objects to prove, to our knowledge, the first non-trivial lower bound for fixed-schedule neighbourhood learning in SINR models (Sect. 7).

2 Models and Definitions

2.1 Network Model

A static ad hoc network consists of n nodes at arbitrary fixed locations. Nodes do not have information about their location, and nodes know the value of n. Each node p possesses a unique identifier number $ID(p)$ from the range $\{1, \ldots, N\}$, where $N \gg n$. All nodes know the value of N. Denote by $p(i)$ the node with identifier i. The topology of a network is represented as an undirected graph, with a vertex for each node and an edge joining each pair of neighbours. The maximum degree of the network is denoted by Δ and we assume that nodes know the value of Δ. For any node p, the set of nodes that are neighbours of p is known as p's *neighbourhood*. We consider the task of *complete neighbourhood learning*, where each node p must produce a set L_p of IDs such that ID $i \in L_p$ if and only if p has a neighbour with ID i.

At any given time, a node can either *transmit* or *listen*, but not both. The signal transmitted by a node p reaches p and all neighbours of p. We consider networks where the nodes share a single radio channel, which means that two signals that reach the same point at the same time interfere with one another. A listening node that receives two or more signals during a single slot t only hears noise, and we say that a *collision* has occurred during slot t. A collision also occurs if a transmitting node receives one or more other signals. In the case when exactly one neighbour of a listening node transmits, we say that the listening node *receives a message* from the transmitting neighbour. In one radio model, we will assume that nodes cannot distinguish between a collision and silence. In another model, we will assume that nodes possess *strong collision detectors*, so that, whether they are listening or transmitting, they can detect that a collision has occurred.

Each node possesses a clock that divides time into equal-length *slots*, $(t_0, t_1, \ldots)$. Each slot is long enough to allow the complete transmission of any message. We consider *synchronous* models, in which it is assumed that all clocks run at the same rate, that slot boundaries coincide across all nodes, and that each node begins its local algorithm at time slot 1. Time slot 0 represents the initial state of the system. The model allows any set of nodes to transmit during a single slot.

We say that a network consists of *weak* nodes if: nodes cannot distinguish between a collision and silence; each node can only send its own ID in each message; and, no node has any knowledge about the schedules of other nodes. A network consists of *strong* nodes if: nodes possess strong collision detectors; nodes can send arbitrary messages; and, each node initially knows the schedule associated with each node ID.

2.2   Schedules and Algorithms

A node's *schedule* T is a $\{0,1\}$-vector that indicates during which slots it will transmit. Entry $T[i]$ of a node's schedule is 1 if and only if the node transmits during time slot i. A *schedule matrix* S is a $\{0,1\}$-matrix with t rows and N columns. Column j is the schedule of the node with ID j. A family of sets $\{S_1, \ldots, S_N\}$ over $\{1, \ldots, t\}$ can be *represented by* a $\{0,1\}$-matrix with t rows and N columns: entry $M_{i,j}$ is 1 if and only if $i \in S_j$. We will use this fact to relate families of sets with node transmission schedules.

A t-slot *fixed-schedule algorithm* for the complete neighbourhood learning task is a deterministic algorithm run by each node p such that: p knows its entire t-slot schedule at time slot 0; p transmits a message during each slot for which the entry in its schedule is 1, and otherwise stays silent; and, at the end of slot t, p outputs a list of all of its neighbours. A *non-adaptive algorithm* is a fixed-schedule algorithm in which the sequence of messages sent by each node is the same in every execution.

2.3   Cover-Free Families

For any set S, an *r-cover for* S is a collection of r sets other than S, whose union contains S. For $r \geq 1$, an *r-cover-free family* $\mathfrak{F}$ is a collection of N subsets of $\{1, \ldots, t\}$ such that, for each $S \in \mathfrak{F}$, there is no r-cover for S consisting of sets from $\mathfrak{F}$. We will say that t is the *length* of $\mathfrak{F}$ and N is the *size* of $\mathfrak{F}$. This reflects the fact that the schedules that we construct from cover-free families will have t time slots and will provide schedules for up to N nodes. Note that it is easy to find cover-free families of small size (e.g., any family consisting of exactly one set is r-cover-free for all $r \geq 1$) and that it is difficult to construct cover-free families of large size but small length.

Many related results in the literature refer to families of sets that are called *strongly-selective families*. A family $\mathcal{G}$ of subsets of $\{1, \ldots, u\}$ is *k-strongly-selective* if, for each $Z \subseteq \{1, \ldots, u\}$ with $|Z| \leq k$ and for each $z \in Z$, there exists a set $G \in \mathcal{G}$ such that $G \cap Z = \{z\}$. Strongly-selective families and cover-free families can be viewed as 'duals' of one another, as remarked in [3]: when represented in matrix form (see Sect. 2.2), the columns of a matrix form an r-cover-free family consisting of N subsets of $\{1, \ldots, t\}$ if and only if the rows of the matrix form an $(r+1)$-strongly-selective family consisting of t subsets of $\{1, \ldots, N\}$. It follows that asymptotic bounds on the size of cover-free families also apply to strongly-selective families (and vice versa). The definition of *strongly-selective families* first appears in Clementi et al. [3], where they were used to solve the multi-broadcast problem in ad hoc wireless radio networks.

2.4   Thick Cover-Free Families

Our new generalization of cover-free families captures the 'thickness' of a cover. Namely, we would like to specify *how many times* the elements of a set S appear in the sets $S_1, \ldots, S_r$. This information is lost if we use the usual definitions of sets and unions. A *multiset* is a generalization of a set in which there can be multiple copies of the same element. For any multisets $F = [f_1, \ldots, f_m]$ and $G = [g_1, \ldots, g_\ell]$, let $F \uplus G = [f_1, \ldots, f_m, g_1, \ldots, g_\ell]$ be the *multiset union* of F and G. For c sets, we denote the multiset union by $\uplus_{i=1}^{c} S_i$. For any set F, denote by $\uplus^b F$ the multiset consisting of b copies of each element of F. Note that, for any multisets F, G, $|F \uplus G| = |F| + |G|$.

Definition 1. *For any set S, an r-cover of thickness b for S is a family of r sets other than S, whose multiset union contains at least b copies of each element in S.*

Definition 2. *A family $\mathfrak{F}$ of sets is r-cover-free for thickness b if, for every set $S \in \mathfrak{F}$, there does not exist an r-cover of thickness b for S consisting of sets from $\mathfrak{F}$.*

Note that an r-cover of thickness 1 is equivalent to a traditional r-cover, and that an r-cover-free family for thickness 1 is equivalent to a traditional r-cover-free family. Also, any cover-free family for thickness b is also a cover-free family for any thickness $b' > b$.

As far as we know, there is no 'dual' definition for thick cover-free families in the literature. So, we propose *k-strongly-b-selective families* with the following definition: for each Z with $|Z| \leq k$ and for each $z \in Z$, there exists a set $G \in \mathcal{G}$ such that $z \in G \cap Z$ and $|G \cap Z| \leq b$. When represented in matrix form, it is not hard to see that the columns form an r-cover-free family for thickness b consisting of N subsets of $\{1, \ldots, t\}$ if and only if the rows form an $(r+1)$-strongly-b-selective family consisting of t subsets of $\{1, \ldots, N\}$.

3   Known Results

3.1   Cover-Free Families

Cover-free families were defined in Erdös, Frankl, and Füredi [12], but the concept was first introduced by Kautz and Singleton [17] in their study of superimposed binary codes. In particular, they were called *zero-false-drop codes of order r*. When the columns of a matrix are taken to be codewords of a zero-false-drop code of order r, the resulting matrix is called *r-disjunct*. These matrices play a central role in non-adaptive solutions to *combinatorial group testing* problems, where the goal is to identify a small number of defective items from a large set by performing tests on groups of items (see Du and Hwang [7] for a survey on this topic). There are also several generalizations of cover-free families (and strongly-selective families) in the literature [4,5,8,10,24].

Erdös et al. [12] provide a non-constructive proof that there exist r-cover-free families consisting of N subsets of $\{1,\ldots,t\}$ such that $t \in O(r^2 \log N)$. Porat and Rothschild [21] provided the first construction of cover-free families that meet this asymptotic bound. Further, Erdös et al. [12] showed that if $N < \binom{r+2}{2}$ (e.g., if $r \geq 2\sqrt{N} - 1$), then $N \leq t$. The family consisting of $\{1\},\ldots,\{t\}$ meets this bound with equality. For any r-cover-free family consisting of N subsets of $\{1,\ldots,t\}$, a proof that $t \in \Omega((r^2/\log r)\log N)$ has been provided by D'yachkov and Rykov [9], Ruszinkó [22], Chaudhuri and Radhakrishnan [1], and Füredi [13].

3.2   Neighbourhood Learning

Previous work about neighbourhood learning has focused mainly on upper bounds. A trivial upper bound of N slots is achieved by a round-robin algorithm where a node with ID i transmits during time slot i. More efficient algorithms use schedules in which some transmissions may be lost due to collisions. In networks of weak nodes, one solution is to use a schedule based on cover-free families: given a Δ-cover-free family $\mathfrak{F} = \{S_1,\ldots,S_N\}$ where each $S_j \subseteq \{1,\ldots,t\}$, the node with ID j transmits its ID during time slot s if and only if $s \in S_j$. Such a schedule guarantees that every node successfully transmits its ID to all of its neighbours. To see why, suppose to the contrary that there is a node p_i whose ID is not received by some neighbour p_j. This means that during every slot that p_i transmitted, it must be the case that either p_j or one of p_j's other neighbours (there are at most $\Delta-1$ of these) transmitted. It follows that the set of p_i's transmission slots is contained in the union of Δ other sets of transmissions slots, that is, there is a Δ-cover for S_i in the family of transmission schedules. The length of the algorithm is at most t, which, using the construction of cover-free families by Porat and Rothschild [21], leads to an algorithm that uses $O(\Delta^2 \log N)$ slots (or N slots if $\Delta \geq 2\sqrt{N} - 1$).

The algorithm based on cover-free families is relatively easy to implement since it is *non-adaptive*: each execution of the algorithm follows the same schedule and sends the same messages, regardless of the network topology. A fully-adaptive algorithm for networks of weak nodes was provided by Gasieniec et al. [14] and uses $O(\Delta^2 \log N)$ slots when $\Delta < \sqrt{n}$, or $O(n \log^2 n \log^2 N)$ slots otherwise. When nodes can distinguish between a collision and silence, Mittal et al. [20] devised a fully-adaptive algorithm for neighbourhood learning that uses $O(n \log N)$ slots. This algorithm works even if nodes do not know an upper bound for Δ (or, for that matter, n). In these fully-adaptive algorithms, nodes decide during which slots they will transmit based on messages that they receive during the execution. Though these two algorithms have better asymptotic running times than the solution based on cover-free families when Δ is large, they are much more difficult to describe and implement.

In networks of weak nodes, several lower bounds for neighbourhood learning algorithms are known. Krishnamurthy et al. [19] showed that the best-case running time of any collision-free algorithm for neighbourhood learning is at least $N - n$ slots. A straightforward adversary argument shows that the worst-case running time of any collision-free algorithm is at least $N - 1$ slots if nodes know their degree in advance and N slots if they do not. It is not too difficult to show that any direct-discovery fixed-schedule algorithm that solves the neighbourhood learning task must use a schedule that corresponds to a Δ-cover-free family. Using the known size bounds from Sect. 3.1, this gives a lower bound of $\Omega((\Delta^2/\log \Delta) \log N)$ time slots.

For general fixed-schedule neighbourhood learning in networks of strong nodes, we know of only one non-trivial lower bound. Note that, for any neighbourhood learning algorithm, we require that each node is able to distinguish between the case where it has 0 neighbours and the case where it has at least 1 neighbour. This means that the schedule used by a fixed-schedule neighbourhood learning algorithm must correspond to a Δ-selective family, as defined in [2]. Clementi et al. [3] show that any such schedule has at least $(\Delta/24) \log (N/\Delta)$ time slots (when $\Delta \leq N/64$). Our main result in this paper gives a better lower bound for general fixed-schedule neighbourhood learning algorithms that matches the $\Omega((\Delta^2/\log \Delta) \log N)$-slot lower bound for direct-discovery algorithms.

4   A Lower Bound for Neighbourhood Learning with Strong Nodes

In this section, we show that any fixed-schedule algorithm for the neighbourhood learning task in UDG networks of strong nodes uses at least $\Omega((\Delta^2/\log \Delta) \log N)$ slots.

First, for any fixed-schedule algorithm for the neighbourhood learning task, we show that the algorithm's schedule matrix represents a Δ-cover-free family for thickness 2. A node p's *history* $h_{p,\alpha}$ is a vector that stores the messages that it has received during the execution α of an algorithm. Entry i of a node p's history, $h_{p,\alpha}[i]$, is equal to: the message that p received during slot i, if at least one neighbour of p transmitted during slot i and no collision occurred at p; $*$, if a collision occurred at p; $\perp$, otherwise. Two executions α, β of an algorithm are *indistinguishable to a node p up to time slot s* if $h_{p,\alpha}[1 \ldots s] = h_{p,\beta}[1 \ldots s]$.

Observation 1. *For any t-slot algorithm, if two executions are indistinguishable to a node p up to time slot t, p outputs the same list of neighbours in both executions.*

For any column j of a schedule matrix $\mathcal{S}$, let S_j be the set of time slots during which the node with ID j is scheduled to transmit. Consider any $k \in \{1, \ldots, N\}$ and any set $\mathcal{C} = \{k_1, \ldots, k_r\} \subseteq \{1, \ldots, N\} - \{k\}$. Then, the family $\mathfrak{F} = \{S_{k_1}, \ldots, S_{k_r}\}$ is an r-cover of thickness 2 for S_k if and only if, for every time slot t' during which the node with ID k is scheduled to transmit, at least 2 nodes with IDs in C are scheduled to transmit during t'. The next result shows that the existence of such a Δ-cover of thickness 2 in an algorithm's schedule means that there are networks in which the algorithm fails to solve the neighbourhood learning task.

Theorem 2. *For any fixed-schedule algorithm $\mathcal{A}$ with schedule matrix S for the complete neighbourhood learning task in any UDG network of strong nodes with maximum degree Δ, the family of sets $\{S_1, \ldots, S_N\}$ is a Δ-cover-free family for thickness 2.*

Proof. To obtain a contradiction, assume that there exists a $k \in \{1, \ldots, N\}$ and a set $C = \{k_1, \ldots, k_\Delta\} \subseteq \{1, \ldots, N\} - \{k\}$ such that the family $\mathfrak{F} = \{S_{k_1}, \ldots, S_{k_\Delta}\}$ is a Δ-cover of thickness 2 for S_k.

Let $\mathfrak{C}$ be a Δ-clique of nodes $\{p(k_1), \ldots, p(k_\Delta)\}$, and let $V = \mathfrak{C} \cup \{p(k)\}$. Construct $G_1 = (V, E_1)$ with $E_1 = \{p(k), p(k_1)\} \cup E(\mathfrak{C})$. Next, construct $G_2 = (V, E_2)$ with $E_2 = \{p(k), p(k_2)\} \cup E(\mathfrak{C})$. We show that all nodes in $V - \{p(k)\}$ output the same list of neighbours in the executions of $\mathcal{A}$ on G_1 and G_2, which contradicts the correctness of $\mathcal{A}$.

Let α_1 be the execution of $\mathcal{A}$ on G_1, and let α_2 be the execution of $\mathcal{A}$ on G_2. We prove, by induction on the length ℓ of the executions α_1, α_2, that $h_{p,\alpha_1}[1 \ldots \ell] = h_{p,\alpha_2}[1 \ldots \ell]$ for all $p \in V - \{p(k)\}$.

In the base case, all nodes possess the same initial information in both networks. As induction hypothesis, assume that, for each $p \in V - \{p(k)\}$, $h_{p,\alpha_1}[1 \ldots \ell] = h_{p,\alpha_2}[1 \ldots \ell]$. We show that, for each $p \in V - \{p(k)\}$, $h_{p,\alpha_1}[1 \ldots (\ell+1)] = h_{p,\alpha_2}[1 \ldots (\ell+1)]$. Since $\mathcal{A}$ is a fixed-schedule algorithm, each node p in V transmits in slot $\ell + 1$ of execution α_1 iff p transmits in slot $\ell + 1$ of execution α_2. Further, by the induction hypothesis, the nodes in $V - \{p(k)\}$ send the same messages in slot $\ell + 1$ of execution α_1 as they do in slot $\ell + 1$ of execution α_2. Each node q in $V - \{p(k), p(k_1), p(k_2)\}$ only has neighbours in $V - \{p(k)\}$, so $h_{q,\alpha_1}[\ell + 1] = h_{q,\alpha_2}[\ell + 1]$.

Finally, consider whether or not $p(k)$ transmits in slot $\ell + 1$. If $p(k)$ does not transmit in slot $\ell + 1$, then $h_{p(k_1),\alpha_1}[\ell+1] = h_{p(k_1),\alpha_2}[\ell+1]$ and $h_{p(k_2),\alpha_1}[\ell+1] = h_{p(k_2),\alpha_2}[\ell + 1]$. If $p(k)$ does transmit in slot $\ell + 1$, then at least 2 nodes in $p(k_1), \ldots, p(k_\Delta)$ transmit during slot $\ell + 1$, since $\{S_{k_1}, \ldots, S_{k_\Delta}\}$ is a Δ-cover of thickness 2 for S_k. Therefore, $h_{p(k_1),\alpha_1}[\ell + 1] = * = h_{p(k_1),\alpha_2}[\ell + 1]$ and $h_{p(k_2),\alpha_1}[\ell + 1] = * = h_{p(k_2),\alpha_2}[\ell + 1]$. Thus, for each node q in $V - \{p(k)\}$, $h_{q,\alpha_1}[\ell+1] = h_{q,\alpha_2}[\ell+1]$. By Observation 1, all nodes in $V - \{p(k)\}$ output the same neighbour list in both executions, contradicting the correctness of $\mathcal{A}$. $\square$

Theorem 2 tells us that the columns of any schedule matrix used by a fixed-schedule neighbourhood learning algorithm must represent a Δ-cover-free family for thickness 2. Therefore, a lower bound on the number of rows of any such matrix gives us a lower bound on the number of slots used by any fixed-schedule neighbourhood learning algorithm. To prove such a lower bound, we actually fix the number of rows t in the schedule matrix, and then prove an upper bound on the number of columns N in terms of t (see Theorem 7 in Sect. 6). Re-arranging the expression for the upper bound on N gives a lower bound on t in terms of

N (see Corollary 8). Thus, we get the following lower bound for fixed-schedule neighbourhood learning algorithms.

Theorem 3. *Any fixed-schedule algorithm for the complete neighbourhood learning task for any UDG network of strong nodes with maximum degree Δ uses at least $\Omega((\Delta^2/\log\Delta)\log N)$ slots.*

5   The Power of Choice

A t-slot *optional-schedule algorithm* for the complete neighbourhood learning task is a deterministic algorithm run by each node p such that: p knows its entire t-slot schedule at time slot 0; p does not transmit during each slot for which the corresponding entry in its schedule is 0, and otherwise can choose to transmit or stay silent; and, at the end of slot t, p outputs a list of all of its neighbours. We present an optional-schedule algorithm for complete neighbourhood learning that uses $O(n + n\log(N/n))$ slots, which beats our $\Omega((\Delta^2/\log\Delta)\log N)$ fixed-schedule lower bound in networks of high degree (e.g. $\Delta \in \Omega(n)$). This algorithm relies heavily on an algorithm for conflict resolution described by Komlós and Greenberg (KG) [18]. In the conflict resolution task, there is a set of k nodes that have all transmitted to a channel simultaneously and we must schedule these nodes such that each node eventually gets to transmit on its own. Their conflict resolution algorithm fixes each node's entire schedule at time slot 0 and each node p follows its schedule until there is a slot t in which p transmits and all others stay silent. This slot t is guaranteed to exist due to the combinatorial properties of the set of schedules. After slot t, p chooses to remain silent for the remainder of the algorithm. This is possible in their network model since node p is able to detect whether or not it was the only transmitter on the channel. The authors prove that there exists a schedule matrix S of length $O(k + k\log(N/k))$ slots such that the conflict resolution task is solved using their algorithm.

We now show how to use the KG algorithm to construct a neighbourhood learning algorithm $\mathcal{A}$ for arbitrary networks. Starting with a schedule matrix S used by the KG algorithm to solve conflict resolution for n nodes, we construct a schedule matrix S' as follows: for each $i \geq 1$, set row $2i - 1$ of S' to be equal to row i of S, and set row $2i$ to be the row of all 1's. We refer to the odd-numbered slots as *KG slots*, and the even-numbered slots as *feedback slots*. In each KG slot s, the nodes behave as they do in the KG algorithm: if node p's schedule has a 1 in slot s, p transmits its ID if and only if there is no previous KG slot in which p transmitted and p's ID was received by all of p's neighbours. In each feedback slot f, a node q transmits its ID if and only if q detected a collision in slot $f - 1$. The number of slots used by $\mathcal{A}$ is twice the number of slots used by the KG algorithm, which is $O(n + n\log(N/n))$.

Recall that nodes possess strong collision detectors, so they can detect collisions whether they are transmitting or listening. Each feedback slot f allows any node p that transmitted in KG slot $f - 1$ to determine if its message was received by all of its neighbours: a collision has occurred at a neighbour of p

during KG slot $f - 1$ if and only p receives a message or detects a collision during feedback slot f. Hence, before the beginning of any KG slot $f + 1$, p can determine if there is a KG slot $f' \leq f - 1$ during which all of its neighbours received its ID, as required by the algorithm. The correctness of the algorithm follows directly from the correctness of the KG algorithm. To see why, we know that the KG algorithm guarantees that each node p in an n-clique transmits on its own, say during slot t_p. When our algorithm is run on a network of n nodes but with fewer edges than the n-clique, p will successfully transmit its ID to all of its neighbours during the KG slot corresponding to t_p, if not before.

6 An Upper Bound on the Size of Thick Cover-Free Families

In this section, we provide an upper bound on the number of sets in r-cover-free families for thickness b. First, we will need a result about shadows due to Sperner, as well as Dilworth's Theorem.

Consider a family $\mathfrak{F}$ of $(h-1)$-element subsets of $\{1, \ldots, t\}$. The *upper shadow* of $\mathfrak{F}$ [11] is the family consisting of all h-element supersets (over $\{1, \ldots, t\}$) of members of $\mathfrak{F}$. Sperner [23] (see [11]) proved the following bound on the upper shadow of $\mathfrak{F}$.

Theorem 4 (Sperner [23]). *If $h \leq (t+1)/2$, then the cardinality of the upper shadow of $\mathfrak{F}$ is at least $|\mathfrak{F}|$.*

Corollary 5. *For any $h \leq \lceil t/2 \rceil$, consider a set family $\mathfrak{G}$ consisting of subsets of $\{1, \ldots, t\}$, and suppose that each set in $\mathfrak{G}$ has cardinality less than h. If $\mathfrak{G}$ is an antichain with respect to inclusion, then $|\mathfrak{G}| \leq \binom{t}{h}$.*

Proof. Suppose that $\mathfrak{G}$ is an antichain, and let ℓ be the cardinality of the smallest set in $\mathfrak{G}$. Construct the family of all sets in $\mathfrak{G}$ of cardinality greater than ℓ and the upper shadow of the subfamily of all ℓ-element sets in $\mathfrak{G}$. By Proposition 4, we know that the cardinality of this family is at least $|\mathfrak{G}|$. We repeat this process a total of $h - \ell$ times to obtain the family $\mathcal{B}$ of all h-element supersets of members of $\mathfrak{G}$. Then, $|\mathfrak{G}| \leq |\mathcal{B}| \leq \binom{t}{h}$. □

Theorem 6 (Dilworth's Theorem [6]). *Let P be a partially ordered set. The cardinality of largest antichain in P is equal to the minimum number of chains needed to cover P.*

Theorem 7. *Consider any family $\mathfrak{F}$ of subsets of $\{1, \ldots, t\}$, and let $h = \lceil (tb(b+1))/(r(r+1) - b(b-1) - 1) \rceil$. For $r \geq 2b+1$, every family $\mathfrak{F}$ that is r-cover-free for thickness b has size at most $(r - b) + 2b\binom{t}{h}$.*

Proof. Note that $0 < h < \lceil t/3 \rceil \leq t/2$ when $r \geq 2b + 1$. Let $\mathfrak{R}$ be the family of all h-element subsets of $\{1, \ldots, t\}$ that are contained in at least one and at most b sets in $\mathfrak{F}$. We partition $\mathfrak{F}$ into three disjoint sub-families. Let $\mathfrak{F}(\mathfrak{R}) = \{S \in \mathfrak{F} : S \supseteq A$ for some $A \in \mathfrak{R}\}$, let $\mathfrak{F}_{<h} = \{S \in \mathfrak{F} : |S| < h\}$, and, let

$\mathfrak{F}' = \mathfrak{F} - \{\mathfrak{F}(\mathfrak{R}) \cup \mathfrak{F}_{<h}\}$. Equivalently, $\mathfrak{F}' = \{S \in \mathfrak{F} \mid |S| \geq h$ and $\forall C \subseteq S$ with $|C| = h$, there exist at least b sets in $\mathfrak{F} - \{S\}$ that contain $C\}$.

To obtain the desired result, we will prove that: (1) $|\mathfrak{F}(\mathfrak{R})| \leq b\binom{t}{h}$, (2) $|\mathfrak{F}_{<h}| \leq b\binom{t}{h}$, and (3) $|\mathfrak{F}'| \leq r - b$.

To prove (1), note that, for each set $T \in \mathfrak{R}$, there are at most b sets in $\mathfrak{F}(\mathfrak{R})$ that contain T. This implies that $|\mathfrak{F}(\mathfrak{R})| \leq b|\mathfrak{R}| \leq b\binom{t}{h}$. To prove (2), consider the partial ordering of the elements of $\mathfrak{F}_{<h}$ by inclusion. Since $\mathfrak{F}$ is r-cover-free for thickness b, every chain of $\mathfrak{F}_{<h}$ has cardinality at most b (otherwise, every element in the minimal set in the chain would appear at least b times in the multiset union of the other sets in the chain). Also, by Corollary 5, every antichain of $\mathfrak{F}_{<h}$ has cardinality at most $\binom{t}{h}$. By Theorem 6, $\mathfrak{F}_{<h}$ can be covered using $\binom{t}{h}$ chains. Therefore, $|\mathfrak{F}_{<h}| \leq b\binom{t}{h}$. Finally, to prove (3), assume that there are at least $w = r - b + 1$ sets $F_1, \ldots, F_w \in \mathfrak{F}'$. We show that the union of these sets contains more than t elements. This is a contradiction, since sets in $\mathfrak{F}'$ are subsets of $\{1, \ldots, t\}$.

Let $Q = \biguplus_{q=1}^{w} [(\biguplus^b F_q) - \biguplus_{i=1}^{q-1} F_i]$. For each $q \in \{1, \ldots, w\}$, the multiset $(\biguplus^b F_q) - \biguplus_{i=1}^{q-1} F_i$ consists of all elements of F_q that do not appear at least b times in $\biguplus_{i=1}^{q-1} F_i$. We proceed by proving upper and lower bounds on the size of Q (Claims 1 and 3, respectively), and, by comparing the two expressions, we will reach the desired contradiction.

Claim 1. $|Q| \leq (b(b+1)/2)|F_1 \cup \ldots \cup F_w|$

Proof. For any fixed $q \in \{1, \ldots, w\}$, let $X_{q,k}$ be the set consisting of all elements $x \in F_q$ such that x is in exactly k of $F_1, \ldots, F_{q-1}$. Then, $(\biguplus^b F_q) - \biguplus_{i=1}^{q-1} F_i$ consists of exactly $b-k$ copies of each $x \in X_{q,k}$, for all $k \in \{0, 1, \ldots, b-1\}$. Hence, we can re-write $|Q| = |\biguplus_{q=1}^{w}[(\biguplus^b F_q) - \biguplus_{j=1}^{q-1} F_j]| = \sum_{q=1}^{w} \sum_{k=0}^{b-1} (b-k)|X_{q,k}| = \sum_{k=0}^{b-1}(b-k) \sum_{q=1}^{w} |X_{q,k}|$. The desired result follows once we show that, for each $k \in \{0, \ldots, b-1\}$, $\sum_{q=1}^{w} |X_{q,k}| \leq |F_1 \cup \ldots \cup F_w|$.

Fix an arbitrary $k \in \{0, \ldots, b-1\}$. Since $X_{q,k} \subseteq F_q$ for $q = 1, \ldots, w$, it suffices to show that, for each element $x \in F_1 \cup \ldots \cup F_w$, there is at most one $q \in \{1, \ldots, w\}$ such that $x \in X_{q,k}$. Choose the largest ℓ such that $x \in X_{\ell,k}$. By the definition of $X_{\ell,k}$, $x \in F_\ell$ and x is in exactly k of $F_1, \ldots, F_{\ell-1}$. Let $i_1, \ldots, i_k \in \{1, \ldots, \ell-1\}$ be such that $x \in F_{i_1}, \ldots, x \in F_{i_k}$. For each $\alpha \in \{i_1, \ldots, i_k\}$, x is in at most $k-1$ of $F_1, \ldots, F_{\alpha-1}$, so $x \notin X_{\alpha,k}$. Further, for each $\beta \in \{1, \ldots, \ell-1\} - \{i_1, \ldots, i_k\}$, $x \notin F_\beta \supseteq X_{\beta,k}$. ∎

Claim 2. For $F \in \mathfrak{F}'$ and $F_1, \ldots, F_z \in \mathfrak{F}$ with $z \leq r - b$, $|(\biguplus^b F) - \biguplus_{i=1}^{z} F_i| > h(r - z)$.

Proof. Assume that $|(\biguplus^b F) - \biguplus_{i=1}^{z} F_i| \leq h(r - z)$. We can cover the elements of $(\biguplus^b F) - \biguplus_{i=1}^{z} F_i$ using $v = r - z \geq b$ subsets of $\mathfrak{F}$ with h elements each. Call these sets $A_1, A_2, \ldots, A_v$. By the definition of $\mathfrak{F}'$, for each A_k with $k \in \{1, \ldots, v\}$, there exists a set $S_k \in \mathfrak{F} - \{F\}$ that contains A_k. However, this means that $F_1 \uplus \ldots \uplus F_z \uplus S_1 \uplus \ldots \uplus S_v \supseteq \biguplus^b F$. In other words, $\biguplus^b F$ is covered by the multiset union of $z + v = z + (r - z) = r$ sets. This contradicts the fact that $\mathfrak{F}$ is r-cover-free for thickness b.   ∎

Claim 3. $|Q| \geq (h/2)[r(r+1) - b(b-1)]$

Proof. Since $w = r - b + 1$ and $q \in \{1, \ldots, w\}$, $q - 1 \leq r - b$, so, by Claim 2 with $z = q - 1$, $|Q| = \sum_{q=1}^{w} |(\biguplus^b F_q) - \biguplus_{i=1}^{q-1} F_i| > \sum_{q=1}^{w} h(r - q + 1)$. Substituting $w = r - b + 1$, this sum is equal to $h\sum_{j=b}^{r} j = h[(r(r+1)/2) - (b(b-1)/2)]$.∎

Claims 1 and 3 imply that $|F_1 \cup \ldots \cup F_{r-b+1}| \geq (h/b(b+1))[r(r+1) - b(b-1)]$. Since $h = \lceil (tb(b+1))/(r(r+1) - b(b-1) - 1) \rceil$, we get that $|F_1 \cup \ldots \cup F_{r-b+1}| > t$, a contradiction. This implies that $|\mathfrak{F}'| \leq r - b$.   □

Fixing the size of $\mathfrak{F}$ and re-arranging the upper bound from Theorem 7 gives us the following lower bound on t, the size of the universe over which the sets in $\mathfrak{F}$ are taken. The proof is omitted due to space constraints.

Corollary 8. *For any $r \geq 2b + 1$ and any r-cover-free family $\mathfrak{F}$ for thickness b consisting of subsets of $\{1, \ldots, t\}$, $t \in \Omega\left(\frac{r^2}{b^2 \log r} \log |\mathfrak{F}|\right)$.*

7 A Lower Bound for Neighbourhood Learning in the SINR Model

In the SINR model of wireless networks [15], a node p receives a message from node q if q's signal is sufficiently stronger than the sum of all other signals received at p (plus some constant amount of background noise). At p's physical location, the strength of q's signal is calculated as $P/(d(q,p)^\alpha)$, where P is q's transmission power, $d(q,p)$ is the Euclidean distance between nodes q and p, and α is the *path-loss exponent* (usually taken to be greater than 2, so that a signal degrades at least quadratically with respect to distance). Assuming that all nodes transmit with the same power P (known as a uniform power assignment), the sum of all other signals received at p is calculated as $\sum_{q' \neq q}(P/d(q',p)^\alpha)$ for all transmitting nodes q' other than q. Formally, if S is a set of transmitting nodes, then node p receives a message from node q if $\frac{P/d(q,p)^\alpha}{N + \sum_{q' \in S - \{q\}}(P/d(q',p)^\alpha)} \geq \beta$, where N represents the constant amount of background noise, and β is a parameter known as the *minimum signal to interference ratio*. We define q to be a *neighbour* of p if a transmission by q alone is received by p, namely, if $P/(Nd(q,p)^\alpha) \geq \beta$. We assume that each node knows a linear upper bound on the number of nodes in the network.

To prove a lower bound for fixed-schedule neighbourhood learning, we first observe that, if too many nodes transmit during the same time slot t, no node in the network receives a message during t, regardless of the network topology. We denote by D_{min} a fixed lower bound on the distance between any two nodes in the network, and we denote by D_{max} a fixed upper bound on the distance between any two nodes in the network.

Lemma 9. *Suppose that all nodes transmit with the same power P. Let $a = 1 + \lceil (D_{max}^{\alpha}/P)((P/\beta D_{min}^{\alpha}) - \mathcal{N}) \rceil$. For any node p and any time slot t, if exactly $b > a$ nodes transmit during time slot t, then p receives no message during time slot t.*

Proof. Suppose that $q_1, \ldots, q_b$ are the transmitting nodes during time slot t, each transmitting with power P. Re-arranging $b > 1 + (D_{max}^{\alpha}/P)((P/\beta D_{min}^{\alpha}) - \mathcal{N})$, we get that $\beta > \frac{P/D_{min}^{\alpha}}{((b-1)P/D_{max}^{\alpha}) + \mathcal{N}}$. Consider any transmitting node q_i. Note that $d(q_i, p) \geq D_{min}$ and that $\sum_{q_j \neq q_i} P/d(q_j, p)^{\alpha} \geq (b-1)P/D_{max}^{\alpha}$. Therefore, $\beta > \frac{P/d(q_i, p)^{\alpha}}{\mathcal{N} + \sum_{q_j \neq q_i} P/d(q_j, p)^{\alpha}}$. Hence, p does not receive q_i's message. $\square$

We prove a lower bound for the neighbourhood learning task in a model SMAX that is at least as strong as the SINR model, which implies that the same lower bound holds for the SINR model. Specifically, SMAX is a model of wireless radio networks that includes a parameter, s_{max}, that specifies the maximum number of simultaneous transmissions that can occur in a single slot without causing collisions at every node. In particular, a collision occurs at every node if greater than s_{max} nodes transmit, and, otherwise, each node receives the message contained in the strongest received signal. Further, we assume that each node can distinguish between a collision and the case where no nodes transmit.

Analogously to Theorem 2, we can show that, in the SMAX model, any fixed-schedule neighbourhood learning algorithm uses a schedule that corresponds to an $(n-1)$-cover-free family for thickness $s_{max} + 1$.

Lemma 10. *For any fixed-schedule algorithm $\mathcal{A}$ with schedule matrix S for the complete neighbourhood learning task for SMAX networks of strong nodes, the family of sets $\{S_1, \ldots, S_N\}$ is an $(n-1)$-cover-free family for thickness $s_{max} + 1$.*

Proof. To obtain a contradiction, assume that there exists a $k \in \{1, \ldots, N\}$ and a set $\mathcal{C} = \{k_1, \ldots, k_{n-1}\} \subseteq \{1, \ldots, N\} - \{k\}$ such that the family $\mathfrak{F} = \{S_{k_1}, \ldots, S_{k_{n-1}}\}$ is an $(n-1)$-cover of thickness $s_{max} + 1$ for S_k.

Let G_1 be any network consisting of nodes $V = \{p(k_1), \ldots, p(k_{n-1})\}$, and let G_2 be any network consisting of nodes $V \cup p(k)$ where $p(k)$ is a neighbour of at least one node in V. We show that all nodes in $V - \{p(k)\}$ output the same list of neighbours in the executions of $\mathcal{A}$ on G_1 and G_2, which contradicts the correctness of $\mathcal{A}$.

Let α_1 be the execution of $\mathcal{A}$ on G_1, and let α_2 be the execution of $\mathcal{A}$ on G_2. We prove, by induction on the length ℓ of the executions α_1, α_2, that $h_{p,\alpha_1}[1 \ldots \ell] = h_{p,\alpha_2}[1 \ldots \ell]$ for all $p \in V - \{p(k)\}$.

In the base case, all nodes possess the same initial information in both networks. As induction hypothesis, assume that, for each $p \in V - \{p(k)\}$, $h_{p,\alpha_1}[1 \ldots \ell] = h_{p,\alpha_2}[1 \ldots \ell]$. We show that, for each $p \in V - \{p(k)\}$, $h_{p,\alpha_1}[1 \ldots (\ell + 1)] = h_{p,\alpha_2}[1 \ldots (\ell + 1)]$. Since $\mathcal{A}$ is a fixed-schedule algorithm, each node p in $V - \{p(k)\}$ transmits in slot $\ell + 1$ of execution α_1 iff p transmits in slot $\ell + 1$ of execution α_2. Further, by the induction hypothesis, the nodes in $V - \{p(k)\}$

send the same messages in slot $\ell + 1$ of execution α_1 as they do in slot $\ell + 1$ of execution α_2.

Finally, we consider whether or not $p(k)$ transmits in slot $\ell + 1$ of execution α_2. If $p(k)$ does not transmit in slot $\ell + 1$ of execution α_2, then, for each $p \in V - \{p(k)\}$, $h_{p,\alpha_1}[\ell + 1] = h_{p,\alpha_2}[\ell + 1]$. If $p(k)$ does transmit in slot $\ell + 1$ of execution α_2, then we know that at least $s_{max} + 1$ nodes in $p(k_1), \ldots, p(k_{n-1})$ transmit during slot $\ell + 1$ in both executions α_1 and α_2, since $\{S_{k_1}, \ldots, S_{k_{n-1}}\}$ is an $(n-1)$-cover of thickness $s_{max} + 1$ for S_k. Therefore, by the definition of s_{max}, no nodes receive any messages in slot $\ell + 1$ in both executions α_1 and α_2, so, for each $p \in V - \{p(k)\}$, $h_{p,\alpha_1}[\ell+1] = h_{p,\alpha_2}[\ell+1]$. Thus, we have shown that, for each $p \in V - \{p(k)\}$, $h_{p,\alpha_1}[1 \ldots (\ell+1)] = h_{p,\alpha_2}[1 \ldots (\ell+1)]$. By Observation 1, all nodes in $V - \{p(k)\}$ output the same list of neighbours in executions α_1 and α_2, contradicting the correctness of $\mathcal{A}$.   □

By Corollary 8, the resulting lower bound for any fixed-schedule algorithm in the SMAX model is $\Omega(\frac{n^2}{(s_{max}+1)^2 \log n} \log N)$. It follows from Lemma 9 that any algorithm that solves neighbourhood learning in the SINR model will also be correct in the SMAX model when $s_{max} = 1 + \lceil (D_{max}^{\alpha}/P)((P/\beta D_{min}^{\alpha}) - \mathcal{N}) \rceil$. Therefore, we can apply the lower bound for the SMAX model to SINR networks.

Theorem 11. *In the SINR model with uniform power assignments, any fixed-schedule algorithm for the complete neighbourhood learning task uses at least $\Omega(\frac{n^2}{(s_{max}+1)^2 \log n} \log N)$ slots, where $s_{max} = 1 + \lceil (D_{max}^{\alpha}/P)((P/\beta D_{min}^{\alpha}) - \mathcal{N}) \rceil$.*

While we are familiar with some existing upper bounds for neighbourhood learning and local broadcast in SINR models [16,25], the above result seems to be the first non-trivial lower bound for complete neighbourhood learning.

8   Conclusions and Future Work

In this paper, we have demonstrated the fundamental limitations of fixed-schedule algorithms for neighbourhood learning in UDG networks. Even if nodes can send arbitrary messages, if they possess strong collision detectors, and if they know every node's schedule, there is no fixed-schedule algorithm that does significantly better than the $O(\Delta^2 \log N)$ algorithm based on cover-free families (equivalently, strongly-selective families.) This was shown by presenting a $\Omega((\Delta^2/\log \Delta) \log N)$-slot lower bound for any such algorithm. However, if nodes are allowed to leave out some transmissions from their schedule, then there is a $O(n + n \log (N/n))$-slot algorithm for neighbourhood learning, which beats our lower bound in networks of high degree. We have also shown how to use thick cover-free families to prove a lower bound in the more realistic SINR model.

Our work leaves open several directions for future research. First, we would like to prove lower bounds for neighbourhood learning algorithms in networks where nodes are allowed to leave out some transmissions from their schedule. Second, we would like to find a non-trivial lower bound for fully-adaptive algorithms, that is, algorithms where nodes do not follow a pre-calculated schedule.

Finally, we are very interested in better understanding the neighbourhood learning task in dynamic versions of our network model, i.e., by allowing the network topology to change over time.

References

1. Chaudhuri, S., Radhakrishnan, J.: Deterministic restrictions in circuit complexity. In: Proceedings of the Twenty-Eighth Annual ACM Symposium on Theory of Computing (STOC '96), pp. 30–36 (1996)
2. Chlebus, B., Gasieniec, L., Gibbons, A., Pelc, A., Rytter, W.: Deterministic broadcasting in unknown radio networks. In: SODA, pp. 861–870 (2000)
3. Clementi, A., Monti, A., Silvestri, R.: Selective families, superimposed codes, and broadcasting on unknown radio networks. In: SODA, pp. 709–718 (2001)
4. De Bonis, A., Gasieniec, L., Vaccaro, U.: Generalized framework for selectors with applications in optimal group testing. In: Baeten, J.C.M., Lenstra, J.K., Parrow, J., Woeginger, G.J. (eds.) ICALP 2003. LNCS, vol. 2719. Springer, Heidelberg (2003)
5. De Bonis, A., Vaccaro, U.: Constructions of generalized superimposed codes with applications to group testing and conflict resolution in multiple access channels. Theor. Comput. Sci. **306**(1–3), 223–243 (2003)
6. Dilworth, R.P.: A decomposition theorem for partially ordered sets. Ann. Math. **51**, 161–166 (1950)
7. Du, D., Hwang, F.: Combinatorial Group Testing and Its Applications. World Scientific, Singapore (2000)
8. D'yachkov, A.G., Vilenkin, P., Torney, D., Macula, A.: Families of finite sets in which no intersection of sets is covered by the union of s others. J. Comb. Theor. Ser. A **99**(2), 195–218 (2002)
9. Dyachkov, A.G., Rykov, V.V.: Bounds on the length of disjunctive codes. Probl. Peredachi Informatsii **18**(3), 7–13 (1982)
10. Dyachkov, A.G., Rykov, V.V., Rashad, A.M.: Superimposed distance codes. Probl. Control. Inf. Theor. **18**, 237–250 (1989)
11. Engel, K.: Sperner Theory. Cambridge University Press, Cambridge (1997)
12. Erdös, P., Frankl, P., Füredi, Z.: Families of finite sets in which no set is covered by the union of r others. Israel J. Math. **51**, 79–89 (1985)
13. Füredi, Z.: On r-cover-free families. J. Comb. Theor. Ser. A **73**(1), 172–173 (1996)
14. Gasieniec, L., Pagourtzis, A., Potapov, I., Radzik, T.: Deterministic communication in radio networks with large labels. Algorithmica **47**(1), 97–117 (2007)
15. Gupta, P., Kumar, P.R.: The capacity of wireless networks. IEEE Trans. Inf. Theor. **46**(2), 388–404 (2000)
16. Jurdzinski, T., Kowalski, D.R.: Distributed backbone structure for algorithms in the SINR model of wireless networks. In: Aguilera, M.K. (ed.) DISC 2012. LNCS, vol. 7611, pp. 106–120. Springer, Heidelberg (2012)
17. Kautz, W., Singleton, R.: Nonrandom binary superimposed codes. IEEE Trans. Inf. Theor. **10**(4), 363–377 (1964)
18. Komlós, J., Greenberg, A.: An asymptotically fast nonadaptive algorithm for conflict resolution in multiple-access channels. IEEE Trans. Inf. Theor. **31**(2), 302–306 (1985)
19. Krishnamurthy, S., Thoppian, M., Kuppa, S., Chandrasekaran, R., Mittal, N., Venkatesan, S., Prakash, R.: Time-efficient distributed layer-2 auto-configuration for cognitive radio networks. Comput. Netw. **52**(4), 831–849 (2008)

20. Mittal, N., Krishnamurthy, S., Chandrasekaran, R., Venkatesan, S., Zeng, Y.: On neighbor discovery in cognitive radio networks. J. Parallel Distrib. Comput. **69**, 623–637 (2009)
21. Porat, E., Rothschild, A.: Explicit non-adaptive combinatorial group testing schemes. In: Aceto, L., Damgård, I., Goldberg, L.A., Halldórsson, M.M., Ingólfsdóttir, A., Walukiewicz, I. (eds.) ICALP 2008, Part I. LNCS, vol. 5125, pp. 748–759. Springer, Heidelberg (2008)
22. Ruszinkó, M.: On the upper bound of the size of the r-cover-free families. J. Comb. Theor. Ser. A **66**, 302–310 (1994)
23. Sperner, E.: Ein satz über untermengen einer endlichen menge. Math. Z. **27**(1), 544–548 (1928)
24. Stinson, D., Wei, R.: Generalized cover-free families. Discrete Math. **279**(1–3), 463–477 (2004)
25. Yu, D., Wang, Y., Hua, Q., Lau, F.: Distributed local broadcasting algorithms in the physical interference model. In: DCOSS, pp. 1–8 (2011)

Optimal Nearest Neighbor Queries
in Sensor Networks

Gokarna Sharma[(✉)] and Costas Busch

School of Electrical Engineering and Computer Science, Louisiana State University,
Baton Rouge, LA 70803, USA
{gokarna,busch}@csc.lsu.edu

Abstract. Given a set of m mobile objects in a sensor network, we consider the problem of finding the nearest object among them from any node in the network at any time. These mobile objects are tracked by nearby sensors called proxy nodes. This problem requires an object tracking mechanism which typically relies on two basic operations: *query* and *update*. A query is invoked by a node each time when there is a need to find the closest object from it in the network. Updates of an object's location are initiated when the object moves from one location (proxy node) to another. We present a scalable distributed algorithm for tracking these mobile objects such that both the query cost and the update cost is minimized. The main idea is that given a set of mobile objects our algorithm maintains a virtual tree of downward paths pointing to the objects. Our algorithm guarantees an $\mathcal{O}(1)$ approximation for *query* cost and an $\mathcal{O}(\min\{\log n, \log D\})$ approximation for *update* cost in the constant-doubling graph model, where n and D, respectively, are the number of nodes and the diameter of the network. We also give polylog-arithmic approximations for both *query* and *update* cost in the general graph model. Our bounds are deterministic and hold in the worst-case. Moreover, our algorithm requires only polylogarithmic bits of memory per node. To our best knowledge, this is the first algorithm that is asymptotically optimal in handling nearest neighbor queries with low update cost.

1 Introduction

In sensor networks, object tracking is an important application where the presence of particular objects (animals, vehicles, etc.) can be detected by nearby sensors [5,7,10,13,17,20,22,23,25,26,35,37,38,40]. We consider in this paper the *nearest neighbor query* problem, denoted as NNQ, in which the objective is to track these mobile objects in such a way that any node in the network can locate the nearest object, among the set of objects available in the network, from it at any time with the minimum cost possible. We mean by locating nearest object the finding of the closest sensor node that has (i.e., detects) the object. For an example, consider a sensor network tracking the movement of taxies (as objects). A pedestrian injects queries into the network from his location, by sending them through nearby sensor nodes using devices, such as PDAs, for the nearest taxi.

P. Flocchini et al. (Eds.): ALGOSENSORS 2013, LNCS 8243, pp. 260–277, 2014.
DOI: 10.1007/978-3-642-45346-5_19, © Springer-Verlag Berlin Heidelberg 2014

The data will then be extracted from the relevant sensor nodes to respond to the user. Note that NNQ is along the lines of the in-network data processing problem for object tracking, e.g. [23,39], and different from the target tracking problem to approximate the trajectory of moving objects, e.g. [3,28].

We model the problem of tracking mobile objects by a weighted graph G, where graph nodes correspond to sensor nodes and graph edges correspond to communication links between the sensor nodes. We assume that each sensor node has its own memory where the objects information reside. A sensor node that currently detects a mobile object is called the *proxy node* for that object at that time. We consider a *nearest-sensor* model such that the sensor that is near to the object becomes its proxy [8,25]. The proxy nodes change over time when objects move. Proxy nodes are subset of sensor nodes which currently have (at least) a detected object in their memory. Proxy nodes store the detected objects and the bookkeeping information about objects in their memory. We sometime denote sensor nodes by users. In NNQ, the objective is to find the nearest proxy node.

Object tracking in NNQ involves two basic operations: *query* and *update*. A *query* operation is invoked when a user is looking for the location of the nearest object (i.e., the nearest proxy node). An *update* operation is initiated when the object moves from one node to another, i.e., it updates the location of the objects. Location queries and updates may be done in various ways. A naive way to query is to flood the whole network. All proxy nodes will reply to the query and the user which issued the request can choose the nearest one among them. Clearly, this approach is inefficient due to energy consumed when the network is large or the query rate is high [25]. Moreover, the time to answer a query is critical when the network query rate is more frequent. Alternatively, if all location information is stored at a specific node (e.g., the sink), no flooding is needed. But, whenever a movement is detected, update messages have to be sent up to that specific node all the time, which may be a major bottleneck. Moreover, when objects move frequently, abundant update messages will be generated. Therefore, we are interested in an approach for NNQ where *query* operations are not required to be flooded and *update* operations are not always required to be sent up to the sink. We assume that each node can issue *query* operations at any time, query rate is frequent, and multiple queries from different nodes can exist.

Cost Model: Consider a set of m mobile objects in a sensor network G. We assume that the cost to send messages between any pair of nodes in G depends only on the distance (i.e., the edge weights) between them.

- **Query cost:** We measure the cost of a *query* operation with respect to the *communication cost*, which is the total number of messages sent in the network G by any NNQ algorithm to reach the closest proxy node from any requesting node. The optimal communication cost for any *query* operation is the shortest distance between the requesting node and its nearest proxy node in G. We compare the communication cost of a NNQ algorithm for a *query* operation to the optimal communication cost for that operation to obtain *query competitive ratio*, which is the approximation of the NNQ algorithm for queries.

– **Update cost:** When an object moves from one proxy node to another proxy node, the optimal communication cost to update the location of that object is at least the distance between these proxy nodes. This is because the distance between the old and new proxy nodes is the actual minimum distance in G the object traverses and hence any algorithm for NNQ needs to send messages at least equivalent to this distance to reflect the location change in its NNQ data structure. We compare the communication cost of a NNQ algorithm for a set of *update* operations to the optimal communication cost for that set of operations to obtain *update competitive ratio*, which is the approximation of the NNQ algorithm for updates.

We look for a class of algorithms for NNQ that hold following two properties: (1) NNQ can be answered with the cost that is proportional to the shortest distance to the closest proxy node from the requesting node in G, and (2) The data structure needed for answering NNQ can be maintained with the cost that is proportional to the minimum distance the objects traverse if they followed shortest paths in G. We present a NNQ algorithm in this work which is optimal for query cost and near-optimal for update cost.

Existing techniques focused on aggregation and join queries, e.g. [18,30], and not much work has been done on nearest neighbor queries [31]. Previous work on nearest neighbor queries has focused only on finding the nearest sensor nodes to a specific query point [11,36]. This is different from our objective of finding the nearest mobile object from any sensor node at any time. Authors in [23,25,26,38] focused on tracking mobile objects but queries are only supported from the specific sink node. These papers [5,10,33–35] tried to solve the NNQ problem but with only one object; for multiple objects the structure developed should be replicated for every object. This replication increases the *query* cost linearly with the number of objects because, to find the nearest object, an algorithm may need to visit all the structures one after another.

Contributions: We propose a scalable query-optimal distributed NNQ algorithm for m mobile objects with near-optimal update cost in constant doubling networks, including low-dimensional Euclidean and growth-restricted networks, which has been used as an appropriate model of sensor network metric in several prior papers [10,16,17,29,40]. Our NNQ algorithm incurs a *query* competitive ratio of $\mathcal{O}(1)$, i.e. if the nearest object is at proxy node u at distance d from the query node v then our algorithm finds that object with the total cost of only $\mathcal{O}(d)$. We consider *query* operations individually and they have always small cost. To our best knowledge, this is the first result that achieves a constant-factor optimal solution to the nearest neighbor queries in realistic sensor networks that holds in the worst-case. In some cases, our algorithm may not return the actual closest proxy u at distance d but a proxy within distance $\mathcal{O}(d)$ from the requesting node. We call the distance ratios of the returned versus the closest proxies as the *query closeness ratio* which is $\mathcal{O}(1)$ in our algorithm.

Our NNQ algorithm provides $\mathcal{O}(\min\{\log n, \log D\})$ competitive ratio on the total communication cost for any set of *update* operations to maintain the required data structure, where n and D, respectively, are the number of nodes and the diameter of the network. This result assumes the *one by one* case where the next *update* operation does not arrive until the current *update* operation is finished. This assumption captures the application scenarios where event inter-arrival times are much greater compared with the message propagation time [17]. The reason for analyzing a sequence of *update* operations is because they provide small amortized update cost compared to the analysis of a single only *update* operation.

Moreover, we analyze the *memory requirement* of our NNQ algorithm through the total number of memory bits it uses in the network nodes to store both the object and the bookkeeping information. Our algorithm has the total memory requirement of only $\mathcal{O}(\log n \cdot \log D)$ bits per node, independent of the number of objects m.

We also extend our results to general networks. For any *query* operation, our NNQ algorithm provides $\mathcal{O}(\log^4 n)$ competitive ratio, and for any set of *update* operations, it provides $\mathcal{O}(\min\{\log^3 n, \log^2 n \cdot \log D\})$ competitive ratio. Also, an object returned by a *query* operation guarantees the closeness ratio of $\mathcal{O}(\log^2 n)$. This *update* competitive ratio is optimal within a poly-log factor in light of the $\Omega(\frac{\log n}{\log\log n})$ lower bound due to Alon et al. [2] in hypercubes and highly expanding graphs. Moreover, the memory requirement is only $\mathcal{O}(\log^2 n \cdot \log D)$ bits per node. All of our bounds are deterministic and hold in the worst-case.

Approach: The main idea is to maintain a hierarchical structure on top of the underlying sensor network metric so that a distributed algorithm that runs on the hierarchical structure can give almost optimal approximation ratio for both *query* and *update* operations. In the literature, various problems were solved in sensor networks using hierarchical structures providing good approximations, e.g. [1, 5, 10, 16, 17, 22, 29, 40].

We extract from G a hierarchical structure, denoted $\mathcal{HS}$, which organizes the nodes into levels (see Fig. 1) using a simple distributed maximal independent set algorithm [27]. The structure $\mathcal{HS}$ has $\mathcal{O}(\log D)$ levels, where at each level there are leader nodes pointing to leaders of lower level. The leaders of higher levels are subsets (refinements) of leaders at lower levels. All the individual nodes in the network are by default leaders in the bottom level (level 0) of $\mathcal{HS}$ and there is a single leader node of the whole network graph at the top level (level $h = \mathcal{O}(\log D)$) which is called the *root*. Only the bottom level nodes can issue *query* and *update* requests for the objects in the network, while the nodes in higher levels are used to propagate the request operations in the graph. The nodes in higher levels store the pointer information for the objects in their vicinity.

The idea now is to maintain a *virtual tree*, denoted as $\mathcal{T}$ (Fig. 1 with only bold edges), on $\mathcal{HS}$ with edges spanning the given set of mobile objects. (We call $\mathcal{T}$ a virtual tree as it may contain

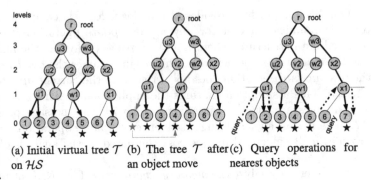

(a) Initial virtual tree $\mathcal{T}$ on $\mathcal{HS}$ (b) The tree $\mathcal{T}$ after an object move   (c) Query operations for nearest objects

Fig. 1. Illustration of mobile object tracking (★ are objects).

cycles when projected to G.) $\mathcal{T}$ consists of *object paths* which are directed paths from the root to the bottom-level nodes that own the objects (for example, the path $r - u_3 - u_2 - u_1 - 1$ in Fig. 1a). Tree $\mathcal{T}$ is *dynamic* and it changes after every *update* operation. The object path pointing to an object is updated whenever the object moves from one node to another (e.g., the paths $u_1 - 1$ and $r - u_3 - v_2 - w_1 - 4$ in Fig. 1b). These path updates are done through corresponding messages, *delete* and *insert* (which together constitute an *update* operation), issued respectively by the nodes that lose and get the objects (nodes 1 and 4, resp., in Fig. 1b). A *delete* goes upward in $\mathcal{HS}$ following the *object path* pointing to it in the reverse order and removes existing downward pointers in its way. An *insert* follows a pre-specified path in $\mathcal{HS}$, which we refer to as *root path*, and may be different than an existing path in $\mathcal{HS}$; while going upward, it sets downward pointers on its way. This upward propagation will be stopped when these *insert* and *delete* messages meet at a common ancestor (u_3 in Fig. 1b). Therefore, our approach is different than the differential one-from studied in [32] for target tracking.

The root path of a node is built by visiting upward the set of limited nearby leader nodes of that node starting from the bottom level up to the top level (details in Sect. 2). The object paths are built upon the root paths of nodes initiating *update* operations in the graph, i.e., object paths are formed from the fragments of root paths of different leaf nodes. In a *query* operation, each requesting node uses its root path to intersect the first object path in its upward propagation and then reaches the object without modifying the existing tree $\mathcal{T}$ in its downward propagation (queries from nodes 1 and 6 in Fig. 1c for nearest objects).

The initial tree $\mathcal{T}$ is obtained by publishing the objects in the proxy nodes they were first detected following their respective root paths towards the root, making each parent pointing to its child. This process is repeated until the root paths intersect with object paths previously set by some other objects, and hence forming the initial object path for each object. While maintaining the object paths, there may be the case that object pointers from two different nodes at some level point to the same immediate child node in $\mathcal{T}$, thus, creating cycles. In this case, we maintain $\mathcal{T}$ by removing appropriately the problematic fragments of object paths that may create cycles (e.g., the fragment $w_3 - w_2 - w_1$ in Fig. 1b).

We borrow some ideas of object and root paths from [35]. The novelty of this work is on how to deal with multiple objects; paper [35] and other closely related work [5,10,20,33,34] dealt with only one object. The extension to multiple objects case is non-trivial. This is because we can create different structures for each object using the approaches of [5,10,20,33–35], but the approximation for a query operation increases by $\mathcal{O}(m)$ factor in the worse case using structure replication. A hierarchically well-separated tree (HST) approach using [6,14] would be probabilistic, giving bounds for the expected cost for both query and update, while our bounds are deterministic.

Related Work: Demirbas and Ferhatosmanoglu [11] use R-trees [19] to locate nearest sensor nodes in sensor networks. Winter and Lee [36] gave an algorithm to locate k nearest sensor nodes. However, they [11,36] focused only on locating nearest sensor nodes which usually do not change over time. Yao et al. [37] focused on moving objects tracked by a sensor network, where the sensor nodes are spreading over a 2-dimensional space, from the observation that the nearest object may not be detected by the nearest sensor node and tried to locate the nearest object to a given query point. Demirbas et al. [9,12] studied the problem of event querying for checking whether a predefined event happened in a region; however, did not consider nearest neighbor queries.

Another line of work [1,15,22,24] studied the problem of providing a location service for ad hoc networks to allow any source node to know the location of any destination node to establish communication sessions between nodes whose location is unknown. Authors in [1,15] presented a routing framework such that each node can send messages to any other node without knowing the position of the destination node. Recently, Kulathumani et al. [22] gave an algorithm *Trail* for distributed object tracking. However, this location service problem is different in the sense that all these work did not consider nearest neighbor queries similar to ours. Moreover, GLS [24] made little effort to handle updates, so that even the update operations for the object at nearby nodes need to reach to the root of the structure to find those nearby nodes.

Paper Organization: The rest of the paper is organized as follows. We discuss the model and the hierarchical structure in Sect. 2. We then present algorithm in Sect. 3 and analyze it in Sect. 4. We then provide extensions to general graphs in Sect. 5. Proofs (and also some related work) are deferred to full version due to space restrictions.

2   Preliminaries

Model: We consider a static sensor network $G = (V, E, \mathfrak{w})$ consisting of n sensor nodes V, where E represents interconnection links between sensor nodes and the weight function $\mathfrak{w} : E \rightarrow \mathbb{R}^+$ corresponds to the latency of communication links. We have m mobile objects $\mathcal{M}$. Objects are tracked by nearby sensor nodes which are proxy nodes. Each node in V has a unique identifier. The weight of a link is equal to the communication cost of sending a message over the link. We assume that $\mathfrak{w}(u, u) = 0$ for any $u \in V$. For simplicity, we assume that the weight of a link is provided according to the distance between two sensor nodes in the network. We take G to be connected, i.e., there is a path of nodes (with respective sequence of edges connecting the nodes) between any pair of nodes in G. Let $\mathrm{dist}_G(u, v)$ be the shortest path length (distance) between nodes u and v with respect to the weight function $\mathfrak{w}$ in G. The k-*neighborhood* of a node v is the set of nodes which are within distance at most k from v (including v). The *diameter* is $D = \max_{u,v \in V} \mathrm{dist}_G(u, v)$, which denotes the maximum shortest path distance over all pairs of nodes in G. We assume that nodes and links do not crash and there is FIFO communication between nodes (i.e. no overtaking of messages occurs).

Hierarchical Structure Construction: We describe how to build a hierarchical structure $\mathcal{HS}$ on the sensor network for later use in the NNQ algorithm. Hierarchical structures were used to solve different problems in sensor networks, e.g. [1,5,10,16,17,22,29,40]. For simplicity, we focus mostly on a constant-doubling network; we describe how to handle other networks in Sect. 5. The doubling dimension graph is defined as follows: Let the space within radius δ of a point be called the ball centered at that point. A point set Γ has doubling dimension ρ if any set of points in Γ that are covered by a ball of radius δ can be covered by 2^ρ balls of radius $\frac{\delta}{2}$. We say that a metric is doubling and has a low dimension if ρ is bounded by a constant and is small. A constant-doubling network has been used as an appropriate model of a sensor network metric in a large body of prior work [10,16,17,29,40]. We select the nodes to include in $\mathcal{HS}$ using a distributed maximal independent set algorithm; we particularly consider the algorithm due to Luby [27]. This structure was used before by Sun and Herlihy in [20] for a problem in the different context.

We now discuss in detail the construction of $\mathcal{HS}$. We define a sequence of connectivity graphs $\mathcal{I} := \{\mathcal{I}_0 = (V_0, E_0), \mathcal{I}_1 = (V_1, E_1), \ldots \mathcal{I}_h = (V_h, E_h)\}$, where 0 is the lowest level where $\mathcal{HS}$ construction begins and $h \leq \lceil \log D \rceil + 1$ is the highest level where the $\mathcal{HS}$ construction ends. We choose to include nodes of the original network G in $\mathcal{I}_i, 0 \leq \ell \leq h$, as follows. At level 0, we include all sensor nodes of G in $\mathcal{I}_0$ such that $V_0 := V$, and for any two nodes $u, v \in V_0$, there is an edge $(u, v) \in E_0$ if and only if $\mathrm{dist}_G(u, v) < 2^1$. We call the nodes in $\mathcal{I}_0$ the leaf nodes. We denote by MIS^0 a maximal independent set of $\mathcal{I}_0$. At level $1 \leq \ell \leq h - 1$, we include only nodes from $MIS^{\ell-1}$ in $\mathcal{I}_\ell$ such that $V_\ell := MIS^{\ell-1}$, and for any two nodes $u, v \in V_\ell$, there is an edge $(u, v) \in E_\ell$ if and only if $\text{dist}_G(u, v) < 2^{\ell+1}$. We call the nodes in $\mathcal{I}_\ell$ the level ℓ nodes. MIS^ℓ is a maximal independent set of $\mathcal{I}_\ell$. At level h, $\mathcal{I}_h$ contains exactly one node, which we call the *root* node.

We define *default parent* and *parent set* for each level ℓ node $w \in \mathcal{I}_\ell$. The default parent of $w \in \mathcal{I}_\ell$ is a node in $\mathcal{I}_{\ell+1}$ that is closest to w, i.e., it is the node at distance at most $2^{\ell+1}$ away from w. The parent set of $w \in \mathcal{I}_\ell$ is a subset of nodes in $\mathcal{I}_{\ell+1}$ that are within distance $4 \cdot 2^{\ell+1}$ of w including the default parent.

$\mathcal{HS}$ described above is a layered node structure on $\mathcal{I}$. The vertex set in it is the level-0 through level-h nodes in the connectivity graphs $\mathcal{I}_0$ to $\mathcal{I}_h$. The edges in $\mathcal{HS}$ are formed by drawing edges between parent-child pairs in two consecutive level connectivity graphs $\mathcal{I}_\ell$ and $\mathcal{I}_{\ell+1}$. (Note that we do not consider the edges we formed in $\mathcal{I}$ construction for the edges in $\mathcal{HS}$.) Therefore, the edges span only between two consecutive level nodes. These edges are logical edges which are simulated by the physical edges in G following the edges in the shortest path between the nodes that are connected in $\mathcal{HS}$. Moreover, the nodes at each level $\ell \geq 1$ are again the logical nodes which are simulated by the physical nodes in the sensor network similar to edges. That is, some nodes in the network G participate as parent nodes for many levels in $\mathcal{HS}$ and some edges participate in connecting many parents in different levels.

We denote by $home^\ell(x)$ the level-ℓ default parent of x. These default parents for a leaf node x for each level are defined recursively up to the root such that $home^0(x) = x$ and $home^\ell(x)$ is the default parent of $home^{\ell-1}(x)$. These $home^\ell()$ nodes are useful in defining parent sets. We denote by $parentset^\ell(x)$ the parent set of $home^{\ell-1}(x)$ (the nodes at level ℓ within $4 \cdot 2^{\ell+1}$ from $home^{\ell-1}(x)$). According to the construction of $\mathcal{HS}$ and the definition of parent sets, all level-$(\ell + 1)$ parents of level ℓ node w can be covered by $2^{2\rho}$ radius-2^ℓ neighborhood balls. Also, different level-$(\ell+1)$ parents of w are at least distance $2^{\ell+1}$ from each other since they are maximal independent sets at level ℓ, i.e. two different parents cannot be from the same radius-2^ℓ neighborhood. Therefore, w has no more than $2^{2\rho}$ level-$(\ell + 1)$ parents.

Observation 1. *In constant-doubling networks, there are at most $2^{2\rho}$ parent nodes at level $\ell + 1$ for any node at level ℓ.*

We now define a path Root_Path(u) for each node $u \in V$ which will refer to as the "root path" of u. Root_Path(u) is formed by connecting the ascending sequence of parent sets of node u in $\mathcal{HS}$ starting from $parentset^0(u) = u$ at level 0 to $parentset^h(u) = r$ at level h (the root level). As there are at most $2^{2\rho}$ nodes in the parent set of any leaf level node at each level ℓ, all the nodes in the parent set $parentset^\ell(u)$ are visited in that level ℓ according to their node IDs starting from the smallest ID node and ending in the highest ID node. Thus, the highest ID node in $parentset^\ell(u)$ at level ℓ is connected to the smallest ID node in $parentset^{\ell+1}(u)$ at level $\ell + 1$ to form Root_Path(u). Nodes in $parentset^\ell(u)$ in each level ℓ for each node u are visited according to this order for both *query* and *update* operations. In Algorithm 1 at Sect. 3, however, we just use the Root_Path(u) formed from considering only one node in $parentset^\ell(u)$ at every level $1 \leq \ell \leq h$ for simplicity in describing *update* operations. We denote the next node towards the root for a current node w in the root path Root_Path(u) of any node u by the *parent* of w and the previous node by its *child*. We say that two paths *intersect* if they have a common node. We also say that two paths intersect at level i if they visit the same leader node at level i. We have the following lemma.

Lemma 1. *For any two nodes* $u, v \in V$, *their root paths* Root_Path(u) *and* Root_Path(v) *intersect at level* $\min\{h, \lceil \log(\text{dist}_G(u,v)) \rceil + 1\}$.

In our algorithm in Sect. 3, we sometime need to examine paths which are obtained from fragments of root paths, which we call "object paths". The root path is a trivial object path. In general, an object path is obtained by concatenating the fragments of root paths of several nodes. This scenario occurs when root paths for the objects are changed by many updates in $\mathcal{HS}$ from different nodes due to object movements. We can bound the lengths of the root paths and object paths as follows.

Lemma 2. *For the root path* Root_Path(u) *from each node* $u \in V$ *(or the object path* Object_Path(u)*) up to level* j, *denoted as* Root_Path$_j$(u) *(or* Object_Path$_j$(u)*),* length(Root_Path$_j$(u)) $\leq 2^{j+2\rho+4}$.

3   NNQ Algorithm

High Level Overview: Consider a set $\mathcal{M} = \{\xi_1, \xi_2, \ldots, \xi_m\}$ of $|\mathcal{M}| = m$ mobile objects. The proxy of each object ξ_i is the sensor node which currently detects it. The main goal of our NNQ algorithm is to maintain a virtual tree $\mathcal{T}$ (Fig. 1 with only bold edges) that spans only the proxy nodes of the objects in $\mathcal{HS}$ (Sect. 2) all the time (even when the objects move). We will argue that $\mathcal{T}$ supports best possible answering of NNQ queries, which are for the nearest objects from the query nodes. The downward pointers pointing to the objects are used in $\mathcal{T}$ formation. This is done by maintaining an object path (for each object ξ_i) which is a directed path from the root node r of $\mathcal{HS}$ to the bottom-level node that is the proxy of ξ_i.

For maintaining $\mathcal{T}$ and facilitating NNQ, we use following three operations: *publish, query,* and *update. publish* operations are used to form the initial object paths in $\mathcal{HS}$ pointing to the objects (this can also be seen the special case of *insert* operations we describe later). These initial objects paths are formed from the root paths of the proxy nodes that detect the objects. The node, say u, becomes the first proxy of ξ_i if it issues a *publish* operation. The *publish* operation is applied for each object only once in the beginning. *publish* makes objects findable in $\mathcal{HS}$ for future *query* and *update* operations. The *query* operation is issued by any sensor node to obtain the nearest object among m objects (from the proxy of that object) from its location. The *update* operation is used when the object is moved from the current proxy node to some new proxy node. This operation is realized through two different operations: (i) *insert* which is issued at the new proxy node and (ii) *delete* which is issued at the current proxy node.

A *query* operation issued by a node x for the closest object ξ_i from its location uses its own Root_Path(x) which intersects with the first object path and then leads to the object ξ_i. Let v be the proxy of ξ_i. Let w be the intersecting node of Root_Path(x) and the object path Object_Path(v) to v. The *query* message is first sent upward $\mathcal{HS}$ along Root_Path(x) towards the root r until it intersects at a node (i.e., w) with

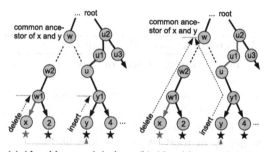

(a) Algorithm to minimize only *update* cost

(b) Algorithm to minimize both *query* and *update* cost

Fig. 2. Maintaining $\mathcal{T}$ (★ are objects)

Object_Path(v), and then the request message follows Object_Path(v) from node w to the object proxy v; then v sends ξ_i (or related information, e.g. its location) to x (along some shortest path in the original network G) as asked by the query node.

An *update* operation for ξ_i is issued when ξ_i is moved from the current proxy x to the new proxy y, y issues an *insert* operation and x issues a *delete* operation, respectively (Fig. 2). Let w be the intersecting node (i.e., common ancestor) of the root path Root_Path(y) of the node y and the object path Object_Path(x) pointing to x. The intersecting node w guaranteed to exist because in the worst-case scenario the paths intersect at the root r. The *update* is implemented by sending *insert* message from y and a *delete* message from x upward in $\mathcal{HS}$, respectively, along Root_Path(y) and Object_Path(x) in the reverse order, until the *insert* and *delete* messages intersect at a node (i.e. node w). While going up, *insert* and *delete* messages modify their respective paths to point toward y. The *insert* operation sets the directions of the edges in the fragment of Root_Path(y) between y and w to point toward y. Whereas, the *delete* operation deletes the downward pointers (or links) in the fragment of Object_Path(x) from w to x. The new object path for ξ_i now points toward y. This process has resulted to a *object path* that consists of two root path fragments, a fragment of x's root path from r to w, and a fragment of y's root path from w to y. Subsequent *update* operations may result into further fragmentation of the object path into multiple root path fragments.

The above description of *query* and *update* operations applies when there is only one object in the network. When there are many objects, the *insert* and *delete* operations for ξ_i, when it moves from x to y, may not always need to reach to the common ancestor w. This is because *insert* operation from y may find an object path pointing to an object at proxy z, before reaching w following Root_Path(y). As shown in Fig. 2a, *insert* from y only needs to set the downward pointers up to node y_1 (not up to the common ancestor w as shown in Fig. 2b), as there exists already an object path pointing to the object at node 4. Similarly, the *delete* operation also may not need to remove the downward pointers up to

---

Algorithm 1: NNQ algorithm

---

1 Publish object at leaf node v_i:
2    $k \leftarrow 1$;
3   **Until** node $p^k(v_i)$ is not a root node **OR** $p^k(v_i).degree = 0$ **do**
4   Set $p^k(v_i).pointer = p^{k-1}(v_i)$; $k \leftarrow k + 1$;
5   Set $p^k(v_i).pointer = p^{k-1}(v_i)$;
6 Query nearest object by leaf node v_i:
7    $k \leftarrow 1$;
8   **Until** $t.degree > 0$ or $t.spcl\_degree > 0$ for any $t \in parentset^k(v_i)$ **do** $k \leftarrow k + 1$;
9   **If** $t.degree > 0$ **then** Go to the proxy node following downward pointers $t.pointer \neq \perp$ and send information to v_i;
10   **Else** Go to the proxy node following downward pointers starting first $t.spcl\_pointer \neq \perp$ and send information to v_i;
11 Insert object by leaf node v_i received from leaf node v_j:
12    $k \leftarrow 1$;
13   **Until** a *delete* message from v_j is not found **do**
14   **If** $p^k(v_i).degree = 0$ **then** Set $p^k(v_i).pointer = p^{k-1}(v_i)$; Set $spcl\_parent(p^k(v_i)).spcl\_pointer = p^k(v_i)$;
15   **If** $t.pointer = p^k(v_i)$ from $t \notin p^{k+1}(v_i)$ **then** $delete\_downward\_fragment(p^k(v_i))$;
16       $k \leftarrow k + 1$;
17   **If** $p^k(v_i).degree = 0$ **then** Set $parent^k(v_i).pointer = p^{k-1}(v_i)$; Set $spcl\_parent(p^k(v_i)).spcl\_pointer = p^k(v_i)$;
18   **If** $t.pointer = p^k(v_i)$ from $t \notin p^{k+1}(v_i)$ **then** $delete\_downward\_fragment(p^k(v_i), t)$;
19 Delete object from leaf node v_j that moved to leaf node v_i:
20    $k \leftarrow 1$;
21   **Until** an *insert* message from v_i is not found **do**
22   **If** *delete* message is of type *erase* **then**
23   Set $p^k(v_i).pointer = \perp$ to $p^{k-1}(v_i)$ in Object_Path(v_i);
24   Set $spcl\_parent(p^k(v_i)).spcl\_pointer = \perp$ to its special child $p^k(v_i)$;
25   **If** $p^k(v_i).degree > 0$ **then**
26   Set $p^k(v_i).pointer = \perp$ to $p^{k-1}(v_i)$ in Object_Path(v_i);
27   Set $spcl\_parent(p^k(v_i)).spcl\_pointer = \perp$ to its special child $p^k(v_i)$;
28   Change *delete* message type from *erase* to *relay*;
29   **If** *delete* message is of type *relay* **then** Forward the *delete* message to $p^{k+1}(v_i)$;
30       $k \leftarrow k + 1$;
31 $delete\_downward\_fragment(x, y)$:
32   Set $y.pointer = \perp$ for x;
33   **If** $y.degree = 0$ **then** $delete\_downward\_fragment(y, y')$ for y' which has $y'.pointer = y$;

---

w; simply deleting the downward pointers in the fragment of object path from x to the node that has downward pointer (w_1 in Fig. 2a) pointing to some other node is sufficient. This approach provides the minimum communication cost for *update* operations because the *insert* and *delete* messages reach to the common ancestor only in the worst-case. However, it may not provide any guarantee on the cost of *query* operations, i.e., the nearest object that is requested by a node may be in its neighboring node but the *query* operation may need to reach up to the root r to find an object path leading to that object. This is due to the possible enormous fragmentation of the object paths after several *update* operations.

We avoid this situation and guarantee efficiently *query* cost along with low *update* cost using the following two ideas: (1) To minimize the *update* cost, our algorithm always matches the *insert* message from y with its corresponding *delete* message from x at the common ancestor node w (Fig. 2b), and (2) To guarantee low *query* cost simultaneously with low *update* cost, our algorithm uses the concept of a *special parent* node, such that whenever a downward pointer is set at a node $t \in parentset^k(.)$ the special parent node of t is also informed about t holding a downward pointer. The special parent is selected in Root_Path(y) for any node t in such a way that any *query* close to t will either reach t or its special parent (details in Definition 1).

Algorithm Description: We now formally describe the algorithm. The algorithm is given in Algorithm 1. We assume that each node in G knows its *parent* and *special_parent* in $\mathcal{HS}$ (both in its root path), except the *root* node, whose *parent* and *special_parent* are both $\perp$ (null). We define *special_parent* node as follows:

Definition 1. *We denote a special-parent node of a level i parent node $p^i(u)$ of a node u in* Root_Path(u) *as* $spcl\_parent(parent^i(u))$ *such that* $spcl\_parent(parent^i(u))$ *is the* $p^k(u)$ *of u in* Root_Path(u) *at level k, where $k = i + 2\rho + 4$, i.e., $spcl\_parent(p^i(u))$ is some parent node of u in* Root_Path(u) *at level k.*

A node might have *pointers* and *spcl_pointers* towards its children (otherwise they are $\perp$); there is at least a *pointer* at the root that is not $\perp$. If we view *pointers* and *spcl_pointers* that are not null as directed edges in $\mathcal{HS}$, they always point down. We denote by *degree* the number of *pointers* and *spcl_degree* the number of *spcl_pointers* that are set at any node in $\mathcal{HS}$ (e.g., the degree of node y_1, denoted as $y_1.degree$ in Fig. 2 is 2 but $w_1.degree$ is 1). We assume nodes process messages sequentially.

For simplicity, assume that there is only one parent node $p^\ell(.)$ at any level ℓ. Therefore, Root_Path(x) of a leaf node x is the concatenation of $p^0(x) = x$, $p^1(x)$ up to $p^h(x) = r$, such that $p^k(x)$ is the level k parent node of x, $p^{k-1}(x)$ and $p^{k+1}(x)$ are its child and parent node both in Root_Path(x), and r is the root (Sect. 2).

Recall that we did not use *parentset^k*(.) in Algorithm 1. When we use *parentset$^k(v_i)$* set then we only need to set the pointer from next parent to the current one in the order of the nodes in $parentset^k(v_i)$. The lowest ID node $t \in parentset^k(v_i)$ has $t.pointer$ that is set to the highest ID node in $parentset^{k-1}(v_i)$, second lowest ID node $t_1 \in parentset^k(v_i)$ has $t_1.pointer = t$, and so on, such that the lowest ID node $t' \in parentset^{k+1}(v_i)$ has $t'.pointer$ set to the highest ID node in $parentset^k(v_i)$. This technique is very useful as it avoids the need of an expensive synchronization mechanism in concurrent situations.

4   Performance Analysis

In this section, we analyze the performance of Algorithm 1 for $query$ and $update$ operations. We assume a synchronous communication model such that the communication link latency is predictable. Note that our algorithm provides correctness without this assumption. We assume that a $time\ unit$ is of duration required for a message sent by a node to reach a destination node that is a unit distance far from it. Assuming an execution starts at time 0, we define a notion of $time\ windows$ similar to [34] which will be helpful in the analysis to obtain upper bounds for the communication cost and also respective lower bounds. As the lower bound is based on actual distances in the original graph G, it will hold also in asynchronous executions where the latency is difficult to predict. We have the sequence of windows for each level k, $0 \le k \le h$, i.e., $\mathbf{W}_k = \{W_k^0, W_k^1, \ldots\}$, where W_k^0 is the first window at level k, W_k^1 is the second window at level k, and so on; each $time\ window\ W_k^i$ is assumed to be of time duration $2^{k+2\rho+4}, 0 \le k \le h$ (Lemma 1). W_k^{j+1} starts immediately after W_k^j expires. Due to synchrony assumption, the windows are aligned in such a way that W_{k-1} starts at the same time with W_k. Thus, for one window at level h, there are exactly two windows at level $h-1$, exactly four windows at level $h-2$, and so on, so that there are exactly 2^{h-k} windows at level k. The requests from level 0 are forwarded to level 1 at the end of the window W_0. Level 1 parents forward requests to level 2 at the end of the window W_1. This proceeds at higher levels and in a similar way to the downward direction. Therefore, at the end of a window, each level k parent node can exchange a message with its parent neighbors at level $k+1$ or level $k-1$.

We start with the cost due to $publish$ operations. The $publish$ cost is the total communication cost that is needed for a leaf proxy node that initially detects the object to insert it in $\mathcal{HS}$. Note that the $publish$ operation adds downward pointers on the publishing leaf proxy node's root path towards the root. This process repeats until the publish operation visits either the root node or there exists a downward pointer, $z.pointer$ at any level k parent node z in the root path of the publishing proxy node. Therefore, the total communication cost for a $publish$ operation is $\mathcal{O}(D)$, which immediately follows from Lemma 2, since the number of levels $h = \lceil \log D \rceil + 1$ in $\mathcal{HS}$. Note that this is one time cost and will be compensated from $query$ and $update$ operations issued thereafter.

Query Cost: We focus on the competitive ratio $C(r)/C^*(r)$, where $C(r)$ is the total communication cost of serving a *query* operation r using Algorithm 1 and $C^*(r)$ is the optimal cost of serving r. If there are no *update* operation in the system, it is trivial to see that a *query* operation r from x finds the first object path to the nearest proxy v, at level $\lceil \log(\text{dist}_G(x,v)) \rceil + 1$ (Lemma 1), following Root_Path(x), where $\text{dist}_G(x,v)$ is the distance of the proxy v from x in G. When there are *update* operations in the system, the *query* r issued by x may not find the object path to v from it at level $\lceil \log(\text{dist}_G(x,v)) \rceil + 1$ because the object path Object_Path(v) to v might have been deformed significantly such that the $2^{(\lceil \log(\text{dist}_G(x,v)) \rceil + 1)}$-neighborhood of x has no information about the proxy v. Nevertheless, we prove Theorem 1.

Lemma 3. *If a node x issues a query operation r for the closest object ξ_i at proxy v at distance $\text{dist}_G(x,v) \leq 2^i$, Root_Path($x$) is guaranteed to find the Object_Path(v) to v at level $k \leq i + 2\rho + 4$.*

Theorem 1. *The competitive ratio of Algorithm 1 for any query operation is $O(1)$.*

Let a node x issues a *query* operation r and there is the closest object ξ at the proxy node v at distance $\text{dist}_G(x,v) \leq 2^i$. Our NNQ algorithm may return an object ξ that is not at v but at some other node u. We bound the ratio $D(r)/D^*(r)$, which we denote as the *query closeness ratio*, where $D(r)$ is the distance to the proxy node of the farthest object that may return by a *query* operation and $D^*(r)$ is the distance to the proxy node of the closest object. This is also due to the fragmentation of root paths caused by *update* operations. In particular, we prove the following lemma.

Lemma 4. *The closeness ratio of Algorithm 1 for any query operation is $\mathcal{O}(1)$.*

Update Cost: We now prove the performance of our NNQ algorithm for *update* operations. We consider a sequence of *update* operations (realized through the combination of *insert* and *delete* operations) on the objects. Lets define a one by one execution of a set $\mathcal{E}$ of l (l does not need to be known) *update* operations $\mathcal{E} = \{r_1 = (u_1, v_1), r_2 = (u_2, v_2), \ldots, r_l = (u_l, v_l)\}$, where $r_i = (u_i, v_i), 1 \leq i \leq l$, are the subsequent *update* operations with old proxy nodes u_i and new proxy nodes v_i. This execution captures the scenarios where event inter-arrival times are much greater than the message propagation times [17]. We define an array $\mathcal{S} = \{S_0, S_1, \ldots, S_h\}$ of counters to keeps track of the levels visited by each operation r_i in $\mathcal{HS}$ when objects move. Initially, each $S_i = 0$. At the end of the execution of $\mathcal{E}$, each S_i have the count of the total number of operations that reached level i. The *peak level* for an operation r_i is the maximum level reached by r_i in $\mathcal{HS}$. According to our assumption, as *publish* operations for each object are executed before the first *update* operation r_1, the peak level reached by r_0 for at least an object is h (the maximum level in $\mathcal{HS}$) and r_0 is counted at all S_i starting from S_0 to S_h. In other words $S_i = 1$ for $0 \leq i \leq h$ after the first *publish* operation. For each *update* operation $r_i \in \mathcal{E}, i > 0$, we keep track of it by registering it at all the counters S_i from 0 to k until *insert* meets its corresponding *delete* operation at its peak level k. That is, after each *update* $S_i = S_i + 1$ for $0 \leq i \leq k$ when k is the peak level for that *update* . We have that $k \leq h$.

Let $C^*(\mathcal{E})$ denote the total communication cost of serving all the operations in $\mathcal{E}$ using the optimal algorithm in G. Let $C(\mathcal{E})$ denote the total communication cost of serving all the operations in $\mathcal{E}$ using Algorithm 1. The total communication cost is measured through the total distance traversed by all the messages. We bound the competitive ratio $CR_{\text{NNQ}} = \max_{\mathcal{E}} C(\mathcal{E})/C^*(\mathcal{E})$ as given in Theorem 2.

Theorem 2. *The update competitive ratio of Algorithm 1 is* $\mathcal{O}(\min\{\log n, \log D\})$.

Memory Requirement: The memory requirement of our NNQ algorithm is the total amount of memory bits it uses in the nodes of the network to track m mobile objects $\mathcal{S}$. We prove the following theorem (details are omitted).

Theorem 3. *The memory requirement of Algorithm 1 is* $\mathcal{O}(\log n \cdot \log D)$ *bits per node.*

5 Extensions to General Networks

We now show the scalability of our NNQ algorithm in general network topologies. We use a $(\mathcal{O}(\log n), \mathcal{O}(\log n))$ *partition scheme* of [4,21,35] to maintain a hierarchical structure $\mathcal{HS}$ in general networks. This partition scheme is based on well-known ideas for clustering the graph to approximate graph distance metrics by distributions over tree metrics [6,14] on a general metric network. A similar scheme is used by [29] to solve the problem of distributing information from a collection of sources to mobile users in a wireless mesh network. This scheme computes a (coarsening) cover $\mathcal{C}$ in deterministic polynomial time [35] such that: (1) for each $v \in V$, $B_r(v) = \{u \in V | \text{dist}_G(u, v) \leq r\}$ is contained in at least one set in $\mathcal{C}$; (2) every vertex v is contained in at most $\mathcal{O}(\log n)$ sets in $\mathcal{C}$; and (3) each set in $\mathcal{C}$ has radius at most $\mathcal{O}(r \cdot \log n)$, for any $r > 0$. This scheme can also be computed through a distributed algorithm in a message efficient manner similar to [17]. We do not use directly the HST tree constructed using [6,14] for NNQ algorithm as the HST tree construction of [6,14] is probabilistic and bounds the cost only in expectation.

There are $h \leq \lceil \log D \rceil + 1$ levels in this $(\mathcal{O}(\log n), \mathcal{O}(\log n))$-partition scheme. At level 0, each node in V belongs to exactly one cluster which consists only of the node itself. At highest level (h), there are $\mathcal{O}(\log n)$ copies of the cluster that contains all the nodes in V; only one copy is kept removing the rest to have just only one cluster at level h in $\mathcal{HS}$. In any level $i, 1 \leq i \leq h - 1$, each node $u \in V$ belongs to exactly $\mathcal{O}(\log n)$ clusters. We select a leader node arbitrary for each cluster from the nodes that are in that cluster so that $\mathcal{HS}$ can be the hierarchical structure of leader nodes of all clusters at all the levels mentioned above.

Similar to Sect. 3, the parent set $parentset^\ell(x)$ of x are all the parent nodes within distance $\mathcal{O}(2^\ell \cdot \log n)$ of x; according to this $\mathcal{HS}$ construction, the total number of nodes in the parent set are $\mathcal{O}(\log n)$. These parent child pairs are connected similarly to Sect. 2. The Object_Path(u) for each node $u \in V$ visits the leader nodes in $parentset^\ell(u)$ at level ℓ that u belongs to according to the level of the clusters (derived from labeling the $\mathcal{O}(\log n)$ hierarchical partitioned hierarchies used in $(\mathcal{O}(\log n), \mathcal{O}(\log n))$-partition scheme construction; details in [35]) starting from the smallest label cluster and ending at the largest label cluster. Object paths are also defined similarly. Moreover, we set the time window of duration $\mathcal{O}(2^k \cdot \log^2 n)$ for each level $1 \leq k \leq h$. We have these following results in general networks.

Lemma 5. *For any two nodes $u, v \in V$ in general networks, their root paths* Root_Path(u) *and* Root_Path(v) *intersect at level* $\min\{h, \lceil \log(\mathrm{dist}_G(u,v)) \rceil + 1\}$. *Moreover, for the root path* Root_Path(u) *from each node $u \in V$ (or the object path* Object_Path(u)*) up to level j, denoted as* Root_Path$_j(u)$ *(or* Object_Path$_j(u)$*),* length(Root_Path$_j(u)$) $\leq \mathcal{O}(2^k \cdot \log^2 n)$.

Lemma 6. *If a node x issues a query operation r for the closest object ξ_i at proxy v at distance* $\mathrm{dist}_G(x,v) \leq 2^i$, Root_Path$(x)$ *is guaranteed to find* Object_Path(v) *to v at level $k \leq i + 2 + 2\log\log n + \log c$, for some positive constant c.*

Theorem 4. *The competitive ratio of Algorithm 1 for any query operation is* $\mathcal{O}(\log^4 n)$ *in general networks.*

Lemma 7. *The closeness ratio of Algorithm 1 for any query operation is* $\mathcal{O}(\log^2 n)$ *in general networks.*

Theorem 5. *The update competitive ratio of Algorithm 1 is* $\mathcal{O}(\min\{\log^3 n, \log^2 n \cdot \log D\})$ *in general networks.*

Theorem 6. *The memory requirement of Algorithm 1 is* $\mathcal{O}(\log^2 n \cdot \log D)$ *bits per node in general networks.*

References

1. Abraham, I., Dolev, D., Malkhi, D.: Lls: a locality aware location service for mobile ad hoc networks. In: DIALM-POMC, pp. 75–84 (2004)
2. Alon, N., Kalai, G., Ricklin, M., Stockmeyer, L.: Lower bounds on the competitive ratio for mobile user tracking and distributed job scheduling. Theor. Comput. Sci. **130**(1), 175–201 (1994)
3. Aslam, J., Butler, Z., Constantin, F., Crespi, V., Cybenko, G., Rus, D.: Tracking a moving object with a binary sensor network. In: SenSys, pp. 150–161 (2003)
4. Awerbuch, B., Peleg, D.: Sparse partitions. In: FOCS, vol. 2, pp. 503–513 (1990)

5. Awerbuch, B., Peleg, D.: Online tracking of mobile users. J. ACM **42**(5), 1021–1058 (1995)
6. Bartal, Y.: Probabilistic approximation of metric spaces and its algorithmic applications. In: FOCS, pp. 184–193 (1996)
7. Can, Z., Demirbas, M.: A survey on in-network querying and tracking services for wireless sensor networks. Ad Hoc Netw. **11**(1), 596–610 (2013)
8. Chen, W.P., Hou, J.C., Sha, L.: Dynamic clustering for acoustic target tracking in wireless sensor networks. In: ICNP, pp. 284– 294 (2003)
9. Demirbas, M., Arora, A., Kulathumani, V.: Glance: a lightweight querying service for wireless sensor networks. Theor. Comput. Sci. **410**(6–7), 500–513 (2009)
10. Demirbas, M., Arora, A., Nolte, T., Lynch, N.A.: A hierarchy-based fault-local stabilizing algorithm for tracking in sensor networks. In: Higashino, T. (ed.) OPODIS 2004. LNCS, vol. 3544, pp. 299–315. Springer, Heidelberg (2005)
11. Demirbas, M., Ferhatosmanoglu, H.: Peer-to-peer spatial queries in sensor networks. In: P2P, pp. 32–39 (2003)
12. Demirbas, M., Lu, X., Singla, P.: An in-network querying framework for wireless sensor networks. IEEE Trans. Parallel Distrib. Syst. **20**(8), 1202–1215 (2009)
13. Demmer, M.J., Herlihy, M.P.: The arrow distributed directory protocol. In: Kutten, S. (ed.) DISC 1998. LNCS, vol. 1499, pp. 119–133. Springer, Heidelberg (1998)
14. Fakcharoenphol, J., Rao, S., Talwar, K.: A tight bound on approximating arbitrary metrics by tree metrics. J. Comput. Syst. Sci. **69**(3), 485–497 (2004)
15. Flury, R., Wattenhofer, R.: Mls: an efficient location service for mobile ad hoc networks. In: MobiHoc, pp. 226–237 (2006)
16. Funke, S., Guibas, L.J., Nguyen, A., Wang, Y.: Distance-sensitive information brokerage in sensor networks. In: Gibbons, P.B., Abdelzaher, T., Aspnes, J., Rao, R. (eds.) DCOSS 2006. LNCS, vol. 4026, pp. 234–251. Springer, Heidelberg (2006)
17. Gao, J., Guibas, L., Milosavljevic, N., Zhou, D.: Distributed resource management and matching in sensor networks. In: IPSN, pp. 97–108 (2009)
18. Gupta, H., Chowdhary, V.: Communication-efficient implementation of join in sensor networks. Ad Hoc Netw. **5**(6), 929–942 (2007)
19. Guttman, A.: R-trees: a dynamic index structure for spatial searching. SIGMOD Rec. **14**(2), 47–57 (1984)
20. Herlihy, M., Sun, Y.: Distributed transactional memory for metric-space networks. Distrib. Comput. **20**(3), 195–208 (2007)
21. Jia, L., Lin, G., Noubir, G., Rajaraman, R., Sundaram, R.: Universal approximations for tsp, steiner tree, and set cover. In: STOC, pp. 386–395 (2005)
22. Kulathumani, V., Arora, A., Sridharan, M., Demirbas, M.: Trail: A distance-sensitive sensor network service for distributed object tracking. ACM Trans. Sen. Netw. **5**(2), 15:1–15:40 (2009)
23. Kung, H.T., Vlah, D.: Efficient location tracking using sensor networks. In: WCNC, vol. 3, pp. 1954–1961 (2003)
24. Li, J., Jannotti, J., De Couto, D.S.J., Karger, D.R., Morris, R.: A scalable location service for geographic ad hoc routing. In: MobiCom, pp. 120–130 (2000)
25. Lin, C.Y., Peng, W.C., Tseng, Y.C.: Efficient in-network moving object tracking in wireless sensor networks. IEEE Trans. Mob. Comput. **5**(8), 1044–1056 (2006)
26. Liu, B.H., Ke, W.C., Tsai, C.H., Tsai, M.J.: Constructing a message-pruning tree with minimum cost for tracking moving objects in wireless sensor networks is np-complete and an enhanced data aggregation structure. IEEE Trans. Comput. **57**(6), 849–863 (2008)
27. Luby, M.: A simple parallel algorithm for the maximal independent set problem. In: STOC, pp. 1–10 (1985)

28. Mechitov, K., Sundresh, S., Kwon, Y., Agha, G.: Poster abstract: cooperative tracking with binary-detection sensor networks. In: SenSys, pp. 332–333 (2003)
29. Motskin, A., Downes, I., Kusy, B., Gnawali, O., Guibas, L.J.: Network warehouses: efficient information distribution to mobile users. In: INFOCOM, pp. 2069–2077 (2011)
30. Nath, S., Gibbons, P.B., Seshan, S., Anderson, Z.R.: Synopsis diffusion for robust aggregation in sensor networks. In: SenSys, pp. 250–262 (2004)
31. Roussopoulos, N., Kelley, S., Vincent, F.: Nearest neighbor queries. SIGMOD Rec. 24(2), 71–79 (1995)
32. Sarkar, R., Gao, J.: Differential forms for target tracking and aggregate queries in distributed networks. IEEE/ACM Trans. Netw. 21(4), 1159–1172 (2013)
33. Sharma, G., Busch, C.: Towards load balanced distributed transactional memory. In: Kaklamanis, C., Papatheodorou, T., Spirakis, P.G. (eds.) Euro-Par 2012. LNCS, vol. 7484, pp. 403–414. Springer, Heidelberg (2012)
34. Sharma, G., Busch, C.: An analysis framework for distributed hierarchical directories. In: Frey, D., Raynal, M., Sarkar, S., Shyamasundar, R.K., Sinha, P. (eds.) ICDCN 2013. LNCS, vol. 7730, pp. 378–392. Springer, Heidelberg (2013)
35. Sharma, G., Busch, C., Srinivasagopalan, S.: Distributed transactional memory for general networks. In: IPDPS, pp. 1045–1056 (2012)
36. Winter, J., Lee, W.C.: Kpt: a dynamic knn query processing algorithm for location-aware sensor networks. In: DMSN, pp. 119–124 (2004)
37. Yao, Y., Tang, X., Lim, E.-P.: In-Network processing of nearest neighbor queries for wireless sensor networks. In: Li Lee, M., Tan, K.-L., Wuwongse, V. (eds.) DASFAA 2006. LNCS, vol. 3882, pp. 35–49. Springer, Heidelberg (2006)
38. Yen, L.H., Wu, B.Y., Yang, C.C.: Tree-based object tracking without mobility statistics in wireless sensor networks. Wirel. Netw. 16(5), 1263–1276 (2010)
39. Zhang, W., Cao, G.: Dctc: Dynamic convoy tree-based collaboration for target tracking in sensor networks. IEEE Trans. Wirel. Commun. 3(5), 1689–1701 (2004)
40. Zhou, D., Gao, J.: Maintaining approximate minimum steiner tree and k-center for mobile agents in a sensor network. In: INFOCOM, pp. 511–515 (2010)

Conflict Graphs and the Capacity
of the Mean Power Scheme

Tigran Tonoyan[✉]

TCS Sensor Lab, Centre Universitaire d'Informatique, Route de Drize 7,
1227 Carouge, Geneva, Switzerland
tigran.tonoyan@unige.ch

Abstract. In this paper the capacity and scheduling problems for wireless networks are considered. It is shown that when using the mean power scheme for the capacity problem, the conflict graph based interference model approximates the SINR model within a constant factor of approximation. Such an approximation does not take place for the uniform and linear power schemes. Using this result, it is also shown that under certain assumptions on the underlying metric space, the mean power scheme yields shorter (asymptotically) schedules than the uniform and linear power schemes, for any set of links. This supports the theoretical evidence that the mean power scheme is more efficient and scalable to different network topologies than more traditional uniform and linear power schemes, and could be a better candidate for MAC protocols.

1 Introduction

Scheduling is one of the fundamental problems in wireless networks: given a set of links, one needs to schedule them into the minimum number of time slots, subject to interference. The subset corresponding to each time slot must be "feasible" with respect to the model of interference adopted. Another related problem is the capacity problem, where it is needed to find a maximum cardinality feasible subset of links. There are a few ways to model the interference. Conflict graph based models are a common way of modeling interference in higher level analysis. In recent years considerable research was conducted for the SINR model, where the cumulative interference is considered as opposed to "binary" interference of graph based models. It has been argued that graph based models can be too optimistic by not taking into account the cumulative interference or path loss. However, these models also have advantages such as simplicity. In fact, in networks with many obstacles and boundaries, the general SINR model seems to be hardly tractable, and graph based models can be the only reasonable option. Hence, it is interesting to find out how well do graph based models approximate the SINR model. In this paper we consider a special conflict graph model based on the SINR formula, and compare it with the SINR model with respect to the capacity problem.

Related Work. The drawbacks of disk graph models have been demonstrated both theoretically and experimentally [1–4]. These drawbacks mainly stem from the

P. Flocchini et al. (Eds.): ALGOSENSORS 2013, LNCS 8243, pp. 278–290, 2014.
DOI: 10.1007/978-3-642-45346-5_20, © Springer-Verlag Berlin Heidelberg 2014

fact that these models do not take the cumulative effect of interference into the account. However, it is known that certain conflict graph models approximate SINR scheduling in doubling metric spaces within a constant approximation factor when the node power levels are uniform and the lengths of the links are "almost equal" [5]. When the link lengths are allowed to be arbitrary, the degree of approximation depends on the power assignment methods used in the network [6]. In [6] three power schemes are considered - the linear, uniform and mean power schemes, and it is shown that when the scheduling problem is considered for links located in doubling metric spaces, then the difference of the solution in a special conflict graph model and in the SINR model is in $O(\log \Delta)$ for the uniform and linear power schemes, and is in $O(\log n)$ for the mean power scheme, where Δ is the ratio between the longest and shortest link lengths and n is the number of links. These approximation bounds hold also for the capacity problem.

Our Contribution. In this paper we show that the optimum SINR capacity of the mean power scheme can be approximated by the maximum independent sets of the corresponding conflict graph within a constant factor, while for the uniform and linear power schemes the approximation factor can still be as large as $\Theta(n)$. This shows that for the capacity problem the gap between the two models is in $O(1)$, which means that a constant factor approximation algorithm for one model can be used to obtain a constant factor approximation for the other model. We use this result and the graph models as a tool to compare optimum schedule lengths for different power schemes. It is known that the mean power scheme yields better solutions for the capacity problem than the uniform and linear power schemes, for any network topology [7]. In this paper we show that under certain assumptions on the metric space where the network is located, the mean power scheme yields shorter *schedules* than the uniform and linear power schemes for any network topology, thus extending the results of [7]. Thus, the mean power scheme is not only scalable in different interference models, but it is also more efficient than the uniform and linear power schemes, from the perspective of the scheduling problem. These results support the theoretical evidence that the mean power scheme could be a better choice in MAC protocol design for wireless networks (see [5] and [8] for more evidence).

2 Models and Definitions

2.1 Capacity and Scheduling Problems

The general theoretical problems that we consider here are as follows. Let $\Gamma = \{1, 2, \ldots, n\}$ denote the set of n links (transmission requests between a sender and a receiver node) in a metric space, where the sender node of each link is assigned a certain power level. Then, using a specific interference model, one can define a *feasible set of links*, i.e. a set of links that can transmit with the given power levels all in the same time slot without interference. *Capacity* problem asks to select a maximum cardinality feasible subset of Γ. A closely related problem, *Scheduling* problem, asks to split Γ into the minimum number of feasible subsets, each of which can be assigned to a different time slot. Each such collection of subsets is called *a schedule*, and the number of subsets is called *the length of the schedule*. Of course, the sizes of feasible sets and schedules depend on the power assignment of the nodes and the interference model. Next we define the SINR model and related conflict graph model.

2.2   The Path-Loss Model of Signal Decay and Cumulative Interference

Let the sender node of each link i be assigned a power level $P(i)$ according to some assignment policy or *power scheme*. According to the path-loss propagation model [9], the receive signal of each link i is $P_i = P(i)/l_i^\alpha$, where l_i is the distance between the sender and the receiver of i and $\alpha > 0$ is the path-loss exponent (typically, $2 < \alpha < 6$ [9]). Similarly, the interference caused by link j at the receiver of link i is $I_{ji} = P(j)/d_{ji}^\alpha$, where d_{ji} denotes the metric distance from the sender node of link j to the receiver node of link i. According to SINR (Signal to Interference and Noise Ratio) model, the transmission of a link i is successful if and only if the SINR is greater than a certain threshold $\beta \geq 1$:

$$\frac{P_i}{\sum_{j \in S \setminus \{i\}} I_{ji} + N} \geq \beta, \tag{1}$$

where the constant $N \geq 0$ denotes the noise and S is the set of links transmitting in the same time slot as i.

In this model, a subset S of Γ is feasible if (1) holds for each link $i \in S$.

We denote the *optimum SINR-capacity* of a set Γ with respect to power scheme P as $OPTC_P(\Gamma)$, and we denote the *optimum SINR-schedule length* by $OPTS_P(\Gamma)$.

Assumptions. The noise coefficient N can be skipped under certain assumptions which are discussed in Sect. 2.5 (see also [7]). From this point on we assume $N = 0$. We also need the assumption that the path-loss parameter α be greater than the *doubling dimension* of the metric space where the nodes are located. The exact definition of doubling dimension can be found in [10]. Here we only need the fact that in a metric space with doubling dimension m, each ball of radius r contains at most $C \cdot (r/r')^m$ disjoint balls of a smaller radius r' where C is an absolute constant. It is known that the m-dimensional Euclidean space has doubling dimension m (see [10]).

By assuming $N = 0$ and $\beta \geq 1$, it is easy to check that the following condition in equivalent to the SINR condition:

$$A_P(S, i) = \sum_{j \in S \setminus i} \min\{1, I_{ji}/P_i\} \leq 1/\beta. \tag{2}$$

This form of SINR condition has been considered in a number of papers (e.g. [11]) because it has the convenient additivity property: if there are two disjoint sets S_1 and S_2 then $A_P(S_1 \cup S_2, i) = A_P(S_1, i) + A_P(S_2, i)$. We can also define $A(S_1, S_2)$ for two sets of links as $A_P(S_1, S_2) = \sum_{j \in S_2} A_P(S_1, j)$.

Following [12], we say that a set S is a *p-signal set* if $A_P(S, i) \leq 1/p$. Similarly, a partition of the set of links is a p-signal partition if each subset is a p-signal set. The following theorem shows that one can vary the threshold of the SINR condition without changing the schedule length much. It holds for any fixed power assignment.

Theorem 1. *[12] There is a polynomial-time algorithm that takes a p-signal schedule and transforms into a p'-signal schedule, for $p' > p$, increasing the number of slots by a factor of at most $\lceil 2p'/p \rceil^2$.*

2.3   The Conflict Graph Model

Graph models incorporate the notion of independence of links. Two links i and j are *q-independent* with respect to a power assignment P if $A_P(\{i\}, j) < 1/q^\alpha$ and $A_P(\{j\}, i) < 1/q^\alpha$, otherwise we say that i and j are *q-adjacent* w.r.t P. When P is clear from the context, we will just say q-independent, q-adjacent etc.

This helps us define the q-adjacency graph $G_q^P(\Gamma)$ of the set of links Γ w.r.t power assignment P, where the set of vertices of $G_q^P(\Gamma)$ is Γ and two links i and j are adjacent in $G_q^P(\Gamma)$ if and only if they are q-adjacent links. In the conflict graph interference model related to $G_q^P(\Gamma)$, we assume that a set of links is feasible iff it is an independent set in $G_q^P(\Gamma)$ in the sense of graph theory. Capacity problem in this model is to find a maximum cardinality independent set in $G_q^P(\Gamma)$. We denote the size of such a set by $\tau(G_q^P(\Gamma))$. A schedule in this model is a coloring of graph $G_q^P(\Gamma)$, so Scheduling problem is to vertex-color the graph with the least number of colors (so that adjacent vertices get different colors). We denote the number of colors used in the optimum coloring by $\chi(G_q(\Gamma))$.

Note that if two links are q-adjacent then at least one of them interferes with the other one "too much", so $G_q^P(\Gamma)$ is a natural approximation of the stricter SINR model.

It follows from the definition of q-independence that each q^α-signal set of links is q-independent. Taking into account Theorem 1, we can state the following relation between schedule lengths of the same set of links in the two models.

Theorem 2. *For any set of links Γ, any power scheme P and a constant $q \geq 1$, $\chi(G_q^P(\Gamma)) \in O(OPTS_P(\Gamma))$.*

2.4   Distance Dependent Power Schemes

The power assignments considered in this paper are made according to generic schemes. For example, the *uniform power scheme* assigns all links the same power levels. By saying power scheme P we will understand the corresponding power assignment for a given network. The motivation for considering power schemes is the fact that they are computable in a localized and general manner and do not depend on the whole network topology, which makes them well-suited for decentralized wireless networks as opposed to complex power control algorithms.

We consider the power schemes which are given by a formula of type $P(i) = cl_i^{t\alpha}$, where $t \geq 0$ is the parameter describing the power scheme and $c > 0$ is a constant (not necessarily small). The well-known *uniform, linear and mean* power schemes are included in this class of power schemes. The uniform power scheme corresponds to the parameter value $t = 0$, i.e. all the links are assigned the same transmit power. We denote this power scheme by $\mathcal{U}$. The linear power scheme, denoted by $\mathcal{L}$, corresponds to the parameter value $t = 1$, i.e. all the links are assigned power levels such that they have a constant receive power. The mean power scheme, denoted by $\mathcal{M}$, corresponds to the parameter value $t = 1/2$. Our analysis concentrates on $\mathcal{M}$ and its relation with the other power schemes.

2.5   Noise Factor

Here we show that the noise factor can be avoided in the SINR formula in asymptotic computations, under certain assumptions.

Let us suppose that there is a noise N and a link i is in a SINR-feasible (w.r.t. some power assignment P) set S that contains at least two links. Then it must be that $P(i)/l_i^\alpha \geq \beta N$. We make a slightly stronger assumption: e.g. $P(i)/l_i^\alpha \geq 2\beta N$.

If a set S is SINR-feasible w.r.t. P *without* the noise factor and with $\beta' = 2\beta$, then we have $P(i)/l_i^\alpha \geq \beta' \sum_{j \in S \setminus i} P(j)/d_{ji}^\alpha$, so

$$P(i)/l_i^\alpha \geq \beta \sum_{j \in S \setminus i} P(j)/d_{ji}^\alpha + P(i)/(2l_i^\alpha) \geq \beta \sum_{j \in S \setminus i} P(j)/d_{ji}^\alpha + \beta N,$$

so S is also SINR-feasible *with* the noise N and with the original β. Thus, if our assumption holds we can do the computations with $N = 0$.

3   The Capacity of $\mathcal{M}$ and the Conflict Graphs

In this section we show that for any set of links, the optimal SINR capacity w.r.t. $\mathcal{M}$ can be approximated by the maximum independent set of the corresponding conflict graph within a constant approximation ratio. Let us start with some lemmas.

We call S an *equilength* set of links if for any pair of links $i, j \in S$, $l_i < 2l_j$.

Lemma 1. *Let i be a link, S be a set of links and $h > 0$ be an integer. We assume:*

(a) S is an equilength set of links,
(b) for all links $j \in S$, $l_j \geq l_i$,
(c) for all $j, k \in S$, $d(s_j, s_k) > \sqrt{l_j l_k}$ and $d(r_j, r_k) > \sqrt{l_j l_k}$,
(d) either $A_\mathcal{M}(j, i) > h^{-\alpha}$ for all $j \in S$ or $A_\mathcal{M}(i, j) > h^{-\alpha}$ for all $j \in S$.

Then $|S| \in O(h^m)$.

Proof. We present the proof for the case when $A_{\mathcal{M}}(j, i) > h^{-\alpha}$ for all $j \in S$. The other case can be handled in a similar way, by replacing senders with receivers and vice versa. Let ℓ denote the length of the shortest link in S. From (d) we have that for all $j \in S$,

$$d(s_j, r_i) < h\sqrt{l_i l_j},$$

and since S is equilength,

$$d(s_j, r_i) < h\sqrt{2l_i\ell} = \sqrt{2}h\sqrt{l_i\ell}.$$

On the other hand, from (b) and (c) we have that for all pairs $j, k \in S$,

$$d(s_j, s_k) > \sqrt{l_j l_k} \geq \sqrt{l_i\ell}.$$

These inequalities can be interpreted in terms of metric balls as follows: for each pair of nodes s_j, s_k, the metric balls of radius $\sqrt{l_i\ell}/2$ centered at these nodes do not intersect each other, and all these balls are contained in the ball of radius $(\sqrt{2}h + 1/2)\sqrt{l_i\ell}$ centered at node r_i. Thus, using the doubling property of the metric space, we can conclude that the number of different balls (and thus, the number of links in S) is bounded by $(2\sqrt{2}h+1)^m$, which proves the lemma for the case considered. □

We will use the following well known results from calculus in the proof of the next lemma. Let $f(x)$ be a real convex function: then for each x, y from the domain of f,

$$f(x) - f(y) \leq f'(x)(x - y). \tag{3}$$

Let $g(x)$ be a real *decreasing* function, integrable over $[1, \infty)$. Then

$$\sum_{k=1}^{\infty} g(x) \leq g(1) + \int_1^{\infty} g(h)\mathrm{d}h. \tag{4}$$

The following lemma is the place where we use the assumption that $\alpha > m$. We use the following notation: for a set of links S and a link i, we denote

$$S_i^+ = \{j \in S : l_j \geq l_i\} \text{ and } S_i^- = \{j \in S : l_j < l_i\}.$$

Lemma 2. *Suppose that $\alpha > m$ and let S be a set of links which is 2-independent w.r.t. mean power. Then for each link $i \in S$,*

$$A_{\mathcal{M}}(S_i^+, i) + A_{\mathcal{M}}(i, S_i^+) \in O(1).$$

Proof. First observe that for each $s_i, s_j \in S$,

$$d(s_i, s_j) > \sqrt{l_i l_j} \text{ and } d(r_i, r_j) > \sqrt{l_i l_j}. \tag{5}$$

Indeed, from 2-independence we have $d(s_i, r_j) > 2\sqrt{l_i l_j}$ and $d(s_j, r_i) > 2\sqrt{l_i l_j}$. If we assume that $l_i \geq l_j$ then, using the triangle inequality, we get:

$$d(s_i, s_j) \geq d(s_i, r_j) - l_j > \sqrt{l_i l_j}.$$

The second inequality follows in the same way from $d(s_j, r_i) > 2\sqrt{l_i l_j}$. Now, let us first fix a link $i_0 \in S$ and show that $A_{\mathcal{M}}(S_{i_0}^+, i_0) \in O(1)$. In order to bound the sum, we first consider the subsets $T_1, T_2, \ldots$ of $S_{i_0}^+$, where

$$T_h = \{j \in S_{i_0}^+ : A_{\mathcal{M}}(j, i_0) > h^{-\alpha}\}.$$

We will first bound $A_{\mathcal{M}}(T_h, i_0)$ for each T_h separately. For doing this, let us fix T_h for some $h \geq 1$ and split T_h into subsets of equilength links:

$$T_h = T_h^1 \cup T_h^2 \cup T_h^3 \ldots,$$

where $T_h^k = \{j \in T_h : l_j \in [2^{k-1}, 2^k)\}$. Let us upper-bound the value $A_{\mathcal{M}}(T_h^k, i_0)$ for fixed h and k. Note that all the conditions of Lemma 1 are satisfied for the set T_h^k and the link i_0 with corresponding values of the parameters; hence, $|T_h^k| \in O(h^m)$. It remains to find out how many subsets T_h^k are there for a fixed h. According to the assumptions, for each link $j \in T_h$, $A_{\mathcal{M}}(j, i_0) > h^{-\alpha}$, so

$$d(s_j, r_{i_0}) < h\sqrt{l_j l_{i_0}}. \tag{6}$$

On the other hand, we have from (5) that

$$d(s_i, s_j) > \sqrt{l_i l_j} \tag{7}$$

for each pair of links $i, j \in T_h$. At last, applying the triangle inequality, we have:

$$d(s_i, s_j) \leq d(s_i, r_{i_0}) + d(s_j, r_{i_0}). \tag{8}$$

Combining (6), (7) and (8), we get that

$$\sqrt{l_i l_j} < h\sqrt{l_i l_{i_0}} + h\sqrt{l_j l_{i_0}} \tag{9}$$

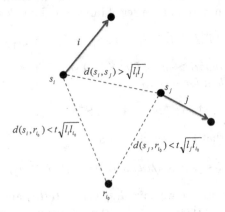

Fig. 1. An illustration of (9).

holds *for all* $i, j \in T_h$ (see Fig. 1). Dividing by the square root of the longer link, this implies that *for all links* $j \in T_h$ *except one*,

$$l_j/l_{i_0} < 4h^2,$$

that one link being the longest link in T_h. Let j_0 denote the *second longest* link in T_h. Then we have that $l_{j_0}/l_{i_0} < 4h^2$. Since $l_j \geq l_{i_0}$ for all $j \in T_h$, we see that the number of subsets T_h^k for a fixed h is upper-bounded by $\log \frac{l_{j_0}}{l_{i_0}} + 1$, where $+1$ counts for the longest link (which might be much longer than the rest of the links). Thus, the number of T_h^k with a fixed h is in $O(\log h)$, and the overall number of links in T_h is in $O(h^m \log h)$. Since $T_1 = \emptyset$ (2-independence) and $T_h \subseteq T_{h-1}$, we have that

$$A_{\mathcal{M}}(S_{i_0}^+, i_0) \leq \sum_{h=2}^{\infty} |T_h \setminus T_{h-1}|(h-1)^{-\alpha} = \sum_{h=2}^{\infty} |T_h|[(h-1)^{-\alpha} - h^{-\alpha}] \leq$$

$$\leq \sum_{h=2}^{\infty} |T_h| \frac{\alpha h^{\alpha-1}}{(h(h-1))^{\alpha}} \leq \sum_{h=2}^{\infty} \frac{ch^m \log h}{h(h-1)^{\alpha}} \leq \sum_{h=1}^{\infty} \frac{c' \log h}{h^{\alpha-m+1}} \in O(1).$$

The explanation follows. We used (3) with $f(h) = h^\alpha$ (which is a convex function for $\alpha \geq 1$). In the last statement we used the fact that $\sum_{h=1}^{\infty} \frac{\log h}{h^{\alpha-m+1}}$ converges: this follows from the assumption that $\alpha > m$. Indeed, since $g(h) = \log h/h^{\alpha-m+1}$ is a decreasing function, we can apply (4):

$$\sum_{h=1}^{\infty} g(h) \leq g(1) + \int_1^{\infty} g(h)dh = \int_1^{\infty} \frac{\log h}{h^{\alpha-m+1}}dh =$$

$$= -(\alpha-m)^{-1} \left(\frac{\log h}{h^{\alpha-m}} \Big|_1^{\infty} - \ln^{-1} 2 \int_1^{\infty} \frac{dh}{h^{\alpha-m+1}} \right) = \frac{1}{\ln 2(\alpha-m)^2}.$$

This concludes the proof of $A_{\mathcal{M}}(S_i^+, i) \in O(1)$. The claim $A_{\mathcal{M}}(i, S_i^+) \in O(1)$ can be proven similarly, by replacing senders with receivers and vice versa in the proof. Note that Lemma 1 is symmetric in this respect and works for both cases. □

The following is a known lemma, e.g. [11].

Lemma 3. *Let S be a set of links such that $A_P(S_i^+, i) + A_P(i, S_i^+) \in O(1)$ for each link $i \in S$, where P is an arbitrary power scheme. Then $OPTC_P(S) \in \Theta(|S|)$.*

Proof. Consider the sum $\sum_{i \in S} A_P(S, i)$. Using additivity of $A(.,.)$ (for both arguments), we can rearrange the sum as follows:

$$\sum_{i \in S} A_P(S, i) = \sum_{i \in S} A_P(S_i^+, i) + A_P(i, S_i^+) \leq c|S|$$

for a number $c \in O(1)$. Using the pigeonhole principle, we can conclude that for at least half of the links $i \in S$, $A_P(S, i) \leq 2c$. This means that there is a subset $S' \subseteq S$ with $|S'| \geq |S|/2$ such that $A_P(S', i) \leq 2c$ holds for each $i \in S'$. Applying Theorem 1 for S', we obtain a *feasible* subset $S'' \subseteq S'$ with $|S''| \in \Omega(|S'|)$, which completes the proof. $\square$

At this point we are ready to prove the main theorem.

Theorem 3. *Suppose that $\alpha > m$. Then for any set of links Γ,*

$$\tau(G_2(\Gamma)) = \Theta(OPTC_{\mathcal{M}}(\Gamma)).$$

Proof. It follows from Theorem 2 that $\tau(G_2(\Gamma)) \in O(OPTC_{\mathcal{M}}(\Gamma))$. It suffices to show that $\chi(G_2(\Gamma)) \in \Omega(OPTC_{\mathcal{M}}(\Gamma))$. Consider a maximum independent set S of $G_2(\Gamma)$. From Lemma 2 we have that for each $i \in S$, $A_{\mathcal{M}}(S_i^+, i) + A_{\mathcal{M}}(i, S_i^+) \in O(1)$. Hence, applying Lemma 3, we have

$$OPTC_{\mathcal{M}}(\Gamma) \geq OPTC_{\mathcal{M}}(S) \in \Theta(|S|) = \Theta(\tau(G_2^{\mathcal{M}}(\Gamma))). \qquad \square$$

This result shows that SINR capacity of $\mathcal{M}$ is well approximated by the capacity of the corresponding conflict graph model. In contrast, it can be shown that the power schemes $\mathcal{L}$ and $\mathcal{U}$ are not so scalable with respect to the corresponding conflict graphs. This is shown in the following theorem (originally from [6]). The network example from this theorem is a generalization of the network example from [2].

Theorem 4. *For any $n > 0$ and $q \geq 1$, there is a network instance Γ, such that $|\Gamma| = n$, Γ is q-independent w.r.t. $\mathcal{L}$ and $\mathcal{U}$, but $OPTC_{\mathcal{U}}(\Gamma) \in O(q^\alpha)$ and $OPTC_{\mathcal{L}}(\Gamma) \in O(q^\alpha)$.*

Proof. In this instance, there are n links $\{1, 2, \ldots, n\}$ sequentially aligned on a straight line in the increasing order of numbers going from left to right. We set $l_i = 2^i$ for $i = 1, 2, \ldots, n$. Each link has its sender on the left side and the receiver on the right side. For each $i > 1$, we define $d_{i,i-1} = ql_i$. Now we can calculate all the other distances. For each $i > 1$, we have:

$$d_{i1} = \sum_{t=2}^{i-1} (d_{t,t-1} + l_t) + d_{i,i-1} = \sum_{t=2}^{i-1} l_t + q \sum_{t=2}^{i} l_t =$$

$$= (q+1) \sum_{t=2}^{i-1} l_t + q l_i = (2q+1)l_i - 4(q+1),$$

where we used the fact that $l_i = 2^i$. Now if $i > j$ then

$$d_{ij} = d_{i1} - d_{j1} - l_j = (2q+1)(l_i - l_j) - l_j \text{ and}$$

$$d_{ji} = d_{i1} - d_{j1} + l_i = (2q+1)(l_i - l_j) + l_i.$$

It is easy to check that this set of links is q-independent with respect to both $\mathcal{U}$ and $\mathcal{L}$.

Suppose that the power scheme $\mathcal{U}$ is used. Consider *any* SINR-feasible subset of links S of size k with the links $i_1 < i_2 < \cdots < i_k$. For the longest link i_k we have:

$$A_{\mathcal{U}}(S, i_k) = \sum_{t=1}^{k-1} \frac{l_{i_k}^{\alpha}}{d_{i_t i_k}^{\alpha}} = \sum_{t=1}^{k-1} \frac{l_{i_k}^{\alpha}}{((2q+1)(l_{i_k} - l_{i_t}) + l_{i_k})^{\alpha}} \geq \frac{k-1}{(2q+2)^{\alpha}}.$$

We should have also that $A_{\mathcal{U}}(S, i_k) \leq 1/\beta$ which allows us bound the number of links in S:

$$k \leq \frac{(2q+2)^{\alpha}}{\beta} + 1.$$

For the case of the power scheme $\mathcal{L}$ a similar argument works if we consider $A_{\mathcal{L}}(S, i_1)$. $\qquad\square$

4   Comparing SINR Schedules Using Conflict Graphs

It is easy to check that q-independence w.r.t. $\mathcal{U}$ or $\mathcal{L}$ implies q'-independence w.r.t. $\mathcal{M}$, where $q' = q - 1$. This remark, combined with Theorem 3 and Theorem 4, implies that there are network instances for which $\mathcal{M}$ can provide much better capacity than $\mathcal{U}$ or $\mathcal{L}$. It is also known that the capacity of $\mathcal{M}$ for any network instance cannot be worse than by a constant factor than the capacity of $\mathcal{U}$ and $\mathcal{L}$ [7]. In this section we discuss the relation between the optimal *schedule lengths* of $\mathcal{M}$ and the two power schemes $\mathcal{U}$ and $\mathcal{L}$, for any set of links. It follows readily from the results presented in the previous section that if $\alpha > m$ then for any set of links, the optimum schedule length w.r.t. $\mathcal{M}$ is not less than the optimum schedule length w.r.t. $\mathcal{U}$ (not counting constant factors).

Theorem 5. *Suppose that $\alpha > m$. Then for any set of links Γ,*

$$OPTS_{\mathcal{M}}(\Gamma) \in O(OPTS_{\mathcal{U}}(\Gamma)).$$

Proof. Let S be a SINR-feasible set of links w.r.t. $\mathcal{U}$. Then S is a q-independent set of links w.r.t. $\mathcal{U}$ for a constant $q \geq 1$. In view of Theorem 1, we can assume that $q \geq 3$. The q-independence condition w.r.t. $\mathcal{U}$ for two links i, j is the following:

$$d(s_i, r_j) > ql_j \text{ and } d(s_j, r_i) > ql_i.$$

Let us assume that $l_i \geq l_j$. Then, using the triangle inequality, we see that the spatial separation between the senders and receivers should be a multiple of l_i, provided that $q \geq 3$:

$$d(s_i, s_j) \geq d(s_j, r_i) - l_i > (q-1)l_i \text{ and } d(r_i, r_j) \geq d(s_j, r_i) - l_j > (q-1)l_i.$$

This means that S is a $q-1$-independent set w.r.t. $\mathcal{M}$; hence, Theorem 3 applies. This implies that for each link $i \in S$, $A_{\mathcal{M}}(S_i^+, i) \in O(1)$. In order to bound $A_{\mathcal{M}}(S_i^-, i)$ it is enough to note that $A_{\mathcal{M}}(S_i^-, i) \leq A_{\mathcal{U}}(S_i^-, i)$. Indeed,

$$A_{\mathcal{M}}(S_i^-, i) = \sum_{j \in S_i^-} \left(\frac{\sqrt{l_i l_j}}{d_{ji}} \right)^\alpha \leq \sum_{j \in S_i^-} \left(\frac{l_i}{d_{ji}} \right)^\alpha = A_{\mathcal{U}}(S_i^-, i).$$

It follows from SINR-feasibility of S that $A_{\mathcal{U}}(S_i^-, i) \in O(1)$. Hence, $A_{\mathcal{M}}(S, i) = A_{\mathcal{M}}(S_i^+, i) + A_{\mathcal{M}}(S_i^-, i) \in O(1)$, and it remains to apply Theorem 1 in order to show that S can be scheduled into a constant number of subsets w.r.t. $\mathcal{M}$. □

A similar result can be achieved for $\mathcal{L}$ as well. However, we were able to prove such a result only with a stricter assumption on α: $\alpha > 2m$. Note that in the case when $m = 2$, i.e. when the nodes are on the Euclidean plane, we require $\alpha > 4$, which is still in the practical range for α (see [9]). For the three dimensional case we require $\alpha > 6$ which might be unpractical.

Theorem 6. *Suppose that $\alpha > 2m$. Then for any set of links Γ,*

$$OPTS_{\mathcal{M}}(\Gamma) \in O(OPTS_{\mathcal{L}}(\Gamma)).$$

Proof. Let S be a SINR-feasible set w.r.t. $\mathcal{L}$. As in the case of $\mathcal{U}$, we can show that $G_{q-1}^{\mathcal{M}}(\Gamma)$ is a subgraph of $G_q^{\mathcal{L}}(\Gamma)$ for $q \geq 2$. Hence, we can assume that we have a 2-independent (w.r.t. $\mathcal{L}$) set S, and $A_{\mathcal{M}}(S_i^+, i) \in O(1)$. We aim to show that under the assumptions of the theorem, $A_{\mathcal{M}}(S_i^-, i) \in O(1)$ holds.

Let us split S_i^- into equilength subsets $T_1, T_2, \ldots$ where $T_t = \{j \in S_i^- : 2^{t-1} l_{min} \leq l_j < 2^t l_{min}\}$. We compute $A_{\mathcal{M}}(T_t, i)$ separately for each t. Let us fix an index t and further split T_t into subsets $T_t^1, T_t^2, \ldots$ where $T_t^r = \{j \in T_t : d(s_j, r_i) \leq l_i + (r-1)\ell\}$, where ℓ is the length of the shortest link in T_t. Note that T_t^1 is empty. Let us fix an $r > 1$ as well. It follows from 2-independence and the construction of T_t that the distance between any two sender nodes s_j, s_k belonging to T_t is at least ℓ. This means that balls of radius $\ell/2$ around those senders do not intersect. Note that for T_t^r, these balls are also contained in the larger ball centered at r_i with radius $l_i + (r-1/2)\ell$. Using the doubling property of the space, we obtain that the number of those sender nodes and the links in T_t^r is upper bounded as follows:

$$|T_t^r| \leq C \cdot \left(\frac{l_i + (r - 1/2)\ell}{\ell/2} \right)^m.$$

Note also that for $r > 1$ and for any link $j \in T_t^r \setminus T_t^{r-1}$, $A_{\mathcal{M}}(j, i) \leq \left(\frac{\sqrt{l_i (2\ell)}}{l_i + (r-1)\ell} \right)^\alpha$; hence,

$$A_{\mathcal{M}}(T_t, i) = \sum_{r \geq 1} A_{\mathcal{M}}(T_t^r \setminus T_t^{r-1}, i) \leq \sum_{r \geq 1} (|T_t^r| - |T_t^{r-1}|) \left(\frac{\sqrt{l_i(2\ell)}}{l_i + (r-1)\ell} \right)^{\alpha} =$$

$$= \sum_{r \geq 2} |T_t^r| \left(\left(\frac{\sqrt{l_i(2\ell)}}{l_i + (r-1)\ell} \right)^{\alpha} - \left(\frac{\sqrt{l_i(2\ell)}}{l_i + r\ell} \right)^{\alpha} \right) \leq$$

$$\leq \sum_{r \geq 2} |T_t^r| (2l_i\ell)^{\alpha/2} \frac{\alpha\ell}{(l_i + (r-1)\ell)^{\alpha+1}} = \sum_{r \geq 2} \frac{(l_i + (r-1/2))^m (2l_i\ell)^{\alpha/2}}{(\ell/2)^m (l_i + (r-1)\ell)^{\alpha+1}} \leq$$

$$\leq 2^{\alpha/2 + 2m} l_i^{\alpha/2} \ell^{\alpha/2 - m} \sum_{r \geq 1} \frac{1}{(l_i + r\ell)^{\alpha+1-m}}.$$

The last sum can be bounded using (4):

$$\sum_{r \geq 1} \frac{1}{(l_i + r\ell)^{\alpha+1-m}} \leq \frac{1}{(l_i + \ell)^{\alpha+1-m}} + \int_1^\infty \frac{dr}{(l_i + r\ell)^{\alpha+1-m}} \leq \frac{2}{\ell(l_i + \ell)^{\alpha-m}}.$$

Thus, $A_{\mathcal{M}}(T_t, i) \leq 2^{\alpha/2 + 2m + 1} \dfrac{l_i^{\alpha/2} \ell^{\alpha/2 - m}}{l(l_i + \ell)^{\alpha - m}} \leq 2^{\alpha/2 + 2m + 1} \left(\dfrac{\ell}{l_i} \right)^{\alpha/2 - m}$. Recall that T_t is an equilength subset of S_i^-, and the minimum link lengths of T_t for different values of t differ at least by a factor 2, so

$$A_{\mathcal{M}}(S_i^-, i) \leq \sum_1^\infty A_{\mathcal{M}}(T_t, i) \leq \frac{2^{\alpha/2 + 2m + 1}}{l_i^{\alpha/2 - m}} \sum_{t=1}^{\lceil \log l_i / l_{min} \rceil} (2^t l_{min})^{(\alpha/2 - m)}$$

Using (4) for the sum in the last term, we can bound it by $c l_i^{\alpha/2 - m}$ with $c \in O(1)$, which will give $A(S_i^-, i) \in O(1)$. $\qquad\square$

5   Conclusion

It was shown in this paper that when the mean power scheme is used, the corresponding conflict graph approximates SINR capacity within a constant factor. This is not the case for the uniform and linear power schemes. It is possible that the constant factor approximation could be extended for scheduling too, so it is an interesting open question whether such an approximation holds.

We also showed that conflict graphs can be used in order to show that the mean power scheme is more efficient than the uniform and linear power schemes in terms of SINR scheduling. These results were obtained under some assumptions. While these assumptions seem to be reasonable, it would certainly be useful to prove the results for more general settings. We believe that the close connection of the two models can yield more results, because it allows to use graph theoretical methods for working with the SINR model.

References

1. Grönkvist, J., Hansson, A.: Comparison between graph-based and interference-based STDMA scheduling. In: 2nd ACM International Symposium on Mobile Ad Hoc Networking and Computing, pp. 255–258 (2001)
2. Moscibroda, T., Wattenhofer, R., Yves, W.: Protocol design beyond graph-based models. In: 5th ACM SIGCOMM Workshop on Hot Topics in Networks (2006)
3. Behzad, A., Rubin, I.: On the performance of graph-based scheduling algorithms for packet radio networks. In: IEEE Global Telecommunications Conference, pp. 3432–3436 (2003)
4. Chafekar, D., Kumar, V.S.A., Marathe, M.V., Parthasarathy, S., Srinivasan, A.: Approximation algorithms for computing capacity of wireless networks with SINR constraints. In: 27th IEEE International Conference on Computer Communications, pp. 1166–1174 (2008)
5. Halldórsson, M.M.: Wireless scheduling with power control. ACM Trans. Algorithms 9(1), 7 (2012)
6. Tonoyan, T.: Comparing schedules in the SINR and conflict-graph models with different power schemes. In: 12th International Conference on Ad Hoc Networks and Wireless, pp. 317–328 (2013)
7. Tonoyan, T.: On the capacity of oblivious powers. In: 7th International Symposium on Algorithms for Sensor Systems, Wireless Ad Hoc Networks and Autonomous Mobile Entities, pp. 225–237 (2012)
8. Halldórsson, M.M., Holzer, S., Mitra, P., Wattenhofer, R.: The power of non-uniform wireless power. In: 24th Annual ACM-SIAM Symposium on Discrete Algorithms, pp. 1595–1606 (2013)
9. Rappaport, T.S.: Wireless Communications: Principles and Practice, 2nd edn. Prentice Hall, Upper Saddle River (2002)
10. Heinonen, J.: Lectures on Analysis on Metric Spaces, 1st edn. Springer, New York (2000)
11. Halldórsson, M.M., Mitra, P.: Wireless capacity with oblivious power in general metrics. In: 22nd Annual ACM-SIAM Symposium on Discrete Algorithms, pp. 1538–1548 (2011)
12. Halldórsson, M.M., Wattenhofer, R.: Wireless communication is in APX. In: 36th International Colloquium on Automata, Languages and Programming, pp. 525–536 (2009)

Rendezvous of Two Robots with Visible Bits

Giovanni Viglietta[✉]

School of Computer Science, Carleton University, Ottawa, Canada
viglietta@gmail.com

Abstract. We study the rendezvous problem for two robots moving in the plane (or on a line). Robots are autonomous, anonymous, oblivious, and carry colored lights that are visible to both. We consider deterministic distributed algorithms in which robots do not use distance information, but try to reduce (or increase) their distance by a constant factor, depending on their lights' colors.

We give a complete characterization of the number of colors that are necessary to solve the rendezvous problem in every possible model, ranging from fully synchronous to semi-synchronous to asynchronous, rigid and non-rigid, with preset or arbitrary initial configuration.

In particular, we show that three colors are sufficient in the non-rigid asynchronous model with arbitrary initial configuration. In contrast, two colors are insufficient in the rigid asynchronous model with arbitrary initial configuration and in the non-rigid asynchronous model with preset initial configuration.

Additionally, if the robots are able to distinguish between zero and non-zero distances, we show how they can solve rendezvous and detect termination using only three colors, even in the non-rigid asynchronous model with arbitrary initial configuration.

Keywords: Rendezvous · Autonomous · Mobile · Robots · Visible

1 Introduction

1.1 Models for Mobile Robots

The basic robot model we employ has been thoroughly described in [1–3,7]. Robots are modeled as points freely moving in $\mathbb{R}^2$ (or $\mathbb{R}$). Each robot has its own coordinate system and its own unit distance, which may differ from the others. Robots operate in cycles that consist of four phases: WAIT, LOOK, COMPUTE, and MOVE.

In a WAIT phase a robot is idle; in a LOOK phase it gets a snapshot of its surroundings (including the positions of the other robots); in a COMPUTE phase it computes a destination point; in a MOVE phase it moves toward the destination point it just computed, along a straight line. Then the cycle repeats over and over.

P. Flocchini et al. (Eds.): ALGOSENSORS 2013, LNCS 8243, pp. 291–306, 2014.
DOI: 10.1007/978-3-642-45346-5_21, © Springer-Verlag Berlin Heidelberg 2014

Robots are anonymous and oblivious, meaning that they do not have distinct identities, they all execute the same algorithm in each COMPUTE phase, and the only input to such algorithm is the snapshot coming from the previous LOOK phase.

In a MOVE phase, a robot may actually reach its destination, or it may be stopped before reaching it. If a robot always reaches its destination by the end of each MOVE phase, then the model is said to be *rigid*. If a robot can unpredictably be stopped before, the model is *non-rigid*. However, even in non-rigid models, during a MOVE phase, a robot must always be found on the line segment between its starting point and the destination point. Moreover, there is a constant distance $\delta > 0$ that a robot is guaranteed to walk at each cycle. That is, if the destination point that a robot computes is at most δ away (referred to some global coordinate system), then the robot is guaranteed to reach it by the end of the next MOVE phase. On the other hand, if the destination point is more than δ away, the robot is guaranteed to approach it by at least δ.

In the basic model, robots cannot communicate in conventional ways or store explicit information, but a later addition to this model allows each robot to carry a "colored light" that is visible to every robot (refer to [2]). There is a fixed amount of possible light colors, and a robot can compute its destination and turn its own light to a different color during a COMPUTE phase, based on the light colors that it sees on other robots and on itself. Usually, when robots start their execution, they have all their lights set to a predetermined color. However, we are also interested in algorithms that work regardless of the initial color configurations of the robots.

In the fully synchronous model (FSYNCH) all robots share a common notion of time, and all their phases are executed synchronously. The semi-synchronous model (SSYNCH) is similar, but not every robot may be active at every cycle. That is, some robots are allowed to "skip" a cycle at unpredictable times, by extending their WAIT phase to the whole cycle. However, the robots that are active at a certain cycle still execute it synchronously. Also, no robot can remain inactive for infinitely many consecutive cycles. Finally, in the asynchronous model (ASYNCH) there is no common notion of time, and each robot's execution phase may last any amount of time, from a minimum $\epsilon > 0$ to an unboundedly long, but finite, time.

Figure 1 shows all the possible models arising from combining synchronousness, rigidity, and arbitrariness of the initial light colors. The trivial inclusions between models are also shown.

Without loss of generality, in this paper we will assume LOOK phases in ASYNCH to be instantaneous, and we will assume that a robot's light's color may change only at the very end of a COMPUTE phase.

1.2 Gathering Mobile Robots

GATHERING is the problem of making a finite set of robots in the plane reach the same location in a finite amount of time, and stay there forever, regardless of their initial positions. Such location should not be given as input to the robots,

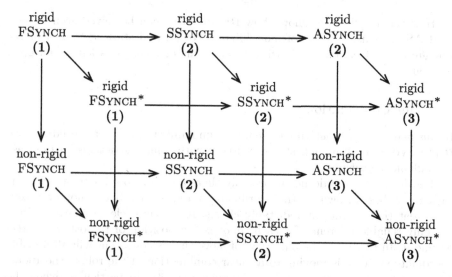

Fig. 1. Robot models with their trivial inclusions. An asterisk means that the initial color configuration may be arbitrary; no asterisk means that it is fixed. The numbers indicate the minimum amounts of distinct colors that are necessary to solve RENDEZVOUS in each model (cf. Theorem 10).

but they must implicitly determine it, agree on it, and reach it, in a distribute manner. Note that this problem is different from CONVERGENCE, in which robots only have to approach a common location, but may never actually reach it.

For any set of more than two robots, GATHERING has been solved in non-rigid ASYNCH, without using colored lights (see [1]). The special case with only two robots is also called RENDEZVOUS, and it is easily seen to be solvable in non-rigid FSYNCH but unsolvable in rigid SSYNCH, if colored lights are not used (see [8]).

Proposition 1. *If only one color is available,* RENDEZVOUS *is solvable in non-rigid* FSYNCH *and unsolvable in rigid* SSYNCH.

Proof. In non-rigid FSYNCH, consider the algorithm that makes each robot move to the midpoint of the current robots' positions. At each move, the distance between the two robots is reduced by at least 2δ, until it becomes less than 2δ, and the robots gather.

Suppose that an algorithm exists that solves RENDEZVOUS in rigid SSYNCH by using just one color. Let us assume that the two robots' axes are oriented symmetrically, in opposite directions. This implies that, if we activate both robots at each cycle, they obtain isometric snapshots, and thus they make moves that are symmetric with respect to their current midpoint. Therefore, by doing so, the robots can never meet unless they compute the midpoint. If they do it, we just activate one robot for that cycle (and each time this happens, we pick a different robot, alternating). As a result, the robots never meet, regardless of the algorithm. □

However, in [2] it was shown how RENDEZVOUS can be solved even in non-rigid ASYNCH using lights of four different colors, and starting from a preset configuration of colors. Optimizing the amount of colors was left as an open problem.

1.3 Our Contribution

In this paper, we will determine the minimum number of colors required to solve RENDEZVOUS in all models shown in Fig. 1, with some restrictions on the class of available algorithms.

Recall that robots do not necessarily share a global coordinate system, but each robot has its own. If the coordinate system of a robot is not even self-consistent (i.e., it can unpredictably change from one cycle to another), then the only reliable reference for each robot is the position of the other robot(s) around it. In this case, the only type of move that is consistent will all possible coordinate systems is moving to a linear combination of the robots' positions, whose coefficients may depend on the colored lights. In particular, when the robots are only two, we assume that each robot may only compute a destination point of the form

$$(1 - \lambda) \cdot me.position + \lambda \cdot other.position,$$

for some $\lambda \in \mathbb{R}$. In turn, λ is a function of $me.light$ and $other.light$ only. This class of algorithms will be denoted by $\mathcal{L}$.

In Sect. 2, we will prove that two colors are sufficient to solve RENDEZVOUS in non-rigid SSYNCH with arbitrary initial configuration and in rigid ASYNCH with preset initial configuration, whereas three colors are sufficient in non-rigid ASYNCH with arbitrary initial configuration. All the algorithms presented are of class $\mathcal{L}$.

On the other hand, in Sect. 3 we show that even termination detection can be achieved in non-rigid ASYNCH with arbitrary initial configuration using only three colors, although our algorithm is not of class $\mathcal{L}$ (indeed, no algorithm of class $\mathcal{L}$ can detect termination in RENDEZVOUS).

In contrast, in Sect. 4 we prove that no algorithm of class $\mathcal{L}$ using only two colors can solve RENDEZVOUS in rigid ASYNCH with arbitrary initial configuration or in non-rigid ASYNCH with preset initial configuration.

Finally, in Sect. 5 we put all these results together and we conclude with a complete characterization of the minimum amount of colors that are needed to solve RENDEZVOUS in every model (see Theorem 10).

2 Algorithms for Rendezvous

2.1 Two Colors for the Non-rigid Semi-Synchronous Model

For non-rigid SSYNCH, we propose Algorithm 1, also represented in Fig. 2. Labels on arrows indicate the color that is seen on the other robot, and the destination of the next MOVE with respect to the position of the other robot. "0" stands for "do not move", "1/2" means "move to the midpoint", and "1" means "move to the other robot". The colors used are only two, namely A and B.

Algorithm 1: Rendezvous for non-rigid SSYNCH and rigid ASYNCH

---

me.destination ⟵ *me.position*
if *me.light = A* **then**
   if *other.light = A* **then**
      me.light ⟵ *B*
      me.destination ⟵ (*me.position + other.position*)/2
   else
      me.destination ⟵ *other.position*
else if *other.light = B* **then**
   me.light ⟵ *A*

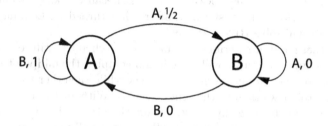

Fig. 2. Illustration of Algorithm 1

Lemma 1. *If the two robots start a cycle with their lights set to opposite colors, they eventually gather.*

Proof. Both robots retain their colors at every cycle, and the A-robot keeps computing the other robot's location, while the B-robot keeps waiting. Hence, their distance decreases by at least δ for every cycle in which the A-robot is active, until the distance becomes smaller than δ, and the robots gather.   □

Theorem 1. *Algorithm 1 solves* RENDEZVOUS *in non-rigid* SSYNCH, *regardless of the colors in the initial configuration.*

Proof. If the robots start in opposite colors, they gather by Lemma 1. If they start in the same color, they keep alternating colors until one robot is active and one is not. If this happens, they gather by Lemma 1. Otherwise, the two robots are either both active or both inactive at each cycle, and they keep computing the midpoint every other active cycle. Their distance decreases by at least 2δ each time they move, until it becomes smaller than 2δ, and they finally gather.   □

2.2 Two Colors for the Rigid Asynchronous Model

We prove that Algorithm 1 solves RENDEZVOUS in rigid ASYNCH as well, provided that the initial color is A for both robots.

Lemma 2. *If, at some time t, the two robots have opposite colors and neither of them is in a COMPUTE phase that will change its color, they will eventually solve RENDEZVOUS.*

Proof. Each robot retains its color at every cycle after time t, because it keeps seeing the other robot in the opposite color at every LOOK phase. As soon as the A-robot performs its first LOOK after time t, it starts chasing the other robot. On the other hand, as soon as the B-robot performs its first LOOK after time t, it stops forever. Eventually, the two robots will gather and never move again. $\square$

Theorem 2. *Algorithm 1 solves RENDEZVOUS in rigid ASYNCH, provided that both robots start with their lights set to A.*

Proof. Let r be the first robot to perform a LOOK. Then r sees the other robot s set to A, and hence it turns B and computes the midpoint m. Then, as long as s does not perform its first LOOK, r stays B because it keeps seeing s set to A. Hence, if s performs its first LOOK after r has turned B, Lemma 2 applies, and the robots will solve RENDEZVOUS.

On the other hand, if s performs its first LOOK when r is still set to A (hence still in its starting location), s will turn B and compute the midpoint m, as well. If some robot reaches m and performs a LOOK while the other robot is still set to A, the first robot waits until the other turns B. Without loss of generality, let r be the first robot to perform a LOOK while the other robot is set to B. This must happen when r is in m and set to B, hence it will turn A and stay in m. If r turns A before s has reached m, then Lemma 2 applies. Otherwise, r turns A when s is already in m, and both robots will stay in m forever, as they will see the other robot in m at every LOOK. $\square$

2.3 Three Colors for the Non-rigid Asynchronous Model

For non-rigid ASYNCH, we propose Algorithm 2, also represented in Fig. 3. The colors used are three, namely A, B, and C.

Observation 3. *A robot retains its color if and only if it sees the other robot set to a different color.*

Lemma 3. *If, at some time t, the two robots are set to different colors, and neither of them is in a COMPUTE phase that will change its color, they will eventually solve RENDEZVOUS.*

Proof. The two robots keep seeing each other set to different colors, and hence they never change color, by Observation 3. One of the two robots will eventually stay still, and the other robot will then approach it by at least δ at every MOVE phase, until their distance is less than δ, and they gather. As soon as they have gathered, they will stay in place forever. $\square$

Algorithm 2: Rendezvous for non-rigid ASYNCH

$me.destination \longleftarrow me.position$
if $me.light = A$ **then**
    if $other.light = A$ **then**
        $me.light \longleftarrow B$
        $me.destination \longleftarrow (me.position + other.position)/2$
    else if $other.light = B$ **then**
        $me.destination \longleftarrow other.position$

else if $me.light = B$ **then**
    if $other.light = B$ **then**
        $me.light \longleftarrow C$
    else if $other.light = C$ **then**
        $me.destination \longleftarrow other.position$

else
    if $other.light = C$ **then**
        $me.light \longleftarrow A$
    else if $other.light = A$ **then**
        $me.destination \longleftarrow other.position$

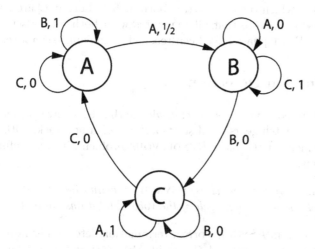

Fig. 3. Illustration of Algorithm 2

Theorem 4. *Algorithm 2 solves* RENDEZVOUS *in non-rigid* ASYNCH, *regardless of the colors in the initial configuration.*

Proof. If the robots start the execution at different colors, they solve REN-DEZVOUS by Lemma 3.

If they both start in A, then let r be the first robot to perform a LOOK. r plans to turn B and move to the midpoint. If it turns B before the other robot s has performed a LOOK, then Lemma 3 applies.

Otherwise, s plans to turn B and move to the midpoint, as well. If a robot stops and sees the other robot still set to A, it waits. Without loss of generality, let r be the first robot to perform a LOOK and see the other robot set to B. r now plans to turn C, but if it does so before s has performed a LOOK, Lemma 3 applies.

So, let us assume that both robots have seen each other in B and they both plan to turn C. Once again, if a robot turns C and sees the other robot still in B, it waits. Without loss of generality, let r be the first robot to see the other robot in C. r plans to turn A, but if it does so before s has performed a LOOK, Lemma 3 applies.

Assume that both robots see each other in C and they both plan to turn A. If a robot turns A and sees the other robot still in C, it waits. At some point, both robots are in A again, in a WAIT phase, but they have approached each other. They both moved toward the midpoint in their first cycle, and then they just made null moves. As a consequence, if their distance was smaller than 2δ, they have gathered. Otherwise, the distance has decreased by at least 2δ. As the execution goes on and the same pattern of transitions repeats, the distance keeps decreasing until the robots gather. As soon as they have gathered, they never move again, hence RENDEZVOUS is solved.

The cases in which the robots start both in B or both in C are resolved with the same reasoning. Note that all the states with both robots set to the same color and in a WAIT phase have been reached in the analysis above.   □

3   Termination Detection

Suppose we wanted our robots to acknowledge that they have gathered, in order to turn off, or "switch gears" and start performing a new task. Although this is not a requirement of the basic RENDEZVOUS problem, it is a useful feature to add as a bonus.

Observation 5. *If the model is* SSYNCH, *termination detection is trivially obtained by checking at each cycle if the robots' locations coincide.*

Unfortunately, in ASYNCH, correct termination detection is harder to obtain. Observe that both Algorithm 1 (for rigid ASYNCH) and Algorithm 2 (for non-rigid ASYNCH) fail to guarantee termination detection. Indeed, suppose that robot r is set to A and sees the other robot s set to B, and that the two robots coincide. Then r cannot tell if s is still moving or not. If s is not moving, it is safe for r to terminate, but if s is moving, then r has still to "chase" s, and cannot terminate yet.

To guarantee correct termination detection in non-rigid ASYNCH, we propose Algorithm 3, represented in Fig. 4. Note that different rules may apply depending on the distance between the two robots, indicated by d in the picture. However, robots need only distinguish between zero and non-zero distances. The colors used are again three, namely A, B, and C.

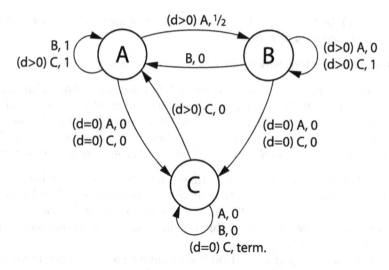

Fig. 4. Illustration of Algorithm 3

Observation 6. *No robot can move while it is set to C.*

Lemma 4. *If some robot ever turns C from a different color, the two robots will gather and their execution will terminate correctly.*

Proof. A robot can turn C only if it performs a LOOK while the other robot is in the same location. If robot r performs a LOOK at time t that makes it turn C, then r stays C forever after, unless it sees the other robot s set to C as well, in a different location. Let $t' > t$ be the first time this happens. Due to Observation 6, r does not move between t and t'. On the other hand, s coincides with r at time t. Then s must turn some other color and move away from r, and then turn C at some time t'' such that $t < t'' \leqslant t'$. But, in order to turn C, s would have to coincide with r, which is a contradiction.

Hence r will stay C and never move after time t. As soon as s sees r set to C, it starts moving toward it (after turning A, if s is also set to C and not coincident with r), covering at least δ at each MOVE phase, until their distance becomes less than δ and s finally reaches r. Then s will turn C as well, and both robots will terminate correctly after seeing each other again.   □

Lemma 5. *If, at some time t, the two robots are set to A and B respectively, and neither of them is in a COMPUTE phase that will change its color, they will eventually gather and terminate correctly.*

Proof. If some robot ever turns C after time t, gathering and termination are ensured by Lemma 4. Otherwise, the two robots keep seeing each other set to opposite colors, and hence they never change color. The B-robot will eventually stay still, and the A-robot will then approach it by at least δ at every MOVE phase, until their distance is less than δ, and they gather. The B-robot then turns C, and Lemma 4 applies again.   □

Let $r(t)$ denote the position of robot r at time $t \geqslant 0$.

Lemma 6. *Let t be a time instant at which both robots are set to A, and neither of them is in a* COMPUTE *phase. Let us assume that robot r will stay still until the end of its current phase (even if it is a* MOVE *phase), and that robot s will either stay still until the end of its current phase, or its destination point is r's current location. Then r and s will eventually gather and terminate correctly.*

Proof. If s is not directed toward r at time t, let d be the distance between $r(t)$ and $s(t)$. Otherwise, let t' be the time at which s performed its last LOOK, and let d be the distance between $s(t')$ and $r(t)$. Furthermore, let $k = \lceil d/\delta \rceil$. We will prove our claim by well-founded induction on k, so let us assume our claim to hold for every k' such that $0 \leqslant k' < k$.

The first robot to perform a LOOK after time t sees the other robot set to A. If they coincide (i.e., if s has reached r or if $k = 0$), the first robot turns C, and Lemma 4 applies. If they do not coincide, the first robot turns B. If it turns B before the other robot has performed a LOOK, then Lemma 5 applies. Otherwise, when the second robot performs its first LOOK after time t, it sees the first robot still set to A. Once again, if they coincide, the second robot turns C and Lemma 4 applies. At this point, if $k = 1$ and s was directed toward r at time t, the robots have gathered and terminated correctly.

Hence, if r and s perform their first LOOK at times t_r and t_s respectively, we may assume that both will turn B, r computes the midpoint m_r of $r(t_r)$ and $s(t_r)$, and s computes the midpoint m_s of $r(t_s)$ and $s(t_s)$. Observe that, if s's destination was not r at time t, then $m_r = m_s$.

Without loss of generality, let r be the first robot to perform the second LOOK. r sees s set to B, hence it turns A. If s performs the second LOOK after r has already turned A, then s necessarily sees r in A (because r keeps seeing s in B), and Lemma 5 applies.

Otherwise, both robots see each other in B, and both eventually turn A. Without loss of generality, let s be the first robot to perform the third LOOK. If $k = 1$ and s was not directed toward r at time t, the robots have indeed gathered in $m_r = m_s$, so s turns C and Lemma 4 applies.

At this point we may assume that $k \geqslant 2$, hence $\delta \leqslant (k-1)\delta < d \leqslant k\delta$. We claim that the distance d' between r and s is now at most $(k-1)\delta$. Indeed, if s was not directed toward r at time t, then each robot has either reached $m_r = m_s$, or has approached it by at least δ. In any case, $d' \leqslant d - \delta \leqslant (k-1)\delta$. Otherwise, if s was directed toward r at time t, then observe that both m_r and m_s lie between $r(t)$ and $m = (r(t) + s(t))/2$. Moreover, s has performed its first LOOK while at distance at most $d - \delta$ from $r(t)$, and subsequently it has further approached $r(t)$. On the other hand, r is found between $r(t)$ and m, thus at distance not greater than $d/2$ from $r(t)$. Hence, $d' \leqslant \max\{d - \delta, d/2\} \leqslant (k-1)\delta$.

Now, s is the first robot to perform the third LOOK, and sees r either already in A or still in B. In the first case, the inductive hypothesis applies, because r

is not in a COMPUTE phase, and its destination is r itself. In the second case, s computes r's location, and it keeps doing so until r turns A. When this happens, the inductive hypothesis applies again.   □

Corollary 1. *If, at some time t, both robots are set to B and are both in a* WAIT *or in a* LOOK *phase, they will eventually gather and terminate correctly.*

Proof. The reasoning in the proof of Lemma 6 also implicitly addresses this case. Indeed, the configuration in which both robots are set to B and in a WAIT or a LOOK phase is reached during the analysis, and is incidentally resolved, as well.   □

Theorem 7. *Algorithm 3 solves* RENDEZVOUS *in non-rigid* ASYNCH *and terminates correctly, regardless of the colors in the initial configuration.*

Proof. If both robots start in A, Lemma 6 applies. If they both start in B, Corollary 1 applies. If one robot starts in A and the other one starts in B, then Lemma 5 applies.

If exactly one robot starts in C, it will stay still forever, and the other robot will eventually reach it, turn C as well, and both will terminate.

If both robots start in C and they are coincident, they will terminate. If they are not coincident, let r be the first robot to perform a LOOK. r will then turn A and move toward the other robot s. If s performs its first LOOK when r has already turned A, it will wait, r will eventually reach it, turn C, and both will terminate. Otherwise, s performs its LOOK when r is still set to C, hence s will turn A as well, and move toward r.

Then, one robot will keep staying A and moving toward the other one, until both have turned A. Without loss of generality, let r be the first robot to see the other one set to A. If they are coincident, r turns C and Lemma 4 applies. Otherwise, r turns B. If this happens before s has seen r in A, then Lemma 5 applies. Otherwise, both robots will turn B. As long as only one robot has turned B, it stays B and does not move. At some point, one robot sees the other in B and Corollary 1 applies.   □

4   Impossibility of Rendezvous with Two Colors

Observe that Algorithms 1, 2 and 3 only produce moves of three types: stay still, move to the midpoint, and move to the other robot. It turns out that, regardless of the number of available colors, any algorithm for RENDEZVOUS must use those three moves under some circumstances.

Proposition 2. *For any algorithm solving* RENDEZVOUS *in rigid* FSYNCH, *there exist a color X and a distance $d > 0$ such that any robot set to X that sees the other robot at distance d and set to X moves to the midpoint.*

Proof. Assume both robots start with the same color and in distinct positions. We may assume that both robots get isometric snapshots at each cycle, so they both turn the same colors, and compute destination points that are symmetric with respect to their midpoint. If they never compute the midpoint and their execution is rigid and fully synchronous, they never gather. □

Similarly, there must exist two states X and Y and a distance d such that, if a robot in X (respectively, Y) sees the other robot in Y (respectively, X) and at distance d, it moves to the other robot's location (respectively, it stays still).

The above observation partly justifies the choice to restrict our attention to a specific class of algorithms: from now on, every algorithm we consider computes only destinations of the form

$$(1 - \lambda) \cdot me.position + \lambda \cdot other.position,$$

where the parameter $\lambda \in \mathbb{R}$ depends only on *me.light* and *other.light*. Similarly, a robot's next light color depends only on the current colors of the two robots' lights, and not on their distance. Recall from Sect. 1 that this class of algorithms is denoted by $\mathcal{L}$. Notice that Algorithms 1 and 2 both belong to $\mathcal{L}$, but Algorithm 3 does not, because it may output a different color depending if the two robots coincide or not.

A statement of the form $X(Y) = (Z, \lambda)$ is shorthand for "if a robot is set to X and sees the other robot set to Y, it turns Z and makes a move with parameter λ", where $\{X, Y, Z\} \subseteq \{A, B\}$ and $\lambda \in \mathbb{R}$. The negation of $X(Y) = (Z, \lambda)$ will be written as $X(Y) \neq (Z, \lambda)$, where as a transition with an unspecified move parameter will be denoted by $X(Y) = (Z, \star)$.

4.1 Preliminary Results

Here we assume that the model is rigid ASYNCH, that only two colors are available, namely A and B, and that the initial configuration is with both robots set to A. All our impossibility results for this very special model are then applicable to both non-rigid ASYNCH with preset initial configuration and rigid ASYNCH with arbitrary initial configuration.

So, let an algorithm that solves RENDEZVOUS in this model be given. If the algorithm belongs to class $\mathcal{L}$, then the following statements hold.

Lemma 7. $A(A) = (B, \star)$.

Proof. If the execution starts with both robots in A, and $A(A) = (A, \star)$, then no robot ever transitions to B, and RENDEZVOUS is not solvable, due to Proposition 1. □

Lemma 8. *If $A(A) = (B, 1/2)$, then $B(A) = (B, \star)$.*

Proof. Let us assume by contradiction that $B(A) = (A, \star)$. If $B(A) = (A, \lambda)$ with $\lambda \neq 1$, we let the two robots execute two cycles each, alternately. As a result, each robot keeps seeing the other robot in A, and their distance is multiplied by $|1 - \lambda|/2 \neq 0$ at every turn. Hence the robots never gather.

If $B(A) = (A, 1)$, we let robot r perform a whole cycle and the LOOK and COMPUTE phases of the next cycle, while the other robot s waits. At this point, their distance has halved, r is set to A, and is about to move to s's position. Now s performs two whole cycles, reaching r's position with its light set to A. Finally, we let r finish its cycle. As a result, the distance between the two robots has halved, both robots have performed at least a cycle, they are in a WAIT phase, and they are both set to A. Hence, by repeating the same pattern of moves, they never gather. $\qquad\square$

Lemma 9. *If $A(A) = (B, 1/2)$ and $B(B) = (A, \star)$, then $B(B) = (A, 0)$.*

Proof. Assume by contradiction that $A(A) = (B, 1/2)$ and $B(B) = (A, \lambda)$ with $\lambda \neq 0$. We let both robots perform a LOOK and a COMPUTE phase simultaneously. Both turn B and compute the midpoint m. Then we let robot r finish the current cycle and perform a new LOOK. As a result, r will turn A and will move away from m. Now let the other robot s finish its first cycle and perform a whole new cycle. s reaches m, sees r still set to B and still in m, hence s turns A and stays in m. Finally, we let r finish the current cycle. At this point, both robots are set to A, they are in a WAIT phase, both have performed at least one cycle, and their distance has been multiplied by $|\lambda|/2 \neq 0$. Therefore, by repeating the same pattern of moves, they never gather. $\qquad\square$

Lemma 10. *If $A(A) = (B, 1/2)$ and $B(B) = (A, 0)$, then $B(A) = (B, 0)$.*

Proof. By Lemma 8, $B(A) = (B, \star)$. Assume by contradiction that $B(A) = (B, \lambda)$ with $\lambda \neq 0$. We let both robots perform a LOOK simultaneously, so both plan to turn B and move to the midpoint m. We let robot r finish the cycle, while the other robot s waits. Then we let r perform a whole other cycle. So r sees s still in A, and moves away from m, while staying B. Now we let s finish its first cycle and move to m. Finally, we let both robots perform a new cycle simultaneously. As a result, both robots are set to A and are in a WAIT phase, both have performed at least one cycle, and their distance has been multiplied by $|\lambda|/2 \neq 0$. By repeating the same pattern of moves, they never gather. $\qquad\square$

Lemma 11. *If $A(A) = (B, 1/2)$ and $B(B) = (A, 0)$, then $A(B) = (A, 1)$.*

Proof. Let us first assume that $A(B) = (B, \lambda)$ with $\lambda \neq 1$. We let one robot perform a whole cycle, thus turning B and moving to the midpoint. Then we let the other robot perform a cycle, at the end of which both robots are set to B. Finally, we let both robots perform a cycle simultaneously, after which they are back to A and in a WAIT phase. Because their distance has been multiplied by $|1 - \lambda|/2 \neq 0$, by repeating the same pattern of moves they never gather.

Assume now that $A(B) = (B, 1)$. We let robot r perform a LOOK and a COMPUTE phase, thus turning B and computing the midpoint. Now we let the other robot s perform a whole cycle, at the end of which it is set to B and has reached r. Then we let r finish its cycle, moving away from s. Finally, we let both robots perform a new cycle simultaneously, which takes them back to A. Their distance has now halved, and by repeating the same pattern of moves they never gather.

Assume that $A(B) = (A, \star)$, and let robot r perform an entire cycle, thus turning B and moving to the midpoint. Due to Lemma 10, $B(A) = (B, 0)$, which means that, from now on, both robots will retain colors. Hence, r will always stay still, and s will never reach r unless $A(B) = (A, 1)$. □

4.2 Rigid Asynchronous Model with Arbitrary Initial Configuration

Lemma 12. *Algorithm 1 does not solve* RENDEZVOUS *in rigid* ASYNCH, *if both robots are set to B in the initial configuration.*

Proof. Let both robots perform a LOOK phase, so that both will turn A. We let robot r finish the current cycle and perform a new LOOK, while the other robot s waits. Hence, r will stay A and move to s's position. Now we let s finish the current cycle and perform a new LOOK. So s will turn B and move to the midpoint m. We let r finish the current cycle, thus reaching s, and perform a whole new cycle, thus turning B. Finally, we let s finish the current cycle, thus turning B and moving to m. As a result, both robots are again set to B, they are in a WAIT phase, both have executed at least one cycle, and their distance has halved. Thus, by repeating the same pattern of moves, they never gather. □

Theorem 8. *There is no algorithm of class $\mathcal{L}$ that solves* RENDEZVOUS *using two colors in rigid* ASYNCH *from all possible initial configurations.*

Proof. Because robots may start both in A or both in B, the statement of Lemma 7, holds also with A and B exchanged. Hence $A(A) = (B, \star)$, but also $B(B) = (A, \star)$. Moreover, by Proposition 2, either $A(A) = (B, 1/2)$ or $B(B) = (A, 1/2)$. By symmetry, we may assume without loss of generality that $A(A) = (B, 1/2)$. Now, by Lemma 9, $B(B) = (A, 0)$. Additionally, by Lemma 10 and Lemma 11, $B(A) = (B, 0)$ and $A(B) = (A, 1)$. These rules define exactly Algorithm 1, which is not a solution, due to Lemma 12. □

4.3 Non-rigid Asynchronous Model with Preset Initial Configuration

Theorem 9. *There is no algorithm of class $\mathcal{L}$ that solves* RENDEZVOUS *using two colors in non-rigid* ASYNCH, *even assuming that both robots are set to a predetermined color in the initial configuration.*

Proof. Let both robots be set to A in the initial configuration, and let $d \geqslant 0$ be given. By Lemma 7, $A(A) = (B, \lambda)$, for some $\lambda \in \mathbb{R}$. If $\lambda \neq 1/2$, we place the two robots at distance $d/|1 - 2\lambda|$ from each other, and we let them perform a whole cycle simultaneously. If $\lambda = 1/2$, we place the robots at distance $d + 2\delta$, and we let them perform a cycle simultaneously, but we stop them as soon as they have moved by δ. As a result, both robots are now set to B, and at distance

d from each other. This means that any algorithm solving RENDEZVOUS with both robots set to A must also solve it with both robots set to B, as well.

Similarly, we can place the two robots at distance $d/|1-\lambda|$ or $d+\delta$, depending if $\lambda \neq 1$ or $\lambda = 1$. Then we let only one robot perform a full cycle, and we let it finish or we stop it after δ, in such a way that it ends up at distance exactly d from the other robot. At this point, one robot is set to A and the other is set to B.

It follows that any algorithm for RENDEZVOUS must effectively solve it from all possible initial configurations. But this is impossible, due to Theorem 8. □

5   Conclusions

We considered deterministic distributed algorithms for RENDEZVOUS for mobile robots that cannot use distance information, but can only reduce (or increase) their distance by a constant factor, depending on the color of the lights that both robots are carrying. We called this class of algorithms $\mathcal{L}$.

We gave several upper and lower bounds on the number of different colors that are necessary to solve RENDEZVOUS in different robot models. Based on these results, we can now give a complete characterization of the number of necessary colors in every possible model, ranging from fully synchronous to semi-synchronous to asynchronous, rigid and non-rigid, with preset or arbitrary initial configuration.

Theorem 10. *To solve* RENDEZVOUS *with an algorithm of class $\mathcal{L}$ from a preset starting configuration,*

- *one color is sufficient for rigid and non-rigid* FSYNCH*;*
- *two colors are necessary and sufficient for rigid* SSYNCH*, non-rigid* SSYNCH*, and rigid* ASYNCH*;*
- *three colors are necessary and sufficient for non-rigid* ASYNCH*.*

To solve RENDEZVOUS *with an algorithm of class $\mathcal{L}$ from an arbitrary starting configuration,*

- *one color is sufficient for rigid and non-rigid* FSYNCH*;*
- *two colors are necessary and sufficient for rigid and non-rigid* SSYNCH*;*
- *three colors are necessary and sufficient for rigid and non-rigid* ASYNCH*.*

Proof. All the optimal color values derive either from previous theorems or from the model inclusions summarized in Fig. 1.

That just one color is (necessary and) sufficient for all FSYNCH models follows from Proposition 1.

Proposition 1 also implies that, for all the other models, at least two colors are necessary. Therefore, by Theorem 1, two colors are necessary and sufficient for all SSYNCH models.

Similarly, Theorem 2 states that two colors are necessary and sufficient for rigid ASYNCH with preset initial configuration. On the other hand, by Theorem 8 and 9, three colors are necessary in the three remaining models, and by Theorem 4 three colors are also sufficient.   □

In the three models in which three colors are necessary and sufficient, it remains an open problem to determine whether using distance information to its full extent would make it possible to use only two colors.

References

1. Cieliebak, M., Flocchini, P., Prencipe, G., Santoro, N.: Distributed computing for mobile robots: gathering. SIAM J. Comput. **41**, 829–879 (2012)
2. Das, S., Flocchini, P., Prencipe, G., Santoro, N., Yamashita, M.: The power of lights: synchronizing asynchronous robots using visible bits. In: 32nd International Conference on Distributed Computing Systems, pp. 506–515 (2012)
3. Flocchini, P., Prencipe, G., Santoro, N.: Distributed Computing by Oblivious Mobile Robots. Morgan & Claypool, San Rafael (2012)
4. Flocchini, P., Prencipe, G., Santoro, N., Widmayer, P.: Gathering of robots with limited visibility. Theor. Comput. Sci. **337**, 147–168 (2005)
5. Izumi, T., Souissi, S., Katayama, Y., Inuzuka, N., Défago, X., Wada, K., Yamashita, M.: The gathering problem for two oblivious mobile robots with unreliable compasses. SIAM J. Comput. **411**, 26–46 (2012)
6. Prencipe, G.: Impossibility of gathering by a set of autonomous mobile robots. Theor. Comput. Sci. **384**, 222–231 (2007)
7. Prencipe, G., Santoro, N.: Distributed algorithms for mobile robots. In: Navarro, G., Bertossi, L., Kohayakawa, Y. (eds.) TCS 2006. IFIP, vol. 209, pp. 47–62. Springer, Boston (2006)
8. Suzuki, I., Yamashita, M.: Distributed anonymous mobile robots: formation of geometric patterns. SIAM J. Comput. **28**, 1347–1363 (1999)

Author Index